The physiology of crop yield

Dedication

This book is dedicated to our mentors:

Michael Kirby
and
John Monteith

to celebrate their contributions in laying the foundations of modern crop physiology

The physiology of crop yield

Second edition

Robert K M Hay
Swedish University of Agricultural Sciences, Uppsala, Sweden

John R Porter
Royal Veterinary and Agricultural University, Copenhagen, Denmark

Blackwell
Publishing

© 2006 by R.K.M. Hay & J.R. Porter

Editorial Offices:
Blackwell Publishing Ltd, 9600 Garsington Road, Oxford OX4 2DQ, UK
  Tel: +44 (0)1865 776868
Blackwell Publishing Professional, 2121 State Avenue, Ames, Iowa 50014-8300, USA
  Tel: +1 515 292 0140
Blackwell Publishing Asia Pty Ltd, 550 Swanston Street, Carlton, Victoria 3053, Australia
  Tel: +61 (0)3 8359 1011

First published 1989 as *An Introduction to the Physiology of Crop Yield* by Longman Scientific & Technical
Second edition published 2006 by Blackwell Publishing Ltd

ISBN-13: 978-14051-0859-1
ISBN-10: 1-4051-0859-2

Library of Congress Cataloging-in-Publication Data
Hay, Robert K. M., 1946–
    The physiology of crop yield / Robert K. M. Hay, John R. Porter. — 2nd ed.
        p. cm.
    Rev. ed. of: An introduction to the physiology of crop yield. 1989.
    Includes bibliographical references and indexes.
    ISBN-13: 978-1-4051-0859-1 (pbk. : alk. paper)
    ISBN-10: 1-4051-0859-2 (pbk. : alk. paper)
    1. Crops—Physiology.   2. Crop yields.   I. Porter, John R.   II. Hay, Robert K. M.,
1946–   Introduction to the physiology of crop yield.   III. Title.

    SB112.5.H38 2006
    633—dc22

                                                        2006005216

A catalogue record for this title is available from the British Library

Set in 10/12.5pt Sabon
by Graphicraft Limited, Hong Kong
Printed and bound in Singapore
by COS Printers Pte Ltd

For further information on Blackwell Publishing, visit our website:
www.blackwellpublishing.com

Contents

Preface

In the preface to the predecessor of this book (Hay and Walker 1989), written in the late 1980s, we noted that the population of the world had risen to 5 billion, but that current food production was sufficient to provide adequate nutrition for 6 billion. The fact that malnutrition was widespread was due primarily to political, sociological and distribution problems rather than deficiencies in agricultural science and practice. Since then, although the rate of increase has slackened significantly, the total population has risen to 6.5 billion and it is unlikely to fall before a peak of 9 billion is reached in the middle of the twenty-first century. Food production has continued to rise in line with demand, but 850 million people, predominantly in poor countries, go hungry each day, according to FAO statistics. Furthermore, balancing the supply and demand of crop products may become even more difficult as emerging countries divert an increasing proportion of their grain production to livestock production, with protein conversion ratios as low as 10%.

There have been other important developments since 1989, notably a general realisation that increased production is not an acceptable goal if the farming systems are not sustainable in the medium to long term. Although there is widespread debate about the concept of sustainability, there is a reasonable consensus that it must involve not only careful husbandry of resources, but also the production of food of appropriate quality for human health. Thus the green revolution, which ensured that food production kept ahead of population growth in the second half of the twentieth century, needs to be succeeded by an 'evergreen revolution', using the many disciplines of agricultural and social sciences to ensure the long-term production of adequate supplies of nutritious food (Swaminathan 1998).

The same period has seen an astonishing expansion in our knowledge of plants at the molecular level. Across the developed world, this has been associated with a general decline in the human and financial resources devoted to research and development at the 'whole plant' level, and the image of crop science has not been helped by the controversial introductions of genetically modified varieties. However, the opening years of the twenty-first century have witnessed a realisation that progress, in terms of improved crop varieties and farming systems, requires collaboration at all levels from the gene to the field, and beyond. We need to ensure the continuation of teams of experts in crop breeding, protection and physiology (and related disciplines) to exploit the many advances at the molecular level.

We hope that this book will contribute to the education, training and, perhaps, the inspiration, of the new generation of crop scientists, who will need to have a broad perspective from the ecosystem through the crop and plant to the gene. Like its predecessor, it is aimed at the advanced undergraduate level, and we have assumed a basic foundation in plant anatomy, biochemistry and physiology. The central core of the earlier book (analysis of yield in terms of the interception of radiation, its conversion into chemical potential energy, and the partitioning of the resulting dry matter) has been conserved in an updated form, and the treatment of crop modelling has been greatly enhanced. However, in meeting the need to consider other resources (principally water and nitrogen) and aspects of crop quality, to equip students to tackle the wider environmental and nutritional aspects of crop production, it has been necessary to sacrifice the 'case histories' approach. Diligent readers will, however, discover that many of the examples in these chapters have been updated and incorporated elsewhere in this book, particularly in the new chapter devoted to crop phenology. Continuing the policy of the first book, we have cited the literature extensively to provide readers with an entry to the scientific literature of the subject.

In planning the structure and content of this book we were assisted greatly by advice from many of those who used its predecessor as a course text; in particular, there is a greater emphasis on maize and soybean, in recognition of their global importance; and some consideration of below-ground aspects. We are very grateful for the support and advice of Andrew Walker, and to those who contributed materials for the text, or read and evaluated chapters as they were completed: Ken Boote, Darleen Demason, Christian Richardt Jensen, David Lawlor, Derrick Moot, Chris Pollock, Graham Russell, Tom Sinclair, Kerr Walker, Dale Walters and the students attending the NOVA postgraduate course in Crop Physiology in March 2005. Nevertheless, we must be responsible for any errors of fact or interpretation, and we should be grateful if any such deficiencies could be drawn to our attention. Although this was a coordinated project, RKMH was *principally* responsible for Chapters 2, 3, 6, 7 and 8; and JRP for Chapters 4, 5, 9 and 10. Finally, John Porter thanks Lincoln University in New Zealand for financial support to spend sabbatical leave there, writing Chapter 4.

Copyright acknowledgements

1981), Figure 6.3 from Figures 2 and 3 (Dale 1985), Figure 6.5 from Figure 6.2 (Evans 1975), Figure 6.17 from Figure 1 (Wurr 1974), Figure 8.1 from Figure 5 (Gooding *et al*. 2002), Figure 9.2 from Figure 1 (Weir *et al*. 1984), Figures 9.3 and 9.4 from Figures 1 and 2 (Porter 1984).

Crop Science Society of America, for: Figure 3.18a from Figure 2a (Williams *et al*. 1965), Figures 4.11 and 4.12 from Figures 3 and 4 (Sinclair & Horie 1989), Figure 4.19b from Figure 4a (Krieg & Hutmacher 1986) Figures 4.23 and 4.24 from Figures 17.6 and 17.7 (Unsworth *et al*. 1994), Figure 6.9 from Figure 10 (Westgate *et al*. 2003), Figure 6.10 from Figure 2 (Uribelarrea *et al*. 2002), Figures 8.3 and 8.4 from Figures 2 and 3 (Cober & Voldeng 2000).

CSIRO Publishing, for: Figure 3.5 adapted from Figure 3 (Williams & Rijven 1965), Figure 4.17 from Figures 2 and 6 (Troughton 1969), Figure 5.2 from Figure 7 (Sale 1974), Figure 6.2 from Figure 1 (Hellman *et al*. 2000), Figure 7.6 from Figures 2 and 3 (Hammer *et al*. 1997).

Darleen Demason, for Figure 3.2.

Elsevier, for: Figure 2.3 adapted from Figure 3 (McMaster 1997), Figure 2.8 from Figure 2 (Bolaños & Edmeades 1993), Figure 3.8 from (Connor & Jones 1985), Figure 3.9 from Figure 2 (Vos & van der Putten 1998), Figure 3.20 from Green 1987, Figure 3.22 from Figure 4 (Jamieson *et al*. 1995b); Figure 4.6 from Figure 2.12 (Fitter & Hay 2002), Figure 6.6 from Figure 2 (Bindraban *et al*. 1998), Figure 6.8 from Figure 1 (Subedi *et al*. 1998), Figure 6.11 from Figure 1 (Bolaños & Edmeades 1996), Figure 6.13 from Figure 1 (Miralles *et al*. 1998), Figure 7.2 adapted from Slatyer 1967, Figure 8.5 from Figure 5 (Triboï & Triboï-Blondel 2002), Figures 9.7 and 9.8 from Figures 1 and 3 (Porter 1993), Figure 10.3 from Figure 7 (Semenov & Porter 1995), Figure 10.6 from Figure 1 (Sinclair *et al*. 2004).

European Association of Potato Research, for: Figure 3.26 from Figure 3 (Khurana & McLaren 1982), Figure 6.16 adapted from (MacKerron & Jefferies 1988), Figure 8.7 from Figure 5 (Schippers 1968).

Haworth Press Inc., for Figure 2.5 from Figure 7.1 (Westgate 2000).

Hodder Education and the author, for Figure 1.1 from Figure 4.5 (Monteith & Unsworth 1990).

International Maize and Wheat Improvement Center, for Figure 7.4 from Figure 1 (Chimenti *et al*. 1997).

International Panel on Climate Change, for: Figure 10.2 from Figure 2.32 (IPCC 2001), Figure 10.5 from Figure 19.13 (IPCC 1995).

International Potash Institute, for Figure 6.4 from Figure 4 (Stoy 1980).

John Warren Wilson, for Figure 3.31a from Figure 3 (Warren Wilson 1959).

Keith Whigham for Figure 2.9 from Figure 4.5 (Evans 1993).

McGraw Hill Education and the author, for Figure 4.14 from Figure 10.11 (Stern *et al*. 2003).

Nottingham University Press, for Figure 4.5 from Figure 2.1 (Long 1994).

Oxford University Press, for: Figure 3.3 redrawn from Figure 6 (Sunderland 1960), Figure 3.15 from Figure 5 (Watson 1947), Figure 3.28 from Figure 1 (Marshall & Biscoe 1980), Figure 4.10 from Figures 6a and 7a (Robson & Parsons 1978), Figure 5.8 from Figure 1 (Cannell & Thornley 2000), Figure 6.23 from Figure 4 (Wachendorf *et al.* 2001b), Figure 7.8 from Figure 2 (Colnenne *et al.* 1998), Figure 7.9 from Figure 5 (Devienne-Barret *et al.* 2000), Figure 7.10 from Figure 2 (Dreccer *et al.* 2000), Figure 8.2 from Figure 5 (Calderini *et al.* 1995), Figures 8.8 and 8.9b from Figures 8.2 and 8.1 (Spedding 1971), Figure 9.5 from Figure 9.6 (Thornley & Johnson 1990), Figure 9.6 from Figure 1 (Marshall & Biscoe 1980).

Paul Biscoe, for: Figure 1.3 from Figure 1 (Biscoe & Gallagher 1977), Figure 3.11 from Figure 6 (Biscoe & Willington 1984).

Roger Sylvester-Bradley, for Figure 7.11 from Figure 15.13 (Scott *et al.* 1994).

The Royal Society of London, for: Figure 4.2 from Figure 1 (Monteith 1977), Figure 7.3 adapted from Raschke 1976.

Springer Verlag, for: Figure 4.18 from Figures 3 and 4 (Krampitz *et al.* 1984), Figure 6.15 from Figure 17.6 (Scott & Wilcockson 1978).

Taylor and Francis Group, for: Figure 3.24 from Rossing *et al.* 1992, Figures 9.9–9.14 from Figures 1, 2, 3.16, 4–7 (Boote *et al.* 1998).

Thomson Publishing Services and the author, for Figures 4.7, 4.13 and 4.21 from Figures 11.2, 12.1 and 13.2 (Lawlor 2001).

Wageningen University Library, for Figure 4.8 from Figure 2 (Uchijima 1970).

Chapter 1

Introduction

> *When an organism interacts with its environment, the physical processes involved are rarely simple and the physiological mechanisms are often imperfectly understood. . . . (environmental) physicists are trained to use Occam's razor when they interpret natural phenomena in terms of cause and effect: they observe the behaviour of a system and then seek the simplest way of describing it in terms of governing variables.*
>
> (Monteith and Unsworth 1990)

This book is about the physiological processes determining the yield that can be harvested from a stand of crop plants; these processes are common to all crop species, including those grown for direct human consumption (cereals, grain legumes, potatoes, vegetables), for indirect consumption *via* livestock (grass, forage legumes, oilseeds), or for industry (fibre, alcohol, fuel). Although crop physiology is founded on plant physiology and biochemistry, in the last fifty years it has emerged as a distinct subject, supporting major advances in food production.

The range of species studied by crop physiologists is a small subset of the plant kingdom, with emphasis being placed on the major crops (wheat, rice, maize, barley, soybean, potato, and a narrow range of pasture grasses), but there is increasing interest in applying the principles of crop physiology to minor and novel species. However, this specialisation is offset by the unique depth of study of individual species at all levels from subcellular biochemistry to field scale agronomy; until the selection of *Arabidopsis* as the model species for plant molecular genetics, wheat was probably the most intensively studied plant species on Earth.

Crop physiology is distinct in that, to provide useful information, it must involve complementary studies of plants growing singly and in stands, under controlled conditions and in the field. This distinction, although important, should not be overstressed as there is much to be learned from autecological and population biological studies of wild plants. The approach has led to concepts such as leaf area index, which considers the crop leaf canopy as a single functional unit, rather than an assemblage of individual plants. This, and the parallel development of robust instruments for sensing the field environment, has resulted in the evolution of new micrometeorological methods for measuring exchanges of mass and energy between atmosphere and canopy (e.g. canopy net photosynthesis).

The primary requirement for the success of a crop in a particular area is that its phenology fits the environment. For example, grain crops that give high yields of biomass but remain vegetative because their reproductive development does not respond appropriately to photoperiodic signals can be as unproductive as crops that set grain but run out of time or water before grain filling is complete. Chapter 2 reviews the approaches to charting and quantifying phenology, and other aspects of development, providing powerful tools for the interpretation of crop yield; for screening new varieties; and for the construction of predictive models.

Growing a crop is an exercise in energy transformation, in which incident solar radiation is converted to more useful forms of chemical potential energy located in the harvested parts (e.g. starch in cereal grains and potato tubers; lipids in oilseeds). In the crop stand, this transformation involves three processes in sequence:

- interception of incident solar radiation by the leaf canopy;
- conversion of the intercepted radiant energy to chemical potential energy (conveniently expressed in terms of plant dry matter); and
- partitioning of the dry matter produced between the harvested parts and the rest of the plant.

The yield (Y) of a crop, per unit area, over a given period of time can, therefore, be expressed by the equation:

$$Y = Q \times I \times \varepsilon \times H \tag{1.1}$$

where:

Q is the total quantity of incident solar radiation received over the period,
I is the fraction of Q that is intercepted by the canopy,
ε is the overall photosynthetic efficiency of the crop (i.e. the efficiency of conversion of radiant to chemical potential energy), commonly expressed in terms of the total plant dry matter produced per unit of intercepted radiation,
H is the harvest index of the crop (the fraction of the dry matter production that is allocated to the harvested parts, normally expressed in terms of above-ground production excluding the root system).

The amount of solar radiation incident upon unit area of cropland (Q) per day, determined by daylength and the diurnal pattern of irradiance, varies regularly with season and latitude, and irregularly with weather factors such as cloudiness. For example, Figure 1.1 shows the diurnal trend in irradiance on three almost cloudless days in January, June and September at Rothamsted in Central England. In Figure 1.2, the upper envelopes of the two datasets show the seasonal trends in maximum daily irradiance and in the total daily receipt of

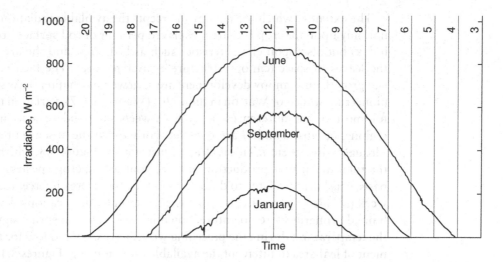

Figure 1.1 Receipts of total solar radiation on three cloudless days at Rothamsted in Central England. The numbers indicate the progression of each day in hours from right to left (from Monteith and Unsworth 1990).

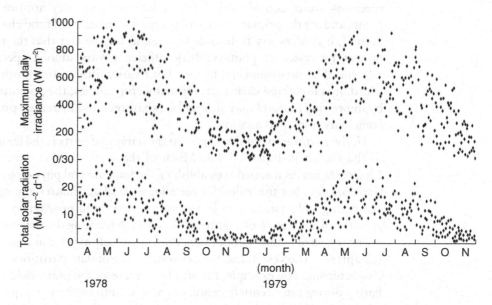

Figure 1.2 Daily records of maximum irradiance and the total quantity of incident solar radiation (0.35–2.5 µm) over two growing seasons in the north-west of England (from Hay 1985).

solar radiation, in England over two growing seasons. These confirm that, even at midsummer, the maximum potential irradiance (approaching 1000 W m^{-2}) can be seriously attenuated by cloud. Of the total solar radiation received, approximately half is photosynthetically active (PAR; 400–700 nm, Chapter 4). Diurnal and seasonal patterns of irradiance for a range of latitudes can be found in sources such as Geiger (1965) and Woodward and Sheehy (1983).

The extent to which the canopy intercepts the available radiation (I) depends not only upon the crop leaf area displayed per unit of soil surface area (leaf area index) but also upon characteristics such as leaf angle and the arrangement of the leaves in space (canopy structure or architecture). The factors controlling leaf growth and canopy development are reviewed in Chapter 3. As shown in the pioneering studies of Watson in the 1940s (Watson 1947), variation in I accounts for most of the differences in yield between sites and seasons in temperate regions, because ε and H are relatively constant in the absence of severe stress (drought, disease etc.). This is clearly illustrated by Figure 1.3, which shows that the total dry-matter production of three contrasting crop species, on a weekly basis, was linearly related to the quantity of radiant energy intercepted during that week (see also Figures 4.2, 6.7). Thus, for example, since canopy development is limited primarily by temperature, potential yield is lost in spring simply because the temperatures during the preceding weeks have been too low for the development of leaf area to intercept the available radiation (e.g. Figures 3.15, 3.25).

Even though ε varies less than I, in the field, many more resources have been expended on studying ε, resulting in a very extensive literature, which is reviewed in Chapters 4 and 5. There are at least three reasons for this concentration of interest upon photosynthesis, respiration and photorespiration. First, the reactions are of considerable intrinsic interest since they are unique to green plants and are the primary source of energy for virtually all food chains on Earth. Second, it is necessary to be able to account for the fact that the rates of component processes of photosynthetic efficiency vary among species, and are influenced by environmental factors, but ε varies little among crops and species. Third, it is becoming clear that, for many crop species, the potential yield gains by improving I and H may shortly be exhausted, and increased production can come only from increases in ε.

H, the partitioning of dry matter to the harvested parts is the least understood of the factors in Equation 1.1. Much of the work on this topic, reviewed in Chapter 6, has been aimed at establishing the fundamental physiology of assimilate translocation, but the 'rules' for the allocation of dry matter among the various competing sinks (storage in leaves, roots, stems and propagules, as well as in growth, maintenance of mature tissues and defence against stress and disease) are now generally understood. Such rules are widely applied in simulation models (Chapter 9). Harvest index, the net result of assimilate partitioning, is a crucial characteristic. For example, the steady increase in the grain yield of wheat and barley during the twentieth century is now seen to be the consequence of a progressive increase in harvest index, with little or no increase in total biomass production. This has important implications for breeding programmes since the indices of the most recent cultivars are approaching the theoretical maximum.

Chapters 3–6 complete the analysis of crops where yield is limited by the interception of solar radiation – a common situation in intensive temperate agriculture. However, in many other parts of the world, yield may be limited by other factors, notably water or nitrogen, and, in the pursuit of 'sustainability', more controlled use of such resources is necessary to prevent both waste and pollution in all countries. In Chapter 7, the approach adopted for crop yields limited by radiation (resolution into resource capture, resource use, and dry-

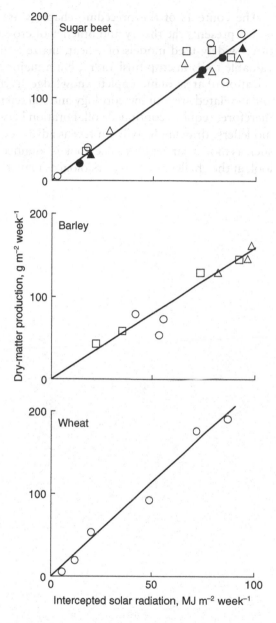

Figure 1.3 Relationships between dry-matter production and intercepted solar radiation, based on weekly measurements, for three crop species grown in the Midlands of England. The different symbols indicate different crops (from Biscoe and Gallagher 1977).

matter partitioning) is applied to water and nitrogen. Although there is a shortage of relevant data, the general conclusion is that resource capture is the most important component of yield. In Chapter 8, physiological aspects of crop quality are considered, including the widely observed inverse relationship between quality and yield; such issues are of considerable importance for the nutritional value and safety of food supplies worldwide, but they also have increasing relevance where customers are applying quality criteria in choosing food items.

The contents of the preceding chapters are drawn together in Chapter 9, which presents the theory and practice of crop modelling, describing, in detail, three widely-used models of wheat, maize and soybean. Crop models can be valuable tools in crop husbandry, but a major benefit of modelling is the identification of areas of incomplete knowledge in crop development and physiology, and in related areas of meteorology and soil science. Progress in crop physiology, therefore, requires continued collaboration between field physiologists and crop modellers, drawing heavily on recent advances in laboratory-based disciplines, such as molecular genetics. The book is rounded off in Chapter 10, by a forward look at the challenges crop physiologists may have to face in coming decades.

Chapter 2
Development and phenology

> *In modern times, plant breeders have sought to improve further the yield, reliability and quality of crops by improving adaptability to different climatic environments and/or management systems. A principal farming objective and a major contribution of the plant breeder to it is therefore to ensure that the life cycles of particular genotypes fit the constraints of the local (or target) environment . . . Annual crops need to be well-adapted to their environments if they are to have the potential to yield well.*
>
> (Summerfield *et al.* 1991)

The major food crops of the world today were first domesticated in eight or more centres of origin, including the Near East (the 'fertile crescent' for wheat, barley, pea); the Americas (Mexico for maize, Peru for potato and *Phaseolus* bean); South Asia (rice); and China (rice, soybean) (Evans 1993). The wild ancestors of these first landraces were adapted to survive and reproduce under the environmental conditions prevailing in the centre of origin (Fitter and Hay 2002). The early cultivators selected plants for yield, quality and other useful characters such as retention of the seed by the plant at maturity and absence of seed dormancy; they also developed technologies for the safe storage of seeds, tubers or other propagules for the next crop, to prevent losses caused by pathogens, predation or premature resumption of growth.

Crop production might have remained a very localised activity had not some of the favoured species shown a remarkable ability to adapt to new environments, as expanding human populations and trade carried seeds and tubers to new areas. Thus wheat and barley, originally adapted to grow in short Mediterranean winters at around 32°N, produce their highest grain yields from crops grown for 10 months at 55°N. Potatoes, originating under the short cool days of the Andes, are now cultivated throughout the world from the long cool days of Northern Europe to the short hot days of the lowland tropics. Similarly, rice and soybean have spread widely across the planet, with the development of a wide range of responses to photoperiod. Much of this adaptation occurred by simple selection before the advent of scientific plant breeding.

Successful cultivation of a crop variety at a given site depends upon two requirements. First, the influence of the prevailing environment (primarily, but not exclusively, photoperiod and temperature) on the development of the crop must result in a life cycle that fits the available growing season: the timing of developmental events (the crop phenology) must be right. It is important for the available time to be partitioned to give an appropriate balance between the generation of 'sources of resources' (leaves and roots) and the filling of 'sinks for resources' (grains, tubers); and the sinks must mature in time for harvest. Second, for a variety that fits its environment, the crop stand must capture and use resources effectively in generating harvestable yield. Thus, a grain crop that produces a high yield of dry matter, but remains vegetative owing to inappropriate photoperiodic responses, is as impractical as a crop that flowers at the right time but is unable, owing to extreme temperature, flooded soil or disease, to capture adequate resources to produce an economic yield. For example, if the timing of anthesis in wheat coincides with very high or very low temperatures, yield can be depressed by ineffective floret fertilisation and grain set (Saini and Aspinall 1982; Subedi *et al.* 1998; Porter and Gawith 1999); here crop phenology has an important influence on sink size (Section 6.5). Some of the most interesting examples of mismatch between genotype and environment have occurred during modern programmes of introduction of novel crops; for example, inappropriate interactions between environment and development resulted in growing seasons of up to 14 months for seed crops of *Oenothera* (evening primrose) in North-West Europe, making the existing varieties impractical to harvest (Russell 1988).

This chapter is about such interactions between development and environment, and the resulting phenologies that make it possible for crops to be cultivated for yield in a given geographic area; a series of crop examples is used to illustrate the important principles. It provides the foundation for subsequent chapters on the capture and use of resources; and an introduction to the quantification of developmental processes, an essential tool in crop modelling (Chapter 9).

2.1   Crop development: concepts and tools

Development is distinct from growth, although they generally proceed simultaneously. In the accepted crop physiological sense, growth refers to increase in plant or crop dry weight, the net result of acquisition and loss of resources (Chapters 4, 5, 7). Development is the sequential production, differentiation, expansion and loss (for example, by orderly senescence) of the structural units of the plant: the phytomers. Although all crop plants, including their reproductive structures, can be broken down into phytomers, primarily made up of a node, an internode, a leaf and an axillary meristem (e.g. Figures 2.1, 2.2, 2.4 and 2.6; Fitter and Hay 2002), there are many variations on the basic theme (Harper 1986) (e.g. the spikelet and its associated section of the wheat ear rachis; Figure 2.7). The distinction between growth and development can be seen clearly in the unfertilised margins of commercial crops of wheat or barley where small single-stemmed plants reach anthesis simultaneously with the much larger multi-stemmed plants of the fertilised crop stand. Any confusion between growth and development can be minimised by the use of the terms expansion or extension rather than growth when referring to increases in dimensions of organs, which are not necessarily associated with increases in dry matter. Development can proceed without growth: the germinating seed or sprouting potato tuber *consumes* stored resources to generate new organs.

The rates of developmental processes such as germination, initiation of phytomers, or unfolding of leaves are determined by temperature, unless the plants are exposed to stresses, such as drought or nutrient deficiency. The temperature relations of the different processes are not identical (Porter and Gawith 1999) but, in temperate crop species, minimum temperatures normally fall in the range 0–5°C, and the increase in rate up to the optimum temperature (within the range 20–30°C), can normally be taken to be linear (e.g. for wheat and maize, Warrington and Kanemasu 1983; Jame *et al.* 1999). Even where more pronounced sigmoid responses are found, the early lag phase can normally be ignored (Shaykewich 1995). Above the optimum, the rate tends to fall more sharply to the maximum temperature, which has rarely been determined precisely, but generally falls between 40 and 50°C (Porter and Gawith 1999).

Within the linear phase, therefore, the plant cannot distinguish between, say, 5 hours at 20°C and 10 hours at 10°C, and this provides the basis for the use of thermal time, in degree days (°days), in charting development. Other terminology includes 'accumulated temperature' or 'growing degree days, GDD' (particularly in crop modelling; Chapter 9), but the related term 'heat units' should be avoided as it, incorrectly, suggests the progressive accumulation of energy. Thermal time is, strictly, the integral over time of environmental temperature above a base temperature (at which the rate of the process falls to zero), but, in practice, a daily mean of maximum and minimum temperatures is normally used. This is a satisfactory approach as long as the diurnal temperature curve is not seriously skewed.

The use of thermal time is central to the study of crop development under fluctuating temperature in the field, permitting the temperature responses of processes to be described by linear relationships in thermal time: degree days accumulated above the relevant minimum, base or threshold temperature (e.g. Figures 2.1b, 2.2b, 2.4b) (Bonhomme 2000; Purcell 2003). The underlying assumption is that the effect of 10 days at a mean temperature of 20°C is equivalent to that of 20 days at 10°C (200 degree days above a base temperature of 0°C). Such relationships tend to depart from linearity only where the temperature rises above the optimum or falls below the base temperature for a significant part of the day (e.g. Jame *et al.* 1999); where the plants are exposed to extreme events (e.g. damaging sub-threshold temperatures) or stresses; or where there is a distinct ontogenic adjustment in the temperature relations (e.g. Hay and Delécolle 1989). The temperature relations of development in a range of species are considered from Section 2.1.2 onwards.

2.1.1   Growth stages and phasic development

Practical agronomists have resolved the problem of describing the dynamics of development by using systems of 'growth stages', which are readily observed in

Table 2.1 Primary and secondary stages used in the Decimal Code for the description of the development of wheat (Tottman and Broad 1987).

0 Germination	**3 Stem extension**
01 water absorption	30 pseudostem (leaf sheath) extension (= 5 cm)
.	31 first node detectable (above ground)
.	.
07–09 coleoptile above ground	.
	.
1 Seedling growth	36 sixth node detectable
10 first leaf through coleoptile	37–39 flag leaf visible
11 first leaf emerged (ligule visible)	
12 2 leaves emerged	**4 Ear in 'boot'**
.	49 tip of ear visible
.	
18 8 leaves emerged	**5 Ear emergence**
19 9 or more leaves emerged	
	6 Anthesis
2 Tillering	
20 main shoot only	**7 Milk development**
21 main shoot and 1 tiller	
.	**8 Dough development**
.	
.	**9 Ripening**
29 main shoot and 9 or more tillers	

the field without the need to bring plants back to the laboratory for dissection, and these are very useful in timing agronomic operations such as spraying for crop protection (see Tables 2.1, 2.2 for wheat and maize; Section 2.1.5 for soybean; Meier (1997) for a wide range of species). The problem for crop physiologists is that these readily visible stages do not necessarily bear simple relationships to the underlying processes of initiation and development of organs, until after the reproductive structures become visible (Landes and Porter 1989). For example, although 'tasselling' is classed as the final stage in the vegetative development of maize (Table 2.2), the transition to reproductive development at the stem apex has already occurred some weeks earlier. Full understanding of crop development, therefore, involves the study of apical meristems, although

Table 2.2 Stages used in the description of the development of maize.

Vegetative	Reproductive
VE Emergence	R1 Silking
V1 First leaf emerged (ligule visible)	R2 Grain 'blister' stage
V2 Second leaf	R3 Milk
V3 Third leaf	R4 Dough
Vn Leaf on nth node emerged	R4 Dent visible (for dent grain types)
VT Tasselling	R6 Physiological maturity

many investigations and crop simulation models (Chapter 9) are based on phasic development (quantifying the duration, in days or degree days, of the intervals between visible growth stages, and the corresponding rates of development; e.g. Summerfield *et al.* 1991). Some models adopt an intermediate approach, including some readily-recorded apical stages: for example, recording double ridges as an approximate index of the start of reproductive development in cereals (Figure 2.7).

The following three sections (2.1.2–2.1.4) introduce the concepts of plant and meristem development using species of increasing complexity as examples. The phasic development approach, which is more practical for crops producing multiple flowering apices on each plant, is taken for soybean (2.1.5).

2.1.2   Events at the stem apex: the leek as a simple model species (Figure 2.1)

Quantitative approaches to crop development can be introduced by considering the leek: a biennial monocotyledonous species, harvested in the first year for its fleshy leaf sheaths (Figure 2.1a), but requiring a period of low temperature vernalisation before it can flower in the second year. Development of the single, unbranched, culm during the first, vegetative, season consists of the successive initiation by the stem apex of simple phytomers which, when mature, have un-extended internodes but long leaf sheaths and blades, and no axillary develop-ment. The stem apical meristem, therefore, remains near to the soil surface, above a closely-packed stack of nodes and internodes (Figure 2.1a).

The first sign of the initiation of a new phytomer is the appearance of a localised outgrowth on the side of the apical dome, detectable only after dissection using a low power microscope: the primordium of the future leaf (Figures 2.7, 3.1, 3.2). The number of phytomers (or leaves) initiated by the apex increases linearly with thermal time (Figure 2.1b), as does the number of visible/appeared leaves, although the rate is lower than that of initiation because of the time elapsing before the tip or ligule can appear above the enclosing sheath of the previous leaf. (The dynamics of individual leaf development from primordium to fully-expanded organ are explained in more detail in Section 3.1, in relation to the generation of the leaf canopy.) There is, however, no accumulation of leaf primordia on the apical meristem (Figure 2.1b), and it is clear that leaf death is also a scheduled and orderly component of plant development.

Because of the linearity of the relationships in Figure 2.1b, it is possible to calculate the plastochron, or interval between the *initiation* of successive leaves (or phytomers), which, here, remained constant at 92°Cdays under fluctuating, and generally increasing, temperatures in the field (base temperature 0°C). The phyllochron, or interval between the *appearance* of successive leaves remained constant at 135 (tip appearance) or 233°Cdays (ligule) (allowing a progressively longer duration of extension for each, successively longer, leaf), and the interval between the death of successive leaves was approximately equal to the ligule phyllochron. The leek plant can, therefore, be viewed as a system for the regular production of vegetative phytomers until it receives the signal to add reproductive phytomers; and the progress of development can be expressed in phyllochrons rather than thermal time.

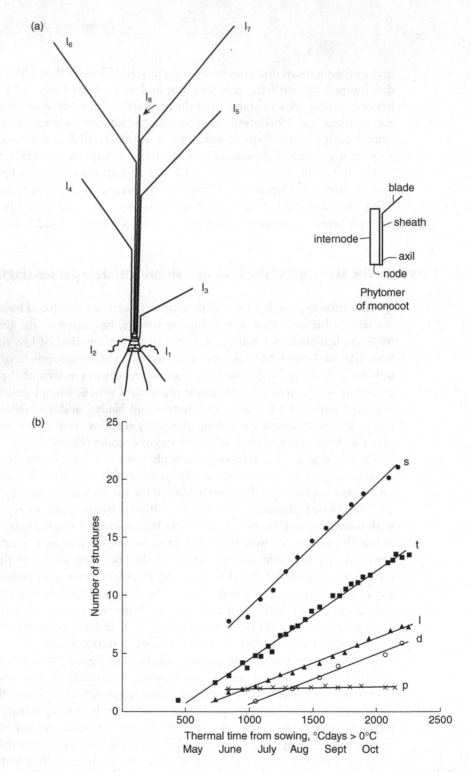

Figure 2.1 Development in leek: (a) schematic diagram of a vegetative leek plant with the tip of leaf 8 just emerging above the ligule of leaf 7; (b) thermal time courses (base 0°C) of the cumulative production of phytomers (s), the appearance of leaf tips (t), the appearance of leaf ligules (l), the death/senescence of leaves (d), and the number of unexpanded primordia on the shoot meristem (p), of leek plants growing in the field in south-west Scotland, 1988 (from Hay and Kemp 1992).

2.1.3 Events at stem apices: branching and reproductive development in wheat (Figure 2.2)

The modular structure of the wheat plant is similar to that of leek, with three principal differences. First, the axillary buds of the lower phytomers of the stem can produce one or more branches (tillers), although many of these die before crop maturity (see Section 2.2.1); Figure 2.2a illustrates a plant in which two

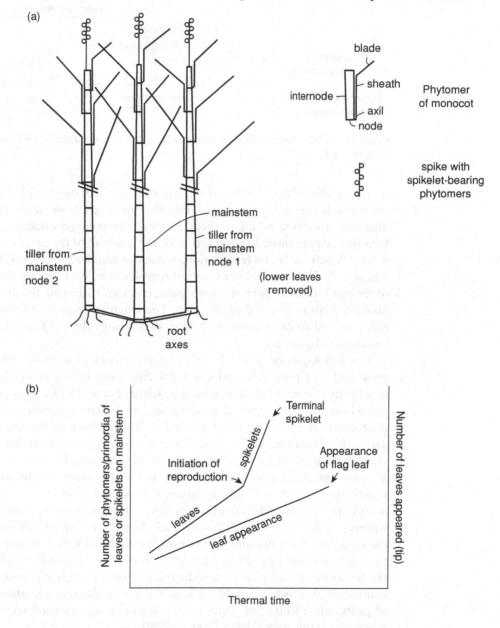

Figure 2.2 Development in wheat: (a) schematic diagram of a wheat plant, with two subsidiary reproductive tillers, at anthesis; (b) model time courses of leaf initiation, spikelet initiation and leaf appearance.

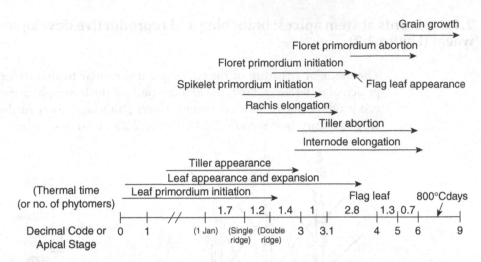

Figure 2.3 Relative timing of the developmental events of a reproductive wheat stem (adapted from McMaster 1997).

surviving tillers have developed to bear ears synchronously with the mainstem (Hay and Kirby 1991). Second, during the growing season, at the apices of the mainstem and principal tillers, there is a switch from the production of phytomers bearing leaves to those bearing spikelets (the modules of the ear, each generating several florets which, on fertilisation, become grains; Kirby and Appleyard 1984; Figure 2.7). Third, in the later stages of reproductive development, the internodes of the stacked lower phytomers elongate, eventually carrying the developing ear above the last or flag leaf of the culm. The schematic representation of a wheat plant around anthesis in Figure 2.2a is, thus, quite different from the vegetative structure in Figure 2.1.

The full sequence of events in the development of a winter wheat crop is presented in Figure 2.3, and corresponding information is available on the synchrony of events below ground (e.g. Klepper *et al.* 1984). After germination, the wheat mainstem apex is programmed to initiate a regular series of leafy phytomers, at plastochrons of around 50°Cdays, base temperature 0°C (Figure 2.2b). However, unlike leek, wheat is a quantitative long-day plant, with reproduction starting earlier as photoperiod increases. There is no evidence of a lower threshold photoperiod under north-temperate conditions; the light signal can be perceived only by green tissues after crop emergence; and only one photoperiodic cycle is required to induce the apex. Consequently, in a spring variety, at least six or seven leaves, including three or four already present in the seed, have been initiated by the time (around 200°Cdays from germination) that the mainstem produces its first reproductive primordium (the collar of the future ear, which is, at first, indistinguishable from a leaf primordium). This is normally, but not invariably, associated with a sharp acceleration in the rate of primordium initiation (Figure 2.2b), resulting in a marked accumulation of primordia at the apex (Delécolle *et al.* 1989).

Primordia cannot be unequivocally identified as reproductive until they achieve the 'double ridge' stage (Figure 2.7), but the timing of this is variable and can be delayed until more than half of the spikelet primordia have been initiated (Jamieson *et al.* 1998a, b; Kirby *et al.* 1999). The only certain method of identifying the collar primordium, and determining the time of its initiation, is therefore, by calculation, once the final number of leaves per stem is known. Plants of winter varieties, in which the transition is delayed until the stem apex has been vernalised by low temperature (Section 2.2.1), produce more leaves per stem. The end of spikelet initiation is readily recognised, after apical dissection, because the terminal spikelet of the ear is initiated centrally (Figure 2.7). Further differentiation of the ear 'in boot', up to the stage of anthesis (pollination), is complete by the time of emergence (Kirby and Appleyard 1984).

As in the leek, leaf appearance proceeds linearly in thermal time (phyllochron range 70–160°Cdays, base temperature 0°C), in coordination with internode extension, such that the flag leaf appears shortly before the emergence of the ear above the canopy (Hay and Kirby 1991; Figure 2.7). In some crops, successful coordination appears to require the resetting of the rate of leaf appearance. Anthesis (pollen shedding) of mainstem and tiller ears is virtually synchronous, and self-pollination occurs within the spikelets of each ear. The phenology of a wheat crop is, thus, determined principally by the timing of onset of reproductive development. This is illustrated in Section 2.2.1 where the influence of crop management on wheat phenology is explored in detail.

Figure 2.2b provides a basis for the quantification of wheat development, and evaluation of the environmental conditions at critical points in the life cycle. For example, the number of leaves produced is the product of the rate and duration of initiation. The number of spikelets produced is, similarly, the product of rate and duration of initiation, but, unlike the case of leaves, some of the spikelets of the future ear do not develop to produce grains (see Sections 2.2.1, 6.5). Modelling of the grain number per mature ear, therefore, requires information on the initiation of florets within spikelets (up to ten), the survival of spikelets and florets up to anthesis, and the success of pollination of florets (Figure 2.7).

2.1.4   Events at stem apices: the consequences of separation of male and female organs in maize (Figures 2.4, 2.5)

The development of maize is a variation on the theme presented for wheat. Here, apart from the fact that the plants are much taller (typically 2 m compared with <1 m for wheat), under intensive cropping there is no axillary activity at the basal nodes: maize plants are effectively uniculm (but see Moulia *et al.* 1999). However, there is axillary activity at higher leaf nodes, and separation of the flowering organs into male and female.

Starting with five in the seed apex, further vegetative phytomers are initiated by the mainstem apex at plastochrons of 20–30°Cdays (base 4°C), approximately half the duration of the corresponding phyllochron (Warrington and Kanemasu 1983), leading to an accumulation of leaf primordia and partly expanded leaves. Leaf appearance is delayed, as in leek and wheat, by the requirement for the tip or ligule to extend beyond the previous sheath (leaf tip phyllochrons in the range 30–60°Cdays, threshold 6–9°C; Vinocur and Ritchie 2001). Maize is a short-day

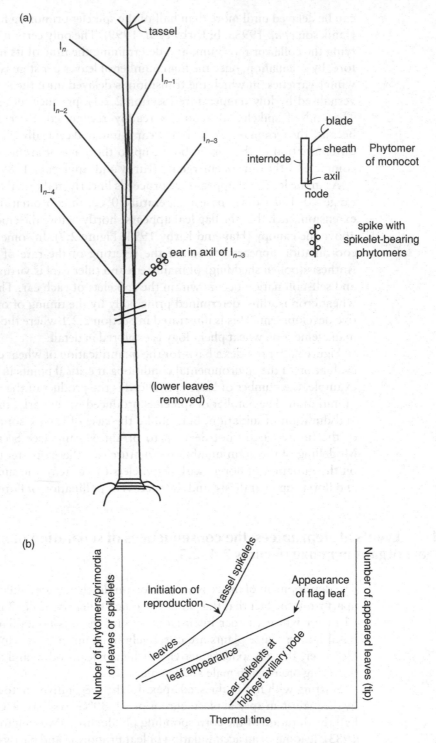

Figure 2.4 Development in maize: (a) schematic diagram of a maize plant at anthesis; (b) model time courses of leaf initiation, spikelet initiation by tassel and ear, and leaf appearance.

species, normally showing a delay in reproductive development in days longer than a threshold of around 13 hours (e.g. Kiniry *et al.* 1983), although some varieties adapted to higher latitudes are insensitive. However, unlike spring wheat, maize stem apices are prevented from responding to the prevailing photo-period at crop emergence by juvenility (Evans and Poethig 1995). Tollenaar and Hunter (1983) found that, under controlled conditions, maize plants became capable of responding to daylength between the four- and seven-leaf stages (Table 2.2); that is, the transition from vegetative to reproductive development at the stem apex did not occur until 13 to 16 leaves had been initiated. Here juve-nility takes the place of the vernalisation requirement in winter wheat in delaying the transition beyond the period immediately after crop emergence. Under the conditions prevailing in the main growing areas of North America, maize plants typically produce 16–23 leaves per stem, and the transition occurs after the appearance of 8–10 leaves.

At the transition from vegetative to reproductive development, as in wheat, the plastochron shortens (i.e. the rate of initiation increases; Figure 2.4b) but the maize mainstem apex generates staminate spikelets only, and stem internode extension results in the emergence of an inflorescence (tassel) above the flag leaf, which is capable of producing pollen only. Ten to twelve days normally elapse after the initiation of the tassel before the axillary meristems at the 5–7th node below the tassel, and at the distal two or three nodes, begin to initiate pistillate spikelets, generating potential ears (Figure 2.4b). At the high population densities

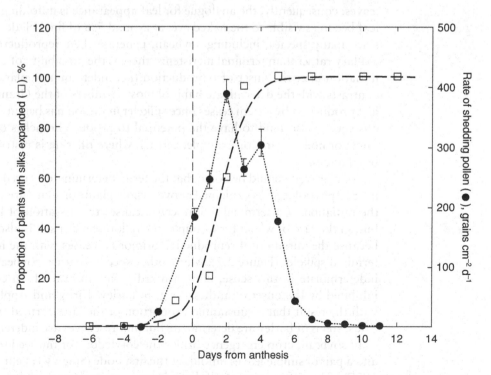

Figure 2.5 Time courses of pollen shedding and silk emergence in maize in relation to anthesis, demonstrating a short anthesis–silking interval (from Westgate 2000).

of modern crops, the distal ears normally cease to develop at an early stage, leaving one per stem.

Shedding of pollen by the tassel (anthesis), which lasts 7–10 days, normally begins before the long styles of the ovules of the ear (the silks) have been fully extruded, a process which may take 4–5 days (Figure 2.5). Once extruded, the silks remain receptive to pollen for around 8 days. Figure 2.5 illustrates a maize crop in which the shedding of pollen and the availability of receptive silks are synchronous. In maize, therefore, the modelling of grain number involves similar information on spikelet dynamics to that of wheat, but, in this case, the focus is on an axillary meristem; the timing of reproductive development at the terminal meristem is critical for successful fertilisation (note the consequences of asynchrony of pollen shedding and silk development in Sections 2.2.2, 6.5), but quantitative analysis of the development of tassel spikelets is not necessary.

2.1.5   Phenology determined by events at axillary meristems: determinate and indeterminate soybean varieties (Figure 2.6)

The development of dicotyledonous crop plants (e.g. legumes, brassicas, potato) differs from that of monocotyledons (leek, cereals, grasses) in several important aspects. They do share the same modular method of construction but, since dicotyledonous leaves are normally petiolate, the production and expansion of the primordia at each node are not obscured by encircling sheaths of older leaves; consequently, the analogue for leaf appearance is unfolding, when a new leaf becomes visible to the naked eye, at around 5% of final blade area. Second, since many species, including soybean, generate their reproductive organs at axillary rather than terminal meristems, there is the possibility of continued leaf initiation after the onset of reproduction (i.e. indeterminate plant habit). This contrasts with the determinate habit of most members of the Gramineae, where leaf production by a stem ceases once spikelet initiation has begun. Dicotyledonous species also tend to have the potential to produce branches over a greater range of nodes than, for example, cereals, where tillering is restricted to a few basal nodes.

For clarity, it should be noted that the term 'determinate' is used in two senses in crop physiology. As explained above, whole plants or stems are determinate if the initiation of a terminal inflorescence causes the cessation of leaf initiation but, in the case of wheat for example, the inflorescence itself is also determinate because the initiation of reproductive primordia ceases with the initiation of a terminal spikelet (Figure 2.7). Many other species (barley, soybean, maize) are indeterminate in this sense; here primordial initiation tends to continue until inhibited by the onset of anthesis (e.g. in barley, Kirby and Appleyard 1984), with the result that a substantial proportion of the later formed structures die. Thus, stems of barley are determinate but inflorescences are indeterminate.

In soybean, crop emergence brings the cotyledons of the seed to the surface, and a pair of simple leaves unfolds at the first node (Stage V1; Fehr and Caviness 1977). Thereafter, single trifoliate leaves are produced at successive nodes (stages V2 to Vn), and branch stems can occur in the axils of lower nodes at low

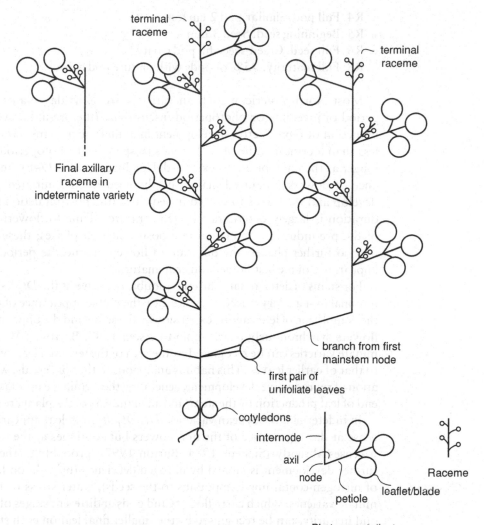

Figure 2.6 Development in soybean: schematic diagram of a determinate soybean plant, with a single branch, at flowering showing the position of the terminal raceme at the apex of each reproductive stem. Inset: the final axillary raceme of the stem of an indeterminate variety.

plant population densities, producing trifoliate leaves according to the same pattern as the mainstem. Flowers are borne in clusters (racemes) at the stem nodes (normally from node 5 or higher), and each fertilised flower produces a pod. Reproductive development can be described by the following growth stages (Fehr and Caviness 1977):

- R1 Beginning bloom. First open flower at any node on the mainstem
- R2 Full bloom. Open flower at one of the two uppermost nodes on the mainstem with a fully developed leaf
- R3 Beginning pod. Pod 5 mm long at one of the four uppermost nodes on the mainstem with a fully developed leaf

- R4 Full pod. Similar pod 2 cm long
- R5 Beginning seed. Seed 3 mm
- R6 Full seed. Green seeds fill pod cavity
- R8 Full maturity. 95% of pods with mature pod colour.

Most North American soybean varieties are short-day plants with a short period of juvenility or photoperiod-insensitivity (the pre-inductive phase). The induction of reproductive development in axillary meristems, under daylengths less than a critical value of 13 h, tends to span 7–11 photoperiodic cycles (but longer as the photoperiod is extended; Upadhyay *et al.* 1994b), and the appearance of flowers is delayed mainly by the need for the initiated phytomers to develop and the leaves to unfold in sequence (the post-induction phase, whose duration is largely determined by temperature). Time to flowering is the sum of the pre-induction, induction and post-induction phases; these are followed by two further phases: the duration of flowering, and the period between the appearance of the last flower and crop maturity.

For stems of determinate varieties (double recessive at the Dt_1 locus), there is a terminal (not axillary) raceme (Figure 2.6); the visible appearance of flowers marks the completion of leaf canopy expansion by the stem; and the upper nodes produce flowers synchronously (Carlson and Lersten 1987; Burton, J W 1997). Determinate varieties can be recognised by the area of the terminal leaf, which is similar to that of earlier leaves. This habit is analogous to that of cereals, where the initiation of reproductive development, generating the terminal ear or tassel, marks the end of leaf production by the stem, and all of the ears of the plant are synchronised.

In indeterminate soybean varieties ($Dt_1 Dt_1 dt_2 dt_2$), leaf appearance does not cease at the appearance of the first flowers but continues at the same rate until it ceases abruptly (Sinclair 1984; Burton 1997), probably at the point where further development is limited by nitrogen deficiency (brought on by the transfer of nitrogen-containing compounds to the seeds), water stress or frost. Indeterminate varieties, which carry flowers and pods at different stages of development and maturity, can be recognised by the smaller final leaf on each stem. They differ from varieties with a determinate habit in the ability to produce yield even if the early flowers or pods are lost owing to stress; in this respect they probably resemble the wild progenitors of cultivated soybean. Indeterminate varieties generally have more nodes and fewer branches than determinate. Semi-determinate varieties have intermediate habits (Burton 1997).

Applying the methods of analysis used in Figures 2.1, 2.2 and 2.4, soybean stems and branches can be viewed as systems for the production of vegetative phytomers at regular intervals until inhibited by the initiation of reproductive development at axillary and terminal meristems (determinate varieties) or by the diversion of resources from leaf production (indeterminate). Constancy of stem phyllochron during the life cycle has been recorded for a range of varieties (range 40–90°Cdays, threshold 5°C; Hofstra *et al.* 1977; Sinclair 1984; Patterson 1992). There is a shortage of information on events at the vegetative stem meristem of soybean, although it can be inferred from data on other legumes that leaf plastochrons are constant and shorter than the corresponding phyllochrons (Jeuffroy and Ney 1997).

The routine monitoring of the initiation of reproductive phytomers, which has proved useful (if arduous) in cereals, is neither practical nor relevant for soybean; up to 35 flowers can be produced by each of up to 20 mainstem nodes in a commercial crop. Depending upon the pattern of assimilate partitioning (Section 6.2), and the incidence of stress at anthesis, a large proportion of the flowers initiated are aborted (e.g. Kokubun *et al.* 2001), resulting in numbers of pods per node varying from 0 to 20 (Whigham 1983). Most studies of flowering in soybean, and crop simulations (for example the widely-used CROPGRO-Soybean; Boote *et al.* 1998), therefore, employ phasic development models, relying on visible growth stages (see Section 2.2.3 and Chapter 9).

2.1.6 Components of yield

Measurement of components of yield, from samples of the mature crop, can provide insight into the effects of environment and management on development and yield. They are particularly important in determining the extent of losses of initiated organs (branches, spikelets, florets); and associated measurements, such as numbers of leafy nodes, can assist in the interpretation of events at stem apices.

For example, the grain yield of a cereal crop can be split into three major components:

Grain yield = ear population density (no of ears/unit area)
  × ear size (no. of grains per ear)
  × individual grain weight

Each of these components can, to a certain extent, vary independently of the others, and their magnitudes are determined at different stages of the crop life cycle (e.g. Figure 2.3). Thus, ear population density is determined primarily by plant population density, but also by tiller initiation (before GS 31; Table 2.1) and loss (around GS 31) in wheat, and by the number of surviving ears per stem in uniculm maize (around GS R1; Table 2.2). The limit to ear size is set during spikelet initiation, but the final magnitude depends upon spikelet and floret survival and the success of floret fertilisation (by GS 69 or R1; Tables 2.1, 2.2; Section 6.5). Finally, mean grain weight, not strictly under developmental control, is determined by the quantity of assimilate available for transport to the ear after anthesis, from current photosynthesis and/or storage within the plant (e.g. Sections 6.6, 6.7).

This approach can be applied usefully to a wide range of grain crops. For example, components of soybean yields can include plant population density, number of flowering nodes per plant, number of pods per node, number of seeds per pod, and individual seed weight. It can also be used to interpret the yields of other crop types including those yielding vegetative organs: in the potato, it has proved useful to partition tuber yield into the number of stems produced per seed tuber, the number of daughter tubers per stem, and mean tuber weight. Each of these is determined at a different period of crop phenology, providing clues to the potential impact of environment and management on yield.

2.2   Case histories: the influence of environment and management on crop development and phenology

Crop development and phenology can, therefore, have profound influences upon the yield of a crop, not least in determining the number and size of sources and potential sinks for assimilate. This section includes five varied case histories, selected to provide a broad rather than a comprehensive view of the possible interactions between development and yield, covering a range of crop types.

2.2.1   Convergence and synchrony: the influence of sowing date on winter wheat in Northern Europe

In temperate areas, where the winter is not sufficiently severe to cause extensive plant damage or death, the yield potential of autumn-sown cereal crops is considerably higher than that of spring-sown crops. A crop stand already established in spring is able to respond immediately to rising temperature and increased receipts of solar radiation; by contrast, since a spring crop cannot be sown until a soil moisture deficit develops, part of the growing season is lost.

In the UK, therefore, there is a preference for winter cereals although, in practice, the sowing date for winter wheat can vary from September to April, depending upon a range of practical aspects of crop management (date of harvesting the preceding crop; soil moisture conditions). The remarkable degree of adaptation of modern winter wheat varieties to climate and management means that the development of crops sown several months apart can converge to give anthesis dates spanning a few days, and effectively the same grain-filling period and harvest date (e.g. anthesis dates between 29 June and 15 July for Maris Hustler wheat crops sown between 9 September and 9 March, representing durations from sowing of 293 to 128 days, respectively; Table 2.3).

Table 2.3 The phenology of winter wheat crops (cv. Maris Hustler) sown on nine dates in 1982–83 in the field in South-West Scotland, expressed in days from sowing, with number of appeared leaves in brackets (from Hay 1986).

Sowing date	Emergence	Double ridge	Terminal spikelet	Flag leaf	Ear emergence	Anthesis
9 Sept	12	161 (9l)	225 (11l)	263 (14l)	284	293 (29 Jun)
5 Oct	16	191 (9l)	220 (11l)	237 (13l)	261	267 (29 Jun)
21 Oct	18	193 (9l)	212 (11l)	229 (13l)	249	257 (5 Jul)
8 Nov	31	179 (8l)	194 (9l)	211 (11l)	231	239 (5 Jul)
10 Dec	35	147 (7l)	162 (8l)	179 (10l)	201	209 (7 Jul)
14 Jan	48	112 (6l)	131 (8l)	154 (10l)	170	178 (11 Jul)
7 Feb	38	95 (6l)	107 (8l)	136 (10l)	148	156 (13 Jul)
9 Mar	29	75 (6l)	90 (8l)	114 (10l)	122	128 (15 Jul)
29 Apr	13	37 (10l)[a]	47 (12l)[a]	53[a]	66[a]	80[a]

[a] Very variable reproductive development (see text).

This fine tuning of the crop to its environment (which relies *primarily* on the partitioning of the season into appropriate durations of vegetative and reproductive development) is a consequence of interactions among plant responses to different environmental factors (temperature, photoperiod; Hay and Kirby 1991; Hay 1999). These can be interpreted using the detailed accounts of wheat development in Figures 2.3 and 2.7. Within the range of conditions experienced in north-temperate areas, leaf initiation is dependent solely upon temperature. There is evidence that soil temperature is the most appropriate measurement in the vegetative stages of development, as the apex is below or near to the soil surface; however, for practical reasons, air temperatures are normally used in the analysis of crop development, as the two measurements are closely correlated. The apex continues to generate leaf primordia at plastochrons of around 50°Cdays (base 0°C) until the signal for flowering is received.

However, winter wheat stem apices do not respond to photoperiod until their low-temperature vernalisation requirement has been met, a process that acts to delay the transition to reproductive development until after the onset of winter, thereby increasing the number of leaves initiated per stem. This tends to protect the more vulnerable reproductive apex against frost damage, as there is a close relationship between apical development and frost tolerance (Prášil *et al.* 2004). The model of Brooking (1996) and Robertson *et al.* (1996), which draws on data from a range of earlier experiments, indicates that, for varieties requiring vernalisation, the rate of saturation of the requirement rises linearly with increasing temperature to an optimum around 11°C, declining to zero at 18°C; the corresponding times to saturation are 70 days at 0°C and 40 days at 11°C. There is, therefore, a complex interaction between leaf initiation and vernalisation, two processes which are effectively in opposition: up to 11°C, increase in temperature promotes leaf initiation (Figures 2.2, 2.7) but also promotes the saturation of vernalisation (and the cessation of leaf initiation when the requirement is satisfied). At higher temperatures, causing devernalisation, leaf initiation becomes increasingly dominant. Subsequent measurements of plants of cv. Winter Batten under controlled conditions have broadly supported the conclusions of this model, indicating a peak rate of vernalisation at 8°C, falling to zero at 16°C under 16-h days, but a relatively constant rate between 8°C and 22°C in short days (Brooking and Jamieson 2002). In summary, the processes of apical development, vernalisation and frost tolerance are associated in a complex manner, and it is likely that they will not be thoroughly unravelled until the molecular basis of each is understood.

Thus, plants sown in the autumn, and prevented from responding to photoperiod until the vernalisation requirement is accumulated (40–70 days of low temperature, or longer if higher temperatures result in devernalisation), initiate more leaves than later sowings before reproductive development starts. For example, plants from the September sowing in Table 2.3 generated 14 leaves per mainstem compared with 10 for the December to March sowings. This mechanism alone led to the effective convergence of the double ridge stage (Table 2.3; albeit an inferior index of the transition to flowering) for the sowings from late October to January. Only in the case of the late April sowing, exposed to generally higher temperatures, was vernalisation ineffective, resulting in the initiation of more leaves (four more than for the March sowing), and irregular flowering.

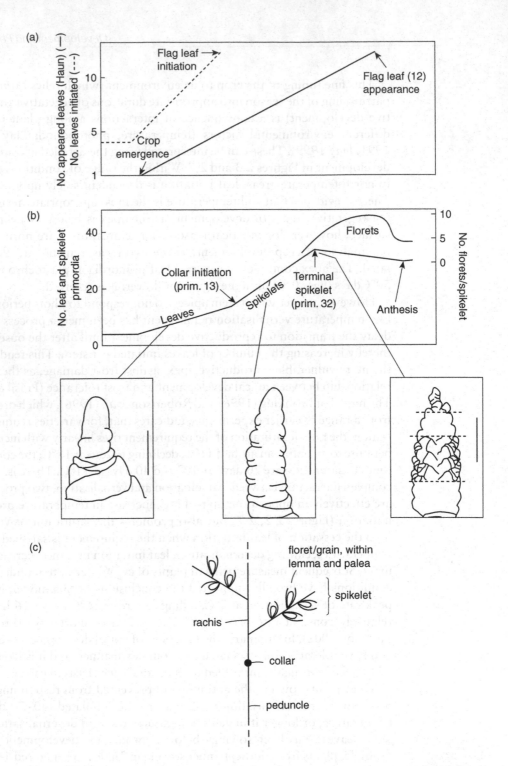

Figure 2.7 Schematic diagram of the thermal time course of development of a model wheat plant (12 mainstem leaves, 19 spikelets per mainstem, 4 grains per spikelet) including relationships between events: (a) mainstem leaf initiation and appearance. (b) mainstem leaf, spikelet and floret initiation and death (note the change in scale for leaf initiation from (a)). The three apex stages illustrated are vegetative (bar 0.25 mm), 'double ridge' (here around 50% spikelet initiation; bar 0.5 mm), and terminal spikelet (bar 1 mm). (c) the first two spikelets of an ear, with four and three viable florets (adapted from Hay and Ellis 1998).

The normal shortening of the plastochron, around the time of initiation of the collar (Figure 2.7), and the more rapid initiation of reproductive phytomers, is intensified by delay in sowing (i.e. by longer photoperiod after midwinter). However, as the increase in rate of initiation of spikelets is associated with a shortening of the total duration of initiation (Stern and Kirby 1979), these effects tend to be mutually compensating so that similar numbers of spikelets per ear are produced for different sowing dates. In the crops shown in Table 2.3, the time from double ridge to terminal spikelet fell from 64 days (September sowing) to 15 days (March). Meanwhile, the mainstem phyllochron also shortened with delay in sowing, such that the appearance of the flag leaf occurred within a period of 4 weeks for all the crops, and further convergence under the influence of photoperiod resulted in the near-synchrony of anthesis. Here the integration of developmental responses to the environment ensured that the flag leaf always preceded the emergence of the ear: generation of the source of assimilates (the leaf canopy) was coordinated with generation of the sink (the ear).

These responses are only part of the complex system of response to the environment, which controls the coordination of the wheat plant (Hay and Kirby 1991). For example, the synchrony of emergence of mainstem and tiller ears involves a parallel acceleration of tiller development with delay in sowing. As explained in Chapters 3, 6 and 7, management can have other effects on the development of cereal crops, notably the effect of plant population density on numbers and dimensions of tillers and ears; and the influence of nitrogen fertilisation on the dimensions and survival of organs (leaves, tillers, spikelets, florets).

2.2.2   Crop improvement and the anthesis–silking interval in maize

The external nature of the pollination of maize crops, which contrasts with that of small grains, has several implications for crop growth and yield. In particular, it facilitates crop improvement by heterosis (the use of hybrid varieties) since the tassels are readily excised in the field; and the development of the ear within, rather than at the top of, the canopy has an influence on assimilate partitioning (Chapter 6) and on crop protection. However, anthesis and silking occur at a time when the maize plant, having just completed an extensive series of internode extensions; carrying its peak leaf area; and with resources being diverted away from the root system; is at its most vulnerable to stress, particularly drought. The relative timing of anthesis and silking (expressed as the anthesis–silking interval, the ASI: the date when 50% of plants have visible silks minus the date when 50% of plants first extrude anthers; Edmeades *et al.* 2000; Figure 2.5) can be affected by water and shading stresses, normally causing delay in silk emergence; pollen shedding can therefore be over before the silks become receptive (e.g. an ASI longer than 10 days), resulting in barren spikelets or whole ears. In several investigations, addition of viable pollen has shown that late-emerging silks tend to be intrinsically sterile, and the corresponding florets barren (Otegui *et al.* 1995).

Many programmes of maize yield improvement involve an unconscious selection for shorter ASI, thus increasing yield stability under stress. For example, the substantial progressive increases in maize yields in the US Corn Belt between 1930 and 1980 (four- to five-fold increase; Russell 1991; Duvick 1992), associated with the introduction of hybrid varieties, were achieved by selecting plant materials which were tolerant of the shading and other stresses associated with much higher plant population densities (e.g. change from 30 000 to 60 000 plants ha^{-1}; see Uribelarrea *et al.* 2002). The associated progressive decrease in ASI (e.g. from 4.1 days for pre-1930s open-pollinated varieties to 1.2–2.0 days for 1970–80s hybrids; Russell 1991) has contributed to increased yield potential by reducing stem barrenness (rather than promoting prolificacy in number of ears per stem), and increasing the number of grains per ear. These effects are considered in more detail in Section 6.5.

An even clearer example is provided by the programme of improvement of tropical maize in Mexico by recurrent selection of the population Tuxpeño Sequía for drought tolerance, using an index that selected for a balance between drought resistance characters and maintenance of yield under well-watered conditions. After eight cycles of selection, the ASI under drought had fallen from over 30 to 10 days, although the corresponding change was 4 to 1 days under well-watered conditions. Thus the associated marked increases in grain yield, a consequence of reduced barrenness of stems and increased floret fertility (Figure 2.8), resulted from changes in crop phenology (Section 6.5).

2.2.3 Adaptation of soybean to different latitudes: phasic analysis of the photoperiodic control of flowering

Probably originating in N. China around 40°N, soybean has a greater range than any other grain legume crop, being grown at latitudes from 0 to 50°N in northern latitudes, under rainfed and irrigated cropping. Clearly, there are soybean phenologies to fit environments with a very wide range of photoperiods. The genetics of photoperiodic response in soybean are complex, involving at least six loci (E_1 to E_5, E_7) (Upadhyay *et al.* 1994a, b; Summerfield *et al.* 1998; Cober *et al.* 2001), whose chromosomal locations have been determined, and isolines have been generated for research purposes. The dominant alleles delay flowering principally by increasing the duration of the floral induction phase, an effect that tends to increase progressively as additional dominant alleles are added. Investigation is complicated by interactions, including epistasis (one gene affecting the expression of another), among loci, and the fact that individual loci can be involved in several phases of development, including the pre-inductive phase. The nature of the controlling genes and their products will become clearer as the mapping of fully-characterised genes for flowering time in *Arabidopsis* onto the soybean genetic map progresses (Tasma and Shoemaker 2003).

The most fruitful approach to understanding the factors determining flowering in soybean has been by varying conditions during each phase of development, demarcated by visible growth stages, under controlled conditions. For example, Upadhyay *et al.* (1994a, b) found that, at 25°C, isolines of the variety Clark that varied at loci E_1, E_2 or E_3 had induction phases ranging from 5 to 12 days in short days (11.5 h) and 25 to 55 days in long days (16–18 h) (Table 2.4).

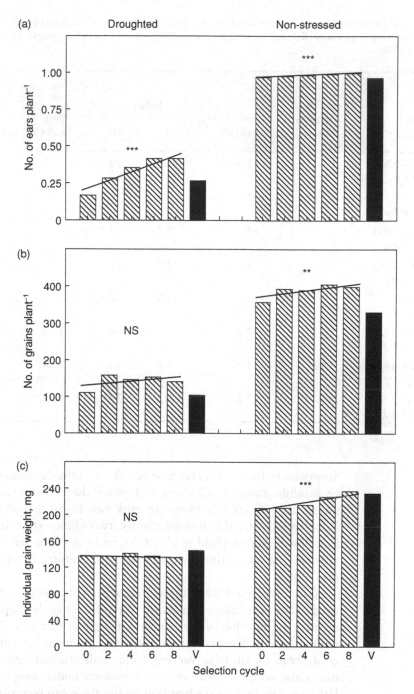

Figure 2.8 The effect of eight cycles of selection for drought tolerance on the components of yield of crops of maize (population Tuxpeño Sequía) grown under well-watered and droughted conditions in Mexico. Measurements of the control variety P21 are included for comparison (V). Effects were significant at $P < 0.01$ (**), $P < 0.001$ (***) or non-significant (NS) (from Bolaños and Edmeades 1993).

Table 2.4 Durations (days) of the phases of pre-flowering development of isolines of soybean cv. Clark with differing combinations of alleles of maturity genes, at 25°C in short (SD) or long days (LD). Data from Upadhyay *et al.* 1994b.

Isoline	Maturity class	Pre-induction	Induction		Post-induction	Sowing to first flower	
			SD	LD		SD	LD
L71-920 ($e_1e_2e_3$)	I	2.8	11.9	24.9	12.7	27.4	40.4
L63-3117 ($e_1e_2E_3$)	II	2.8	10.9	28.0	13.9	27.6	44.7
L63-2404 ($e_1E_2e_3$)	II	3.5	9.7	27.7	14.7	27.9	45.9
L80-5914 ($E_1e_2e_3$)	IV	5.4	12.0	41.0	13.7	31.1	60.1
Clark ($e_1E_2E_3$)	IV	3.7	11.5	38.7	12.6	27.8	55.0
L66-432 ($E_1e_2E_3$)	IV	7.8	9.0	46.6	15.9	32.7	70.3
L74-441 ($E_1E_2e_3$)	IV	11.7	4.8	43.7	17.4	33.9	72.8
L65-3366 ($E_1E_2E_3$)	V	11.2	5.5	54.9	18.2	34.9	84.3

Surprisingly, there were also genetic effects on the duration of the pre-induction or juvenile (range 3–12 days) and post-induction (induction to first flower) phases (13–18 days). Subsequent work has shown that alleles at the first three loci also influence the durations of the two phases after the appearance of the first flower (Summerfield *et al.* 1998); and that there are marked temperature/photoperiod interactions, with higher temperatures delaying maturity (Cober *et al.* 2001).

As for other grain crops, therefore, soybean phenology is determined primarily by the date of initiation/induction of reproductive development, which largely dictates the duration of the crop and its date of maturity (Table 2.4). Selection within this rich genetic background in the Americas has led to the introduction of a series of varieties varying from: indeterminate photoperiod-insensitive materials, suited to northern environments under long photoperiods, where late frost can damage the first flowers but these can be replaced at higher nodes (e.g. southern Canada and the northern US states); to determinate materials with alleles prolonging the inductive phase over several weeks (late varieties), for southern states and in the sub-tropics. Here a balance must be struck between delay in flower induction to ensure that the leaf area index of the crop (Section 3.3) is adequate to optimise the capture of solar energy; and a phenology that matches the available growing season, ensuring grain filling and crop maturity.

There is also a need for indeterminate short-duration varieties in environments subject to water stress. The classification of these responses to the environment, based on a network of agronomic trials, has resulted in the system of maturity groups which guides soybean growing throughout North and Central America (Figure 2.9).

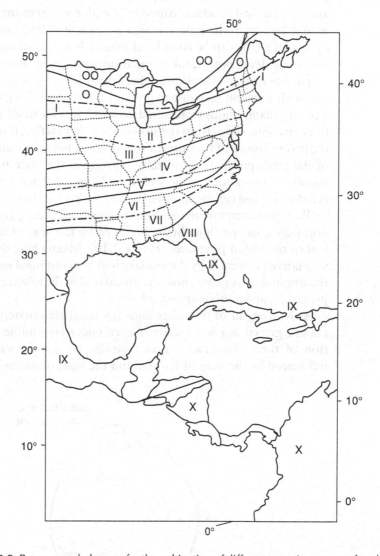

Figure 2.9 Recommended zones for the cultivation of different maturity groups of soybean varieties in North and Central America. In the northern states of the USA and Canada, the use of varieties which respond weakly to photoperiod (Groups 00 to I, with recessive genes; see Table 2.4) is appropriate for short growing seasons, limited by low temperature. Varieties with greater response to photoperiod would not mature within the growing season. At lower latitudes, where the growing season is less constrained by low temperature but photoperiod is progressively shorter, the inclusion of dominant genes extends the time to flowering; this increases the potential yield of the crop, if other resources such as water are not limiting (from Evans 1993).

2.2.4   Development in storage: physiological age and tuber initiation in the potato

The case histories considered so far have involved crops whose yield depends upon reproductive development. Some, but not all, potato varieties flower and set seed, but the resources partitioned to sexual reproduction are normally negligible in the economy of the crop; vegetative tubers, borne on underground stems, are the principal product. Among the major world crops of temperate zones, a unique feature of the potato is that a substantial fraction of the developmental cycle of the crop can be completed in store before planting, permitting the times of canopy development, tuber initiation and harvest to be manipulated.

The life history of a potato tuber begins when an underground, diageotropic (i.e. with a neutral response to gravity) stolon, branching from a normal stem of a potato plant, begins to accumulate starch in a localised swelling at about eight to twelve internodes from the stolon apex. In practice, it is difficult to determine the precise timing of the onset of tuberisation, but once initiated, the development of the tuber proceeds acropetally: starting at the node furthest from the apex, successive internodes of the stolon bud expand longitudinally and radially, by cell division and cell expansion. Each of the eyes of the mature tuber (the original axillary meristems of the stolon, arranged according to a spiral phyllotaxis), normally contains three buds, bounded by the scar of the original (potential) leaf of the stolon phytomer (Figure 2.10). Meanwhile, the initiation of further vegetative phytomers by the axillary bud has continued so that, at tuber harvest, the original stolon apex, now the apical bud of the tuber, contains at least 12 leaf primordia and partly-expanded leaves.

At harvest, all of the tuber buds (sprouts) are innately dormant and cannot resume growth for several weeks even under favourable conditions. The duration of tuber dormancy varies considerably among varieties, and it is also influenced by the date of harvest and the state of maturity of the tubers when

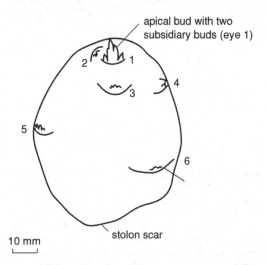

Figure 2.10 A sprouting seed potato tuber cv. Pentland Javelin, showing apical dominance.

they enter storage. Once dormancy has been broken, the rate and pattern of sprout development are determined by the temperature and duration of storage. A short period of warm storage tends to favour apical dominance, giving seed tubers with a single long apical sprout, whereas prolonged cool storage favours the slower development of sprouts from several eyes (although the response varies among varieties; Hay and Hampson 1991).

Sprout development in storage is a component of the annual life cycle of the crop; in extreme cases, tubers stored for prolonged periods can actually produce sprouts with stolon branches and a new cycle of tuber initiation without being planted (a condition termed 'little potato'). In common with other indices of plant development (e.g. plastochrons, phyllochrons), the developmental stage or 'physiological age' of seed tubers at planting can be quantified in terms of the thermal time (degree days) above a threshold (normally taken to be 4°C), which the stored tuber has experienced since the breaking of dormancy. Thus, in an extensive series of investigations by O'Brien *et al.* (1983), linear relationships were established between sprout extension (and, in particular, the length of the longest sprout on the tuber) and accumulated temperature above 4°C.

Physiologically-older tubers tend to produce an earlier (albeit less extensive) leaf canopy, earlier tuber initiation but also earlier crop senescence; yield potential is, therefore, sacrificed in favour of speed to market (Figure 2.11). This pattern of development is exploited in the cultivation of (high-value) early potato crops: the tubers of varieties whose development is intrinsically rapid ('early' varieties which initiate tubers at an early stage of canopy generation) can be preconditioned in storage to give a range of physiological ages, thus spreading the dates of tuber initiation and harvest over the growing season (e.g. Figure 2.12). This effect can also be illustrated by a comparison of the contrasting phenologies, and yield potentials, of early- and main-crops of potatoes in the

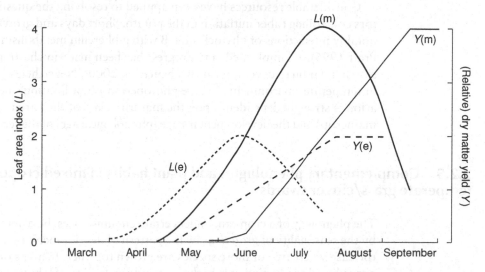

Figure 2.11 Model time courses of canopy leaf area (see Section 3.3 for explanation of leaf area index) and dry matter production for early (– – –) and main (——) crops of potatoes in a north-temperate environment.

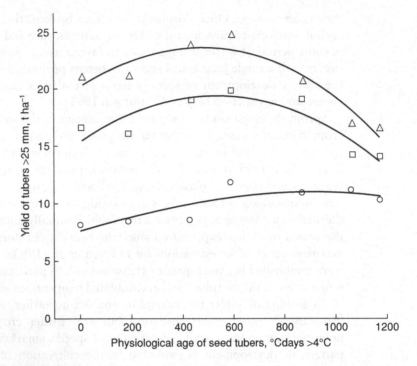

Figure 2.12 The influence of the physiological age of the planted seed tubers on the fresh weight yield of potato tubers (>25 mm) at three harvest dates (o 15 June; □ 30 June; △ 18 July). Data are from early crops of cv. Home Guard grown in Pembrokeshire, Wales, in 1977 (from O'Brien *et al.* 1983).

UK (Figure 2.11). The details of leaf and canopy development in the potato are considered in more detail in Chapter 3.

Considerable resources have been applied to resolving the question of the factors controlling tuber initiation in the potato. Short days and growth substances, notably interactions of phytochrome B with gibberellin metabolism (Jackson and Pratt 1996), are implicated, and progress has been made in charting gene expression in relation to development (Bachem *et al.* 2000). Nevertheless, it is clear that in temperate environments, where photoperiod is not limiting, the time of initiation is strongly dependent upon the maturity class of the variety grown (early, maincrop) and the developmental stage (physiological age) of the seed tubers.

2.2.5   Complementary phenologies and plant habits in mixed cropping: temperate grass/clover swards

The phenology of a crop can, under certain circumstances, be determined as much by the seasonality of *growth* as by the pattern of development; this is particularly the case where two or more species are grown together. In areas of less intensive agricultural production, where the cost and availability of labour are not limiting, mixed cropping is common. For example, in the highly productive traditional agriculture of the New Guinea Highlands (Pimental and Pimental 1979), tubers and fruit are harvested continuously from a range of species grown on the same plot, thus ensuring that bare soil is not exposed to the erosive action of heavy rain. Complementary phenologies and canopy structures, and active management, ensure that each component of the system has appropriate supplies of solar radiation and inorganic nutrients. There is a vast array of other systems ranging from the intercropping of *Phaseolus* beans and maize (the beans raising the nitrogen status of the crop; the maize providing a living beanpole, ensuring the supply of radiation to bean leaves) to agroforestry (for example, combination of grazed pasture with timber production).

Temperate pastures provide the leading example of mixed cropping in intensive temperate agriculture, and there has been a renewed interest in highly productive grass–clover swards in the pursuit of more sustainable systems and safer food (reduced application of fertiliser nitrogen; less reliance on animal protein in ruminant diets; increased digestibility of fodder; e.g. Wachendorf *et al.* 2001a). The productivity and quality of grass–clover swards depend upon the management of the phenologies and habits of the two components. The following analysis considers a sward cut at heading (flowering of the grass component) followed by two or more cuts, depending upon the level of soil fertility; the patterns of growth and development are broadly similar for grazed swards, although the influence of shading tends to be less.

In the competitive relationship between perennial ryegrass (*Lolium perenne*) and white clover (*Trifolium repens*), the grass component has several advantages over the legume in spring and early summer. More rapid leaf expansion at low temperatures, an erectophile canopy (Section 3.4.3), and a high potential for branching (tillering) mean that, early in the season, the grass component predominates in the dry matter production of the sward (Figure 2.13), intercepting most of the incoming solar radiation. As explained in Section 6.10, this effect is enhanced by the higher susceptibility of clover plants to winter stresses, so that they tend to start the new season with little leaf area; and by their very prostrate habit, with stem (stolon) extension restricted to the soil surface.

However, the early peak in growth of the grass component, culminating in ear emergence of reproductive tillers, is followed by a trough in production, typically in June in Northern Europe (Figure 6.21). This depression, following the first harvest, occurs because of the death of young vegetative tillers, developing in deep shade at the base of the reproductive canopy. Furthermore, because the leaves of the surviving vegetative tillers, exposed after the first cut, have developed in shade, they have a low photosynthetic potential (Wolege 1977).

This midsummer depression in grass production, a consequence of both growth and developmental effects, shifts the competitive balance towards the clover component, assuming that water supply is not limiting. White clover is intrinsically tolerant of shade, diverting resources from leaf production to the extension of petioles (Thompson 1993). This response permits the clover component to *survive* during the early peak of grass growth by raising leaves into the upper parts of the canopy. However, with the removal of the grass canopy at the first harvest, the intact clover stolons can rapidly generate a planophile canopy (Section 3.4.3), of leaves carried on shorter petioles, which tends to dominate radiation interception in areas of the sward. The peak of dry matter production

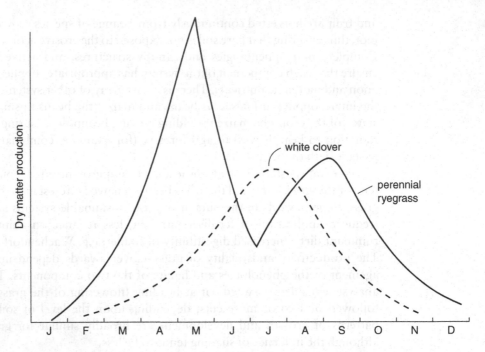

Figure 2.13 Model time courses of dry matter production by the components of a temperate grass–clover sward harvested by cutting two or three times per season.

by the legume, therefore, coincides with the trough in grass production (Figure 2.13). The ramification of clover stolons below cutting height also facilitates the colonisation and exploitation of gaps in the sward.

Later in the season, production of the grass component recovers by the proliferation of new vegetative tillers with (less shaded) leaves of higher potential. However, there is normally a marked seasonal asymmetry in production because of the more rapid rates of photosynthesis of reproductive tillers in spring (vernalised by low temperature) than of vegetative tillers in late summer and autumn (Parsons and Robson 1981a, b). The contribution of clover to sward dry matter production during the second peak of grass growth will depend upon management, particularly nitrogen fertilisation; under low fertility, the clover plants, securing independent supplies of nitrogen by fixation, will compete more effectively with the, generally shorter, grass sward. The challenge to the grass farmer is to manage the sward, including crop protection, to ensure that the clover component is present to exploit the depression in grass production (Hay and Walker 1989).

The tools described in this chapter are used throughout the remainder of this book in the interpretation of the physiology of crop yield. They are also widely applied in the simulation of crop development, growth and yield (Chapter 9) and in prediction of the effects of global climate change (Chapter 10). For example, it should be clear from the analysis of crop development presented here that one important effect of global warming will be to disrupt the interactions between the effects of temperature and photoperiod, with important implications for the phenologies of crop and wild plants.

Chapter 3

Interception of solar radiation by the canopy

> It is probable ... that the development of successful systems of crop husbandry largely depends on discovering empirically the optimum conditions for leaf area production in a particular environment.
>
> (Watson 1952)

Most of the solar radiation absorbed by a crop canopy is intercepted by its leaf blades, although leaf sheaths, petioles, stems and reproductive structures can make important contributions. Because of this, it has become customary to express the capacity of a crop to intercept solar radiation by its leaf area index (L) – the area of leaf lamina (one surface only) per unit of soil surface area. The more precise measurement, green area index (GAI) is also employed, to emphasise that interception is useful only if the leaf area is photosynthetically active. The fact that L is dimensionless (e.g. m^2 of leaf area m^{-2} of land area) has suggested to some that it should be thought of as the number of layers of leaves in a crop. Whichever definition is used, L is a property of the crop rather than of the individual plants, but full understanding of the genetic, environmental and management factors controlling L can come only from consideration of the development, growth and senescence of individual leaves and plants.

3.1 The life history of a leaf

Leaves begin life as a regular series of primordia, localised outgrowths on the sides of the apical dome of a vegetative shoot (Section 2.1.2; Figures 2.7, 3.1 and 3.2). The genetic and environmental interactions controlling the relative positions of these primordia (which also determine the three-dimensional arrangement of the fully developed leaves on the stem, the phyllotaxis), and the thermal time interval between the initiation of successive primordia (the plastochron; Section 2.1.2), are complex, and are only now beginning to be disentangled using the tools of molecular genetics (Lyndon 1994; Sinha 1999; Golz and Hudson 2002). For example, anatomical studies of gene expression in the apical

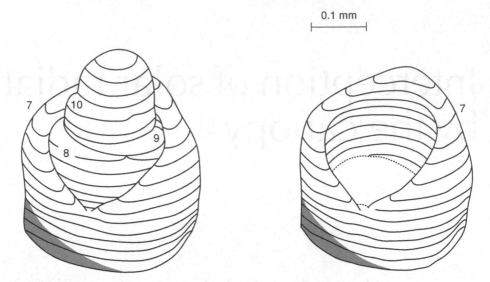

0.1 mm

Figure 3.1 Three-dimensional reconstruction of the vegetative stem apex of a wheat plant which has just initiated the tenth leaf. The first six fully- and partly-developed leaves have been removed to reveal the expanding blade and sheath of leaf 7. The horizontal lines, which have no structural significance, indicate spacings of 20 μm (from Williams 1975).

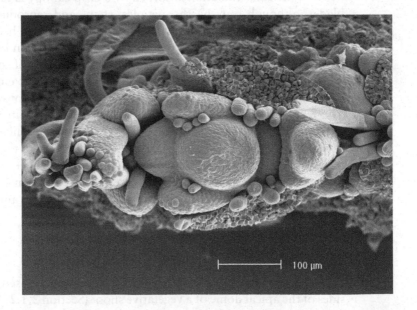

100 μm

Figure 3.2 Scanning electron micrograph of the apex of a soybean (cv. A01-2 Williams) mainstem at the stage (V2) when the first trifolate leaf was fully expanded. To the left of the apical dome, the primordium of the youngest leaf is flanked by the primordia of two stipules. The preceding leaf (to the right, flanked by broken surfaces of the two removed stipules) has just begun to develop as a trifoliate structure. The next leaf, to the left of the youngest, is clearly trifoliate and trichomes are forming on its abaxial surface. All older tissues have been removed for clarity. (Unpublished micrograph kindly provided by Professor Darleen DeMason.)

meristem of maize have shown that the *terminal ear 1* gene plays a primary role in determining the position and timing of leaf primordium initiation (Veit *et al.* 1998); auxin carrier proteins play a major part in determining phyllotaxis in *Arabidopsis* (Reinhardt *et al.* 2003); and the expression of *expansin* genes in the founder cells of the primordium leads to loosening of the cell walls such that lateral cell expansion can proceed (Fleming *et al.* 1999; Cosgrove 2000). Leaf initiation begins in the developing seed when it is still attached to the parent plant; for example, at least three (wheat) or five (maize) leaf primordia are already present in the embryos of cereal grains at sowing. In contrast, because the sprouts of seed potatoes can contain 20 or more primordia and partly developed leaves, much of the development of a potato crop leaf canopy after planting consists of the expansion of already initiated leaves (Section 2.2.4).

After initiation, the developing leaf enters a phase dominated by cell division which, in dicotyledons, continues at least up to leaf unfolding (5% of final blade area; Dennett *et al.* 1978). During this phase, an increasing proportion of the cells of the leaf do not take part in division because of differentiation, for example into vascular tissues. In spite of this, detailed observations, mainly under controlled conditions (e.g. Figures 3.3, 6.3), have shown that the rates of increase of cell number, leaf length, area and fresh weight are approximately exponential. Leaf development in the fluctuating environment of the field will be less regular, unless expressed in terms of thermal time.

In dicotyledons, leaf unfolding tends to be associated with a decline in cell division which, although subsequently more localised in plate and marginal meristems (e.g. Figure 3.4), does persist up to a late stage of leaf expansion, when the leaf blade has achieved 30–50% of its final size. The unique shapes of the leaves of each species are determined by the position and activity of these meristems, under the control of a suite of genes (Sinha 1999). Because of the very large numbers of cells involved, these relatively localised areas of cell division produce a high proportion of the final number of cells in the mature leaf (Dale 1982). Nevertheless, the great increase in leaf area during laminar expansion (Figures 3.3, 6.3) is caused primarily by cell expansion (10- to 35-fold increase in mean cell volume).

The pattern of leaf expansion in cereals and grasses is rather different, generating simpler, strap-like, laminae. The phase during which most of the leaf cells are involved in division is short, and cell division, thereafter, is restricted to a basal intercalary meristem. The founder cells of these meristems can be identified at a very early stage in the life of each leaf (e.g. in maize, Freeling 1992), and they are situated below, or near to, the soil surface (except for the later leaves, whose intercalary meristems are carried upwards as a consequence of stem internode extension following the onset of reproductive development; see Sections 2.1.3, 2.1.4; Figures 2.1, 2.2, 2.4). Cell expansion, predominantly in one dimension, begins at an early stage, with the result that, when the tip of a cereal or grass leaf emerges from the encircling sheath of the previous leaf (leaf appearance; Section 2.1.2, Figures 2.1–2.4), its cells are already fully expanded. Mature, fully functional leaf tissue is, thus, pushed up into view by cell division in the intercalary meristems and expansion in the leaf extension zone, whose length, in wheat, can be 35–60% of the length of the encircling sheath (Kemp 1980; Muller *et al.*

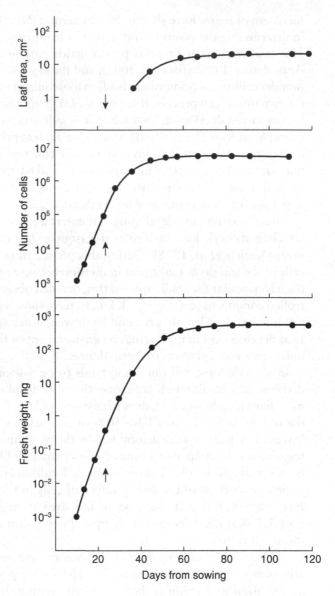

Figure 3.3 Time courses of fresh weight, cell number and laminar area of one of the second pair of leaves of a sunflower plant growing under constant controlled conditions. The arrows indicate the time of leaf unfolding. Note that cell division is completed before the achievement of maximum fresh weight and leaf area (adapted from Sunderland 1960).

2001). The timing of initiation of the cells that will form the ligule, and their passage through the leaf extension zone, appear to be determined either at a very early stage in the primordium (e.g. in maize, Freeling 1992) or around the time of leaf tip appearance (e.g. in pasture grasses, Skinner and Nelson 1994). The relative rate of increase in the laminar area of a leaf tends to decrease after tip appearance, but this later phase of expansion can account for as much as 80% of the final laminar area (Figure 3.5; note the logarithmic scale).

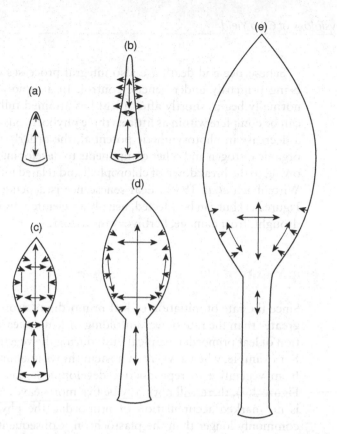

Figure 3.4 The sites and directions of expansion at different stages ((a) to (e)) in the development of a model dicotyledon leaf. The final shapes of the lamina and petiole are determined by localised cell division and expansion in the directions indicated by the arrows (from Dale and Milthorpe 1983).

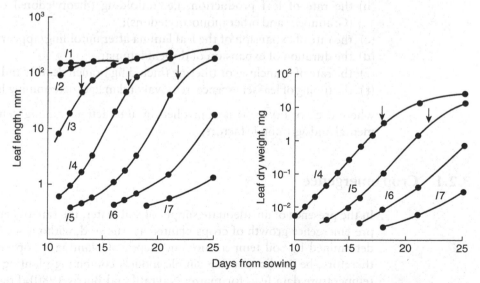

Figure 3.5 The length (stem apex to leaf tip) and dry weight of successive leaves (1 to 7) of a wheat plant grown under constant controlled conditions. The arrows indicate the time of leaf appearance (leaf tip visible). Note the pronounced decline in the relative rate of leaf expansion after appearance (adapted from Williams and Rijven 1965).

Senescence and death are also integral processes in the life history of a leaf, being primarily under genetic control. In a rapidly growing crop, senescence normally begins shortly after a leaf has attained full size (e.g. Figure 3.13) and can be complete within as little as three phyllochrons (e.g. Figure 2.1). It involves a decrease in photosynthetic potential, the orderly mobilisation and export of organic nitrogen and other components to developing leaves, and a loss of colour owing to the breakdown of chlorophyll and related ultrastructure (e.g. in soybean; Wittenbach *et al.* 1980). Leaf senescence is dependent upon thermal time (e.g. Figure 2.1) but can be affected, usually accelerated, by shading, nutrient deficiency, drought, frost damage, herbivory or disease.

3.2   The components of plant leaf area expansion

Since the rate of initiation of leaf primordia of crop plants is almost invariably greater than the rate of leaf unfolding or leaf appearance, there is an accumulation of leaf primordia in apical buds during the vegetative phase of development. For example, when a wheat mainstem shows unequivocal signs of the transition from vegetative to reproductive development (i.e. at the double ridge stage; Figure 2.7), there will normally be five more leaves to appear. Even where there is no marked accumulation of primordia, the phyllochron (Section 2.1.2) is commonly longer than the plastochron. Consequently, the rate of initiation of leaf primordia is not normally a factor in determining the leaf area of a plant.

In growing crops, therefore, the leaf area of a plant at a given time is determined by:

(a) the date of plant emergence;
(b) the rate of leaf production, i.e. unfolding (dicotyledons) or appearance (Gramineae and other monocotyledons);
(c) the rate of expansion of the leaf lamina after unfolding/appearance;
(d) the duration of expansion of the leaf lamina;
(e) the rate of branching or tillering (including branch death); and
(f) the timing of leaf senescence, removal or damage (determining leaf longevity);

where each of (a) to (f) is controlled by interactions among genetic, environmental and agronomic factors.

3.2.1   Crop emergence

In the presence of an adequate supply of soil water, the rate of germination and pre-emergence growth of crops planted as true seed, without seed dormancy, is determined by soil temperature and depth of planting. Crop emergence can, therefore, be predicted using simple models combining planting date and soil temperature data (e.g. for maize; Navratil and Burris 1980). Prediction can be more difficult for vegetatively propagated crops, for example the potato, where the developmental age of the sprouts of the seed tuber (Section 2.2.4) can have a significant effect on time to crop emergence (Firman *et al.* 1992).

3.2.2 Leaf production

The rate of leaf production by the stems of crop plants depends primarily on the temperature of the expanding leaves, strictly the zones of cell expansion. For vegetative cereal plants before stem internode extension, the relevant temperature is that of the surrounding soil or the air near the soil surface (e.g. Jamieson *et al.* 1995a; Vinocur and Ritchie 2001; although there is conflicting evidence: McMaster *et al.* 2003). In studies of dicotyledonous and reproductive monocotyledonous plants, it is common to use the temperature of the air at the appropriate position in the canopy, although, under certain conditions, there can be substantial gradients in temperature between leaf and surrounding air (Fitter and Hay 2002).

The linearity or near-linearity of the temperature response curve for leaf production below the optimum temperature (Section 2.1) and, under many circumstances, the constancy of the response (number of leaves appearing or unfolding per unit of thermal time) throughout leaf production (cereals: Figures 2.1b, 2.2b, 2.4b; potatoes: Firman *et al.* 1995), mean that the process can be characterised by a single phyllochron for the mainstems of the crop (Table 3.1). In crops where a significant proportion of the canopy is generated by branches, corresponding phyllochrons for each node can be evaluated; for example, in barley, phyllochron increases with the node of insertion of the tiller (Kirby *et al.* 1985). Where the constancy of stem phyllochron has been observed to fail, with a new but constant value established (e.g. in wheat, Hay and Delécolle 1989; Calderini *et al.* 1996), this may reflect a readjustment to ensure the coordination of canopy and ear development (Hay and Kirby 1991).

Table 3.1 Typical leaf appearance/unfolding characteristics for mainstems of a range of crop species.

Species	Phyllochron (°Cdays)	Base temperature (°C)	Source
Wheat (tip)	70–160	0	Hay and Kirby 1991
Maize (tip)	25–55	8	Birch *et al.* 1998; Vinocur and Ritchie 2001
Leek (tip)	135	0	Hay and Kemp 1992
Potato	28–31	0	Kirk and Marshall 1992; Vos and Biemond 1992
Soybean	39–79	5	Patterson 1992
Pea	46	2	Truong and Duthion 1993
Brassica spp.	45–50	5	Morrison and McVetty 1991; Nanda *et al.* 1995
Sunflower	20–25	4	Villalobos and Ritchie 1992
Sugar beet	27–31 (early) 34–68 (late)	1	Milford *et al.* 1985a

A constant mainstem phyllochron implies that the response to temperature is determined early in the life of the crop, probably around the time of crop emergence, and that it is unaffected by subsequent variation in the environment. Studies of temperate cereals have shown that, although the magnitude of the phyllochron is highly heritable (Mosaad *et al.* 1995; Dofing 1999), it can be profoundly affected by sowing date (McMaster 1997), but not normally by other management or environmental factors, in the absence of stress (see below). For winter wheat, the rate of change of photoperiod at crop emergence, which increases progressively from autumn to spring (Baker *et al.* 1980), remains the best available predictor of phyllochron; it is used widely in simulation models (Bindi *et al.* 1995; McMaster and Wilhelm 1995; Chapter 9), even though experimental approaches have not supported the involvement of photoperiod (e.g. Hotsonyame and Hunt 1997; Slafer and Rawson 1997). Among others, Jame *et al.* (1998, 1999) have taken a different approach, interpreting the slight curvilinearity of the temperature response curve of leaf appearance as evidence of the continuing influence of photoperiod during canopy development. In summary, the strong influence of sowing date on phyllochron in maize and wheat can be described adequately for the purposes of crop modelling but the underlying biology remains to be explained fully.

The unresponsiveness of the phyllochron to variation in the environment after crop emergence can be striking; for example, there was no effect of an 18-fold variation in plant population density on the rate of leaf appearance of winter wheat in an experiment in southern England (Whaley *et al.* 2000). Similarly, nutrient levels in arable and grassland soils do not generally influence leaf production. Nevertheless, under stressful conditions (e.g. plant water potential lower than -0.6 MPa, Maas and Grieve 1990; Krenzer *et al.* 1991 (wheat); nitrogen deficiency, Vos and Biemond 1992 (potato); Longnecker and Robson 1994 (wheat); phosphorus deficiency, Plénet *et al.* 2000 (maize)), slowing of the expansion of the new leaf can lead to longer phyllochrons.

The *number* of leaves produced per stem is the product of the rate and duration of the process of leaf appearance/unfolding. For crops such as sugar beet, potato and leek, which tend to remain vegetative or flower at a late stage in stem development, the process can continue uninterrupted throughout the growing season, generating as many as 40 or 50 leaves per mainstem (Milford *et al.* 1985a; Cao and Tibbitts 1995; Figure 2.1). In grain crops, the cessation of leaf initiation, and of subsequent leaf appearance or unfolding, normally occurs in response to a photoperiodic signal transmitted from the leaves to the stem apex, after it has been rendered susceptible to induction by vernalisation (effects of low temperature or short days, Section 2.1.3; Hay and Kirby 1991) or loss of juvenility (maize, Section 2.1.4; pre-induction phase of soybean, Section 2.1.5). In many cases, the transition will occur eventually, even under non-inducing conditions. Thus the flag leaf of a wheat crop can vary from leaf 6 in a spring variety under long days to leaf 18 or more in a winter variety held in short days. In indeterminate crops, leaf production tends to continue until the exhaustion of resources for leaf development, principally nitrogen (e.g. Section 2.1.5).

3.2.3   Leaf expansion

Although the relative rate of expansion of leaves decreases after appearance/
unfolding (e.g. Figures 3.3, 3.5), this later phase (which can involve 80% of the
total length, 95% of dry weight and 99% of blade area; Dale and Milthorpe
1983) determines the rate of increase of photosynthetic leaf area. For this reason,
most field studies of leaf area expansion have been restricted to the post-
unfolding/appearance phase.

For each crop species there is a distinct ontogenetic trend in leaf size but,
because the number of leaves produced by a given stem can vary with growing conditions (Section 3.2.2) and with variety, the dimensions of the leaf at a given node
are not invariable. For example, maize varieties with very contrasting phenologies,
grown under uniform conditions, produced between 16 and 30 leaves per plant,
with leaf area at a given node varying by up to 100% (Dwyer *et al.* 1992). However, when leaf sizes were normalised (largest leaf set at unity, irrespective of node
of insertion), the ontogenetic trends of the varieties converged, with the largest
leaf produced by the ear-bearing node, or one of the adjacent nodes (Figure 3.6;

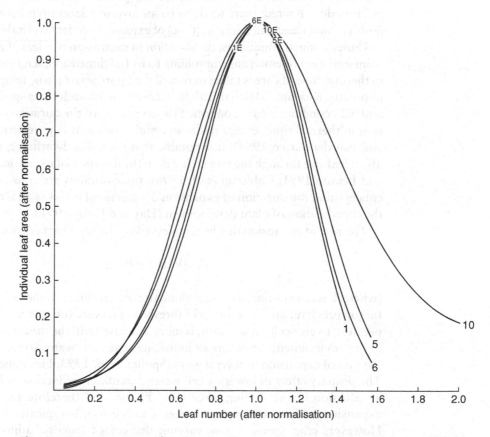

Figure 3.6 Ontogenetic trends in the area of leaves of a series of maize varieties (numbered 1,
5, 6, 10). The data were normalised for comparison among varieties by setting the area of the
largest leaf of each at unity. E indicates the position of the leaf node carrying the ear (from Dwyer
et al. 1992).

see also Figure 3.10). The progressive increase in leaf size from emergence to the ear leaf and the subsequent decrease to the flag leaf could be described by a near-symmetric bell curve in most cases. Wheat and barley have simpler ontogenetic trends, with leaf size normally increasing progressively throughout the development of a stem, apart from the last two or three leaves, particularly the flag leaf, which can be much smaller in barley (e.g. Kirby *et al.* 1982; Calderini *et al.* 1996). Several dicotyledonous crop species have trends similar to that of maize, with a peak in leaf size followed by a decline (e.g. in sunflower, as shown by the irrigated treatment in Figure 3.8; sugar beet, Milford *et al.* 1985b). There are significant differences in the ontogeny of soybean varieties, with indeterminate types showing a trend towards smaller leaves at higher nodes, and determinate types showing greater uniformity (Fehr and Caviness 1977).

For a given crop variety, therefore, the number of leaves produced per stem is an important determinant of the upper limit of leaf size at a particular node; environmental factors act to determine whether the leaf achieves its potential dimensions. Actual leaf size depends upon both the rate and duration of expansion. Cardinal temperatures (base, optimum and maximum) for the rate of leaf expansion are, naturally, similar to those described for leaf production and, since leaf size at a given node is limited, there tends to be an inverse relationship between the rate (area per unit time) and duration (time) of expansion under favourable conditions.

Under optimal conditions, the *duration* of expansion of a leaf at a given node, from initiation or emergence/unfolding to its full dimension, tends to be constant in thermal time, if care is taken to record the appropriate tissue temperature (e.g. in potato, Kirk and Marshall 1992). Measurements under a range of controlled and field conditions have confirmed the constancy of the duration of leaf expansion in thermal time, except under stressful conditions (e.g. Jefferies 1993; Vos and van der Putten 1998). It is notable that even *Rht* dwarfing genes in wheat affect leaf size through the rate rather than the duration of expansion (Appleford and Lenton 1991; Calderini *et al.* 1996); these findings are consistent with the concept that the duration of expansion of individual leaves is an integral part of the coordination of plant development (Hay and Kirby 1991).

The *rate* of expansion of a leaf, governed by the Lockhart equation:

$$1/V.\mathrm{d}V/\mathrm{d}t = \phi(P - Y) \tag{3.1}$$

[where V is cell volume, ϕ is the (volumetric) extensibility of the cell wall, P is cell turgor pressure, and Y is the yield threshold pressure (i.e. the minimum turgor pressure to give cell expansion)], is more sensitive than the *duration* to variation in the environment. Exposure of immature tissues to water stress can influence the rate of expansion in several ways (Spollen *et al.* 1993; Fitter and Hay 2002). The primary effect of lowering cell water potential is reduction in P; this causes a reduction in the driving force $(P - Y)$ for, and therefore the rate of, cell expansion; and expansion ceases when P falls below Y (typically 0.2–0.3 MPa). However, crop species show varying degrees of osmotic adjustment under drought, with P being at least partly maintained by the secretion of solutes into the cell (Section 7.1.1). Furthermore water stress can also affect cell wall extensibility or yield threshold (Cramer and Bowman 1991; Spollen *et al.* 1993).

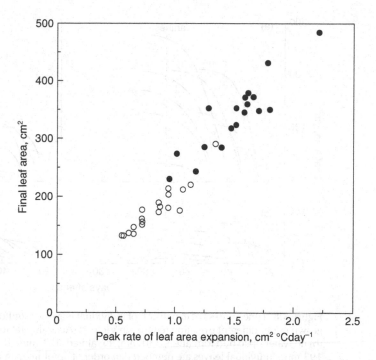

Figure 3.7 Relationships between final area and peak rate of expansion of individual leaves of potato plants grown under irrigation (●) or drought (○) in the East of Scotland. The data include measurements of 19 varieties (from Jefferies 1993).

Leaves that have expanded under water stress, therefore, tend to be smaller than the potential area at that node, owing to a reduction in the rate of expansion, but the actual dimensions will depend upon the severity of the stress and the degree of osmoregulation achieved. Thus, in potato, a species showing weak osmoregulation (approximately 0.1 MPa; Jefferies 1993), there was a clear relationship between area and rate of expansion under moderate drought (Figure 3.7).

Detailed study of sunflower, which has a much greater potential for osmotic adjustment (up to 0.6 MPa), has confirmed that the effects of water stress on the rate of leaf expansion are dominated by cell turgor, whereas duration is relatively unaffected (Takami *et al.* 1981, 1982). Figure 3.8 shows how these responses affected leaf area expansion in sunflower plants grown under varying irrigation treatments in a dry environment. For leaves at a given node, the rate but not the duration of expansion declined with increased severity of water stress up to at least leaf 15. Thus, the fully expanded area of leaf 3 fell from 69 cm^2 (fully irrigated) through 62 cm^2 (dry to flower bud formation) and 38 cm^2 (rainfed) to 31 cm^2 (no irrigation before anthesis). The corresponding values for leaf 12 were 207, 200, 62 and 54 cm^2. However, as the degree of water stress intensified, the duration of leaf expansion also began to be reduced as cell turgor fell below the yield threshold for expansion during an increasing fraction of each day (Fitter and Hay 2002).

Apart from sowing date and, where appropriate, irrigation, the management factor with the most influence upon leaf size is nitrogen fertilisation. The

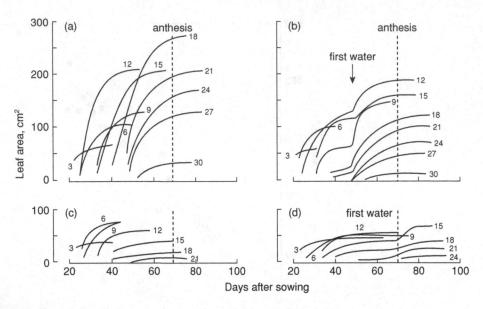

Figure 3.8 Time courses of expansion of individual leaves of sunflower cv. Sungold plants grown under different irrigation regimes in Victoria, Australia: (a) irrigated weekly, 450 mm; (b) no water up to bud formation, 330 mm; (c) rainfed, 121 mm; (d) no water up to anthesis, 193 mm. Individual leaves are numbered in order of unfolding, and the curve for leaf 12 is displaced, for clarity, in (a) and (b) (from Connor and Jones 1985).

influence of nitrogen supply in increasing leaf dimensions, which is most marked when relieving nitrogen deficiency, cannot be interpreted in terms of the Lockhart equation because cell division can be stimulated, generating a higher population of expanding cells per leaf (see the classic study of sugar beet by Morton and Watson 1948). Figure 3.9 demonstrates the incremental influence of nitrogen supply on individual leaf area of potato plants whose fertility was closely controlled in pot culture, indicating saturation at the highest level of application for leaf 10. The upper part of Figure 3.10 shows the interaction of nitrogen application and water supply in determining individual leaf area in a field-grown maize crop, where both factors affected cell turgor (Wolfe *et al.* 1988a). As in other field crops (e.g. wheat; Hotsonyama and Hunt 1998), nitrogen supply did not affect the dimensions of the first few (here six) leaves, showing that, as potential leaf size increased (Figure 3.6), the supply of nitrogen by the less fertilised soil was unable to meet the need of the expanding organ.

The primary effect of increasing the plant population density of a crop is to increase competition between adjacent plants, although the impact can be modified by variations in branching (Section 3.2.4). The resulting shading of plant tissues, involving alterations in the quantity and spectral composition of radiation incident upon shaded leaves, has a profound influence upon the balance of plant growth regulators. In general, there are increases in tissue levels of gibberellins, mediated by phytochrome, whose overall effects are the promotion of leaf expansion and the acceleration of all developmental processes. Thus, closer spacing of cereal plants is associated with larger and more rapidly expanding leaf

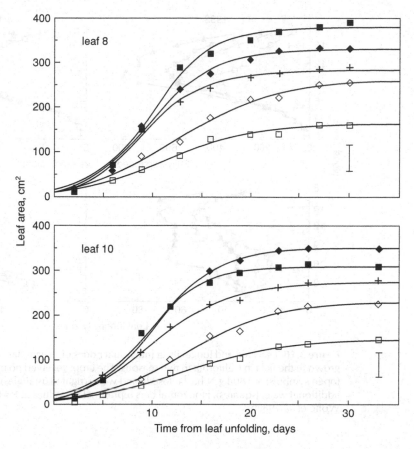

Figure 3.9 Time courses of expansion of leaves 8 and 10 of potato plants cv. Vebeca grown in pots in a glasshouse at five levels of nitogen supply (□ 250; ◇ 500; + 1000; ◆ 1500; ■ 2000 mg N plant^{-1} week^{-1}). Vertical bars represent LSD for treatment means at $P = 0.05$ (from Vos and van der Putten 1998).

canopies (Figures 3.18, 3.19), although later leaves tend to be smaller (e.g. in wheat, Whaley *et al.* 2000; in maize, Edmeades and Daynard 1979a). Again, these effects are predominantly the consequence of changes in the rate of expansion of individual leaves. The agronomic significance of the related promotion of leaf expansion in pasture grasses under cool temperatures and very long days is still to be explored in detail (Hay 1990).

3.2.4 Branching

The modular structure of crop plants allows for the development of secondary and higher-order stems at each node, thereby increasing the potential of the individual plant to generate leaf area. The extent of branching, which is largely determined by management (choice of species or variety; plant population density; and level of nitrogen fertiliser application), ranges from the normally uniculm maize and leek to pasture grasses, whose prolific tillering is stimulated

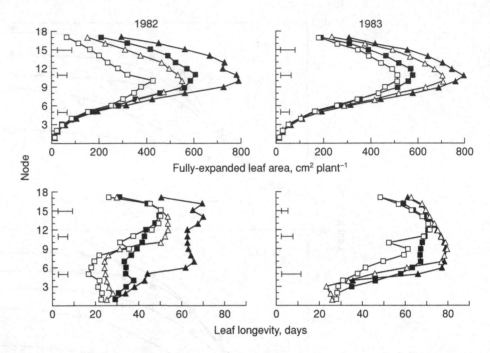

Figure 3.10 Leaf area and longevity at individual nodes of maize plants cv. Dekalb XL25A grown in the field in California in two seasons. The crops received no nitrogen application (open symbols) or 180 kg N ha^{-1} (closed), and were irrigated (triangles) or received no additional water (squares). Horizontal bars represent LSD values at $P = 0.05$ (adapted from Wolfe *et al.* 1988b).

by cutting and grazing (e.g. 30 000–35 000 tillers m^{-2} in continuously grazed perennial ryegrass swards in the UK; Jones *et al.* 1982). However, for most seed crops, any variation among varieties in the genetic potential for branching is normally masked by the effects of high plant-population density. Competition among adjacent plants tends to ensure that a large proportion of potential branches does not develop beyond the bud stage, or dies prematurely (e.g. Figure 3.11).

Table 3.2 and Figure 3.11 illustrate the demography of winter wheat crops grown at a high level of management in Northern Europe. Widely spaced plants, growing without competition, showed the full potential for branching, generating nearly 30 tillers per plant, of which more than 80% survived to produce ears (Table 3.2). At the other extreme, the intense competition among the plants grown at 800 plants m^{-2} ensured that only one branch was formed per plant, and all of these died before harvest. Commercial crops, within the range 200–300 plants m^{-2}, normally produce two to three fertile tillers per plant. Increasing the supply of nitrogen fertiliser can stimulate tiller production but its effect on tiller mortality at the start of internode extension, just after the terminal spikelet stage, is more crucial. Figure 3.11 shows that increasing nitrogen supply from 90 to 330 kg ha^{-1} had little effect on tiller numbers produced at each node but it increased the number of tillers that survived to bear ears by 35%.

Table 3.2 The influence of plant population density on the components of grain yield of a crop of winter wheat cv. Lely grown in the Netherlands (adapted from Darwinkel 1978).

Plants (m⁻²)	Ears (m⁻²)	Ears/plant	Maximum no. of tillers/plant	Tiller mortality (%)
5	118	23.6	29.0	18.6
25	272	10.9	19.7	44.7
50	322	6.4	13.8	53.6
100	430	4.3	9.8	56.1
200	490	2.5	4.7	46.8
400	582	1.5	3.1	51.6
800	777	1.0	2.2	54.5

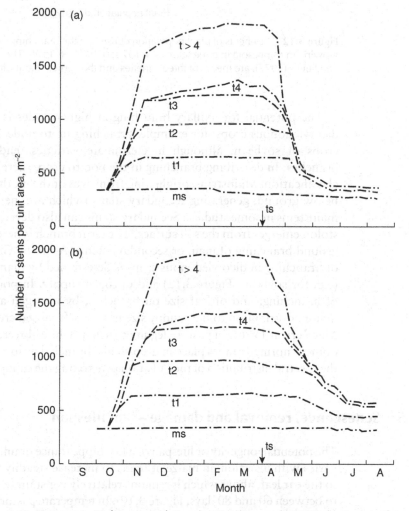

Figure 3.11 Time courses of stem population density of crops of winter wheat cv. Avalon grown in SE England at different levels of nitrogen fertilisation: (a) 90 kg N ha⁻¹; (b) 330 kg N ha⁻¹. Ms refers to the mainstem, t1 to t4, tillers at successive nodes; t > 4, tillers at higher nodes; and ts indicates the timing of the mainstem terminal spikelet stage. The final numbers of fertile tillers were (a) 389 and (b) 526 m⁻² (adapted from Biscoe and Willington 1984).

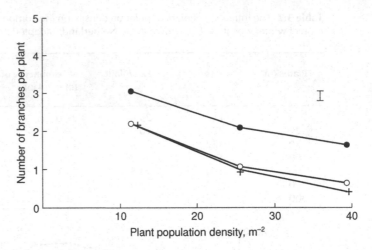

Figure 3.12 The effects of plant population density on the branching of soybean plants growing in Minnesota in three seasons: ● 1973; ○ 1974; + 1975. The data, from Lueschen and Hicks (1977), are means of three varieties and the vertical bar indicates SEM.

The potential for axillary branching at higher nodes is generally greater in dicotyledonous crops; for example, branching up to node 10 can occur in field crops of soybean, although indeterminate varieties tend to generate fewer branches. In describing branching in the potato crop, there is a need for careful classification. Mainstems, originating from eyes of the mother tuber, can branch below ground, generating secondary stems (which are, nevertheless, classed as mainstems in some studies). Secondary stems can also develop if an underground stolon emerges from the soil surface. The term branch is here reserved for above-ground branching of main or secondary stems. As in the Gramineae, the degree of branching in dicotyledonous crops is determined by plant population density (e.g. for soybean, Figure 3.12) and nitrogen supply. In potato, the stimulation of branching, and of leaf size on branches, by nitrogen application, makes a major contribution to the canopy size in highly fertilised crops (e.g. Millard and MacKerron 1986). In pasture legumes such as white clover, where stolon development normally takes place in deep shade, branch development is controlled by the spectral distribution of the radiation penetrating the canopy (Thompson 1993).

3.2.5   Senescence, removal and damage – leaf lifespan

The potential longevity or lifespan of a leaf (appearance or unfolding to senescence) varies with insertion; in maize, it tends to increase steadily with leaf number up to the ear leaf, above which it remains relatively constant (e.g. rising from 20–30 to between 60 and 80 days, Figure 3.10). In temperate pasture grasses and cereals, the regular pattern of leaf appearance, extension and senescence means that each mainstem or tiller tends to carry three to six green leaves during vegetative development (e.g. wheat, Figure 3.13; pasture grasses, Duru and Ducrocq 2000), although other species show a steady increase in green leaf number (e.g. leek,

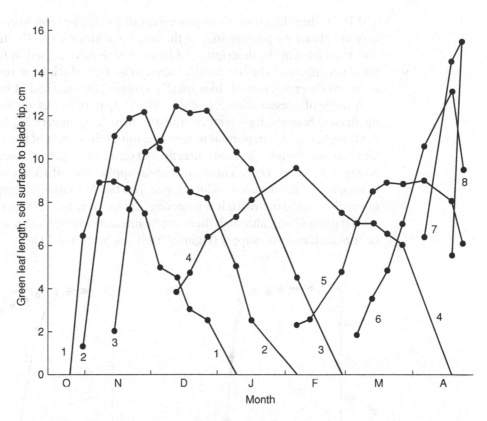

Figure 3.13 Appearance, extension and senescence of successive leaves (1 to 8) of a wheat plant cv. Maris Hustler growing in the field in N. England, 1978–9. The falling section of each curve was constructed from visual assessments of green leaf area not affected by chlorosis, senescence or physical damage (from Hay and Tunnicliffe Wilson 1982).

Figure 2.1). In dicotyledonous crops there can also be a progressive increase in the number of leaves per stem in the first half of the growing season; for example, Vos and Biemond (1992) measured mainstem leaf spans in potato crops increasing from around 30 days (approximately 450°Cdays) at node 1 to over 100 days (1500°Cdays) at nodes 12 to 18. The lifespan of later leaves of grain legumes, particularly indeterminate varieties, tends to be cut short by the remobilisation of protein from canopy to grains during senescence (e.g. Secor *et al.* 1983).

In general, therefore, the ontogeny of a crop plant involves the prompt death, and recycling of the resources of, the leaves formed during the expansion of the canopy; but later leaves, the major source of assimilate for grain or tuber filling, tend to live longer. Furthermore, there is evidence that selection and breeding for increased grain yield, under more intensified cropping conditions, has tended to enhance the functional life of later leaves. For example, in comparisons of maize varieties introduced to Ontario between the 1950s and the 1990s, the newer hybrids, selected for yield under higher plant population densities, had lower rates of senescence of the upper leaves (Tollenaar 1991; Valentinuz and Tollenaar 2004). Similarly, using a 'leaf stay-green' score, Duvick and Cassman (1999) showed that, for maize varieties introduced to the North-Central United States between 1931 and 1991, there had been an improvement of 29%. The term 'stay-green' is used here to indicate the prolongation of the functional life of a leaf. Confusingly, it has also been used in the description of leaves which have ceased to function, with the dismantling of the biochemical apparatus, but which have remained green owing to the protection of chlorophyll pigments (Thomas and Howarth 2000).

A range of stresses can act to reduce the lifespan of leaves, including drought, nitrogen deficiency, high temperature and frost (e.g. Figures 3.10, 3.13) but, in field crops, it is also important to take account of the effects of pests and diseases. Grazing pests remove leaf area directly, but other pathogens can act more subtly, accelerating senescence. Thus the routine application of fungicides to control biotrophic fungal diseases (mildews, rusts) on the flag leaves of temperate cereals normally leads to increased leaf longevity (e.g. by more than 20 days, Figure 3.14) during grain filling, although there can be interactions with other environmental factors such as water supply (Figure 3.14a) (see Section 6.11).

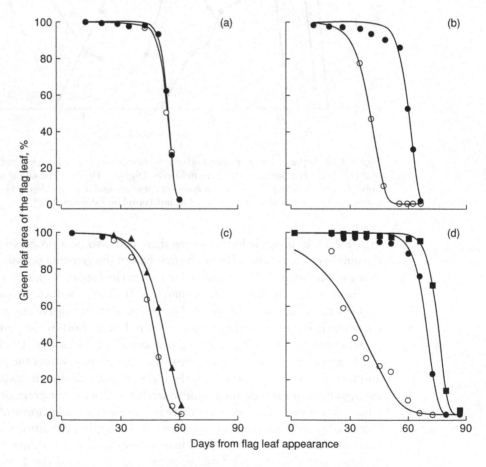

Figure 3.14 The effects of application of fungicides on the time course of green area of flag leaves of wheat plants growing in Central England (o control, untreated, plants; ● plants treated with triazole; ▲ triazole and morpholine; or ■ strobilurin). Plants were (a), exposed to water stress (cv. Avalon); infected with (b) *Septoria tritici* (cv. Longbow); (c) *Erisiphe graminis* (cv. Mission); or (d) *Puccinia striiformis* (cv. Cockpit) (from Gooding *et al.* 2000).

3.3   The development of the crop canopy: leaf area index

Drawing on the review of leaf development in Section 3.2, it can be concluded that the leaf area of a crop in an intensively farmed area depends primarily upon the interactions of crop ontogeny with three or four factors:

- *temperature* – determining the basic dynamics of crop emergence, leaf and branch production, leaf expansion, and senescence;
- *nitrogen status* – determining the size and longevity of leaves, and the development and survival of branches;
- *plant population density* – influencing early-season leaf area (i.e. low in widely spaced crops), and later-season leaf area through the effects of competition (development and survival of branches; leaf size and longevity);
- *water supply* – modifying leaf size and longevity (particularly in areas where the crop experiences moderate to severe water stress, and the associated loss of turgor);

with secondary constraints imposed by environmental stresses and hazards (frost, high temperature, wind, herbivory, disease). The interplay of these factors can be illustrated by time courses of leaf area index under a range of management systems.

3.3.1   Seasonal development of leaf area index

Seasonal patterns of leaf area index in annual temperate crops are well illustrated by Watson's (1947) classic measurements of winter wheat, spring barley, maincrop potatoes and sugar beet growing at Rothamsted (Figure 3.15). In the cereal crops, leaf expansion was depressed by low temperature until April/May, when there was a rapid increase in L, associated with higher temperatures and larger individual leaf sizes, up to a pronounced peak in June/July. After ear emergence, the canopies underwent rapid senescence such that L had fallen to zero by harvest in August. The wheat crop showed an early advantage over barley, and

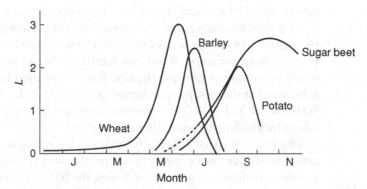

Figure 3.15 Seasonal patterns of leaf area index of crops grown at low levels of fertilisation at Rothamsted in the 1940s (from Watson 1947).

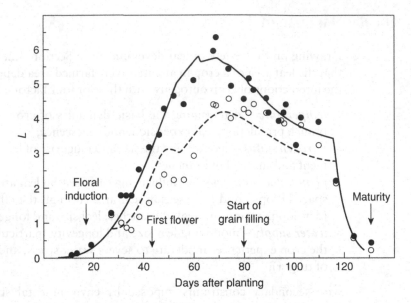

Figure 3.16 Seasonal patterns of leaf area index for crops of soybean cv. Cobb (MG VIII, determinate; see Figure 2.9) planted on 26 June 1981 in Florida, and either irrigated throughout growth (●) or subjected to drought towards the end of the vegetative phase (○) (from Boote *et al.* 1998).

an earlier and larger peak L, mainly because the crop, established in the autumn, was able to respond more rapidly to favourable spring temperatures, irrespective of other soil conditions; in contrast, the spring barley could not be sown until a soil moisture deficiency had developed. Due to the low levels of nitrogen fertiliser applied at the time (1940s), the values of L achieved were very modest (up to 3), and insufficient to intercept all of the incoming solar radiation, apart from a few days in summer (see Section 3.4).

Compared with the cereals, the expansion of the leaf canopy in the dicotyledonous crops was delayed as a consequence of later planting, to prevent frost damage in potatoes, and cold-induced bolting in beet. For these essentially vegetative crops, there was no abrupt leaf senescence related to flowering and, when harvested in October and November, they still carried a significant green leaf area.

Relationships between L and phenology for grain legumes can be illustrated by Figure 3.16, which shows that, in a determinate soybean variety grown in Florida, floral initiation at axillary meristems began at a very early stage in the development of the canopy; the first flower appeared before the peak in L was achieved; but the canopy was declining throughout the phase of grain filling (see Sections 2.1.5, 2.2.3). This contrasts with the seasonal pattern for the potato, where tuber filling begins earlier in the life of the canopy (Figure 2.11).

The time course of L actually achieved by a crop depends upon the specific genotype–environment interactions of the variety used. Thus, under similar growing conditions in south-west Wales, the potato maincrop variety Maris Piper produced canopies of greater extent and longevity than Désirée (Figure 3.17). The finding that the early variety Home Guard consistently generated less leaf area,

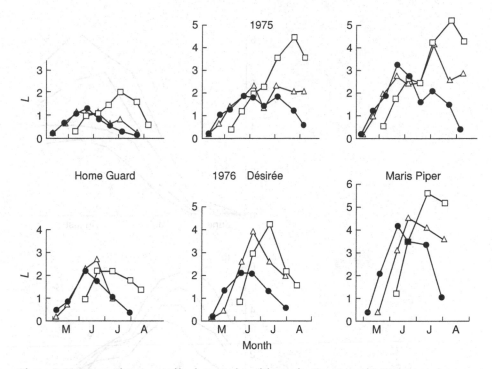

Figure 3.17 Seasonal patterns of leaf area index of three plantings (● mid-March; △ end March/beginning April; □ end April) of three varieties of potato in West Wales, in two contrasting seasons (adapted from Jones and Allen 1983).

particularly with later planting, was, however, principally a consequence of the greater physiological age of the tubers planted (see Section 2.2.4; Figure 2.11).

3.3.2   Leaf area index and crop management

Planting date can have a profound influence upon the course of development of L. In addition to reducing radiation interception early in the season, delayed sowing of temperate cereals causes an acceleration or telescoping of crop development (Table 2.3; Hay and Kirby 1991), resulting in a tendency for lower peak L values. The opposite tends to hold for later-planted potato crops, which, under higher temperatures, develop much larger and longer-lasting canopies (e.g. Figure 3.17). The net effects of variation in dates of planting or emergence on canopy generation in a given species or variety will depend upon the seasonal distribution of other factors such as frost, drought, flooding, and pest or disease incidence.

The effects of plant population density on L tend to be most marked during the early stages of crop development before the canopies of adjacent plants have had time to adjust (e.g. Figure 3.18b, spring barley; Figure 3.19, potato); and in crops such as maize, where there is no possibility of filling gaps between plants by branching (Figure 3.18a). Thus the penalty to pay for low population density or gaps in the canopy, in terms of radiation interception during grain filling, can

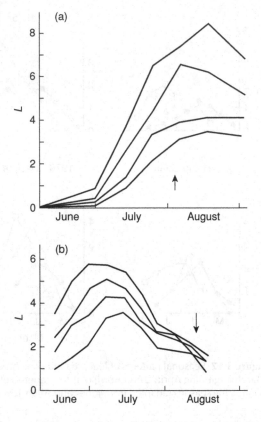

Figure 3.18 The influence of plant population density on the seasonal pattern of leaf area index of crops of (a) maize, California and (b) spring barley, Cambridge, UK. The target densities for maize were 1.8, 3.5, 7 and 13×10^4 plants ha^{-1}, and for barley 1, 2, 4 and 8×10^6 plants ha^{-1}. The arrows indicate tassel (maize) or ear (barley) emergence (adapted from Kirby 1967; Williams *et al.* 1968).

be much higher for maize than for a tillering cereal (compare Figure 3.18a with 3.18b). Because of the branching pattern of some species, notably soybean, spacing between rows and plant population density can influence L independently (Willcott *et al.* 1984).

Increases in the size of later leaves, stimulation of branching, and enhanced survival of branches, caused by increased supply of nitrogen, result in higher peak L and greater canopy duration in a wide range of crops (e.g. wheat, Figure 3.20; maize, Figure 3.21), with the risk of over-investment in leaf area under high levels of application (but see Section 3.4). There are important interactions with other factors, notably water supply; for example, in the extreme case presented in Figure 3.21, the withholding of irrigation from a maize crop in Nebraska in an exceptionally dry year meant that the crop was unable to use the applied nitrogen for canopy expansion.

Generally, the effects of water stress differ from those of other factors affecting L in that timely rainfall or irrigation can lead to some degree of recovery of the canopy. For example, Figure 3.22 shows a range of time courses of L generated in barley crops in New Zealand by varying the scheduling of irrigation; although

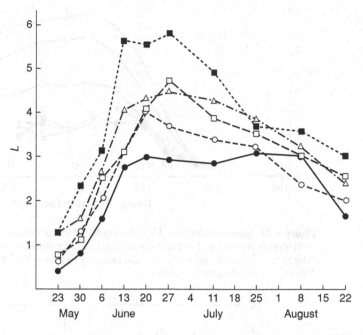

Figure 3.19 Seasonal patterns of leaf area index of maincrops of potatoes (cv. Désirée and Maris Piper) grown at different plant population densities in Wales in 1974 (where ●, ○, □, △ and ■ indicate 25, 30, 37.5, 50 and 75×10^3 plants ha^{-1}; from Allen and Scott 1980).

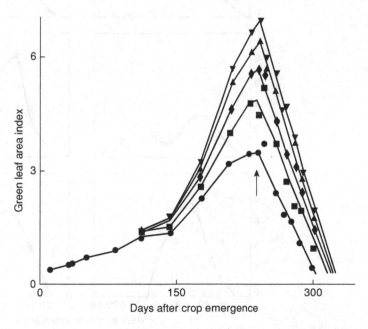

Figure 3.20 Seasonal patterns of leaf area index of wheat crops (cv. Armada) grown at different levels of nitrogen fertiliser application in central England in 1980/81 (where ●, ■, ♦, ▲ and ▼ indicate 0, 50, 100, 150 and 200 kg N ha^{-1}; the arrow indicates the approximate date of anthesis; from Green 1987).

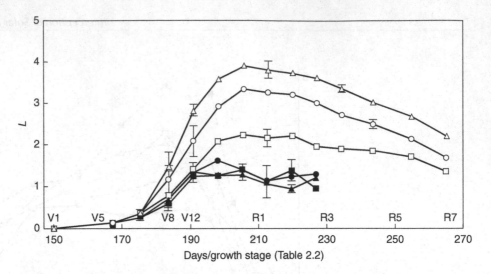

Figure 3.21 Seasonal patterns of leaf area index of maize crops (cv. Pioneer 3394) grown with (open symbols) and without irrigation (closed), at different levels of nitrogen application: 0 kg N ha^{-1} (squares), 68 kg N ha^{-1} (circles), and 135 kg N ha^{-1} (triangles) in Nebraska in 1994 (from Boedhram *et al.* 2001).

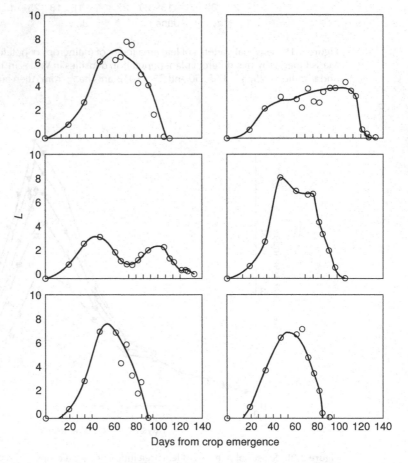

Figure 3.22 Seasonal patterns of leaf area index of barley crops (cv. Triumph) grown under a range of irrigation treatments at Lincoln, New Zealand (times of application indicated by vertical bars; from Jamieson *et al.* 1995b).

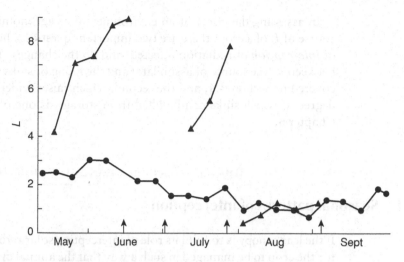

Figure 3.23 The influence of the method of defoliation on the seasonal pattern of leaf area index of swards of S23 perennial ryegrass growing in Southern England (▲ cut four times per season at the dates indicated by arrows – note that the data for the first regrowth are omitted; ● continuously grazed by sheep; from Jones 1981).

there were particularly sensitive periods, the resumption of irrigation was associated with renewed leaf area expansion and/or prolongation of the life of the canopy. Similarly, the soybean crop in Figure 3.16 showed signs of canopy decline and recovery during two periods of drought, resulting in an overall reduction in leaf area duration.

Finally grazers, pests and diseases can cause a wide range of systematic and random variations in the time course of *L*. Figure 3.23 illustrates the gross differences in *L* that can be generated by variation in the method of harvesting a productive grass sward; whereas Figure 3.24 shows more subtle effects on an arable crop resulting from infection by virus diseases.

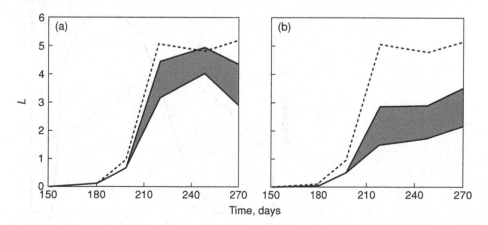

Figure 3.24 Seasonal patterns of leaf area index of sugar beet crops infected with (a) beet mild yellow virus or (b) beet yellow virus, in the Netherlands, 1989. The shaded areas indicate the contribution of yellow leaves, and the broken line is the leaf area index of control, uninfected, plants (from Rossing *et al.* 1992).

In assessing the effect of an environmental or agronomic factor on the time course of L of a crop, there are two important questions: how is the time course of *interception* of radiation affected; and do the changes affect the relationship between L (the source of assimilate) and the filling of sinks? The first question is covered in Section 3.4, and the second, which raises wider questions about the degree to which sinks can be filled from storage, is one of the main themes of Chapter 6.

3.4   Canopy architecture and the interception of solar radiation

3.4.1   Seasonal patterns of interception

If the leaf canopy is to fulfil its role of intercepting solar radiation, it is important for the crop to be managed in such a way that the annual cycle of L matches that of incident radiation. If they do not match, potential yield will be lost as a consequence of unintercepted solar energy or wasteful investment of dry matter in excessively large leaf canopies during periods of low irradiance.

In north-temperate regions such as the UK, the annual fluctuation in incident total solar radiation (Q) follows a broad peak, with highest daily inputs in June, and substantial receipts from April to September (Figure 3.25; Chapter 1). Assuming that, for most species, at least three layers of leaves are necessary ($L = 3$) for the complete interception of all incoming photosynthetically active radiation (PAR) (Section 3.4.2), it is clear that the major crops of temperate agriculture are not particularly well adapted to maximise interception. In wheat and barley, the highest values of L do tend to coincide with the seasonal maximum in Q (e.g. Figures 3.15, 3.18) but individual leaf lifespan is short, such that the functional life of the canopy is curtailed and substantial quantities of PAR are

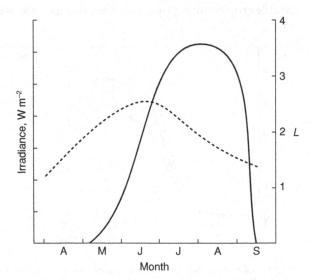

Figure 3.25 Seasonal patterns of mean daily irradiance (- - - - -) and leaf area index (———) of a model potato crop growing in England (adapted from Moorby and Milthorpe 1975).

not utilised. Measures such as heavy application of nitrogen fertiliser can result in very high peak values of L, but they do not necessarily enhance the duration of full interception (e.g. Figure 3.20). Peak L in maize does not normally coincide with the midsummer peak in Q, but the long lifespan of leaves above the ear node tends to ensure that interception is complete during grain filling (e.g. Figures 3.10, 3.18, 3.21).

In temperate zones, potato crop canopies are better adapted, by their longer duration, to intercept a high proportion of Q from midsummer onwards (Figures 3.15, 3.17, 3.25), but as with cereals, management must be appropriate to avoid wasteful investment in leaf area; for example, Figure 3.19 shows that a three-fold increase in stem population density resulted in a doubling of peak L, but little effect on the duration of an L greater than 3. Attempts to increase interception during April and May can be hampered by restricted canopy expansion if physiologically older tubers are planted (Figures 2.11, 3.17), and there is always the risk of defoliation by late frosts.

The ontogeny of most annual crops, therefore, makes it difficult to increase the duration of a canopy with a leaf area index greater than 3, without exposing the canopy to environmental stress. The potential for increased interception by arable crops, outside the period of peak area, is limited by low temperatures, environmental hazards such as frost, flooding, drought or leaf diseases, and by the need for the crop to be mature at a time when soil physical conditions permit the necessary harvest operations. In warmer climates the situation can be reversed, with the period of high Q being avoided because of the hazards of water and thermal stresses.

The pattern of interception can be very different for perennial species. For example, under appropriate cutting or grazing management of temperate grassland swards, a continuous canopy can be maintained throughout the year, resulting in the interception of a high proportion of the available PAR (e.g. Figure 3.23). Here, limits to production are set more by the efficiency of utilisation of intercepted energy than by interception (Chapters 4, 5).

3.4.2 Optimum and critical leaf area indices

In an imaginary crop with large, thick, undissected and horizontal leaves with little overlap, complete interception of incoming PAR would be achieved at an L value close to 1. However, since real crops display a wide range of leaf sizes, shapes, thickness, and dispositions, an L of 3–5 is generally required for the interception of more than 90% of incoming energy, in a range of species and varieties (e.g. for potatoes, Figure 3.26; wheat, Hipps *et al.* 1983; sunflower, Sadras and Trápani 1999; soybean, Wells 1991). An L of more than 6 is, however, normally necessary for maize crops (e.g. Williams *et al.* 1965; Gallo and Daughtry 1986) owing largely to the wider spacing of the large unbranched plants.

It might, therefore, be assumed that investment in leaf area beyond the minimum L to ensure full interception of incoming radiation would be wasteful, with older, shaded, leaves in lower strata of the canopy reaching their light compensation point and becoming sinks rather than sources of current assimilate. For this

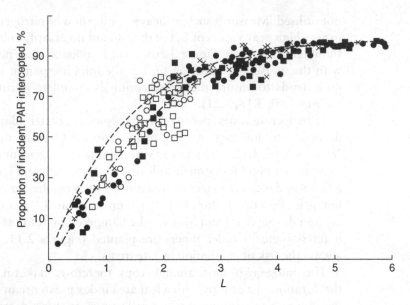

Figure 3.26 The interception of incident PAR by potato (cv. Pentland Crown) canopies of differing leaf area index. The different symbols indicate varying agronomic treatments in two seasons (1979, 1980) in the Midlands of England (from Khurana and McLaren 1982).

model of interception, there would be a clear optimum L, at which crop growth rate (C) would reach its highest value: below such an optimum, C would be dependent upon L, and would be depressed owing to incomplete interception of the incident solar energy, whereas above the optimum, it would be depressed owing to increased respiratory losses (e.g. Figure 5.1a).

In practice, field crops do not generally show such an optimum L. It is more common for C to increase up to a critical value of L at which interception is complete (L_{crit}, which is normally in the range 3–5, but which can be higher, for example in highly erectophile canopies; see below), above which a relatively constant, maximal, value of C is maintained (e.g. for wheat, Figure 3.27). The reasons for such a pattern of response are explored in greater detail in Chapter 5. Briefly, there is little evidence to support the hypothesis that shaded leaves constitute important respiratory sinks for current assimilate; indeed, in Section 3.2.5, it was established that the early leaves of an annual crop canopy senesce promptly whereas the lifespans of those involved in grain or tuber filling are longer-lived. It is generally assumed that rates of whole crop respiration are more closely related to the rate of gross photosynthesis than to biomass or leaf area. For example, in classic experiments on stands of white clover, McCree and Troughton (1966) demonstrated that the increase in crop respiration with increasing L was small beyond L values of 2–3 (Figure 5.1b).

Although such observations are consistent with the existence of critical rather than optimal L values, they do not explain why values of L_{crit} vary considerably among crop species and among varieties of the same species. To do this, it is necessary to consider the rate of net photosynthesis of individual leaves receiving different levels of irradiance up to and beyond saturation, as well as the

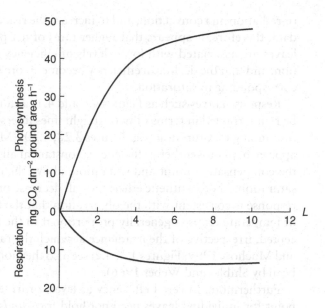

Figure 3.27 Relationships between the rates of respiration and gross photosynthesis (expressed per unit ground area), and leaf area index, of wheat crops receiving 400 W m^{-2} of total solar radiation. (Model curves constructed from measurements in the field and under controlled conditions; adapted from Evans *et al.* 1975.)

architecture of the crop canopy. In particular, it is important to know how the leaf area of a crop is arranged vertically, bearing in mind that the individual leaves will not be illuminated uniformly, with the topmost tending to receive much more solar energy than those lower in the canopy.

3.4.3   Leaf photosynthesis and canopy properties

In studies of individual leaves of C$_3$ species (wheat, soybean, sugar beet, potatoes; Section 4.4), the rate of net photosynthesis increases with irradiance, reaching an asymptote at the irradiance corresponding to light saturation. For example, the response curve in Figure 4.5 indicates that, under the conditions of the experiment, the net photosynthesis of a typical C$_3$ leaf, arranged horizontally, would be saturated at around 500 μmol m^{-2} s^{-1} (equivalent to approximately 100 W m^{-2} PAR, less than half the maximum midday irradiance in midsummer in north temperate zones). Thus, if the leaves at the top of a wheat canopy were held at right angles to the direction of the incoming solar radiation, they would be saturated for a large part of the day during much of the growing season.

In contrast, in Figure 4.6, the photosynthetic efficiency of the leaves (the quantum yield; Section 4.2), calculated from the gradient of the response curve, tended to decrease with increased irradiance. This effect is most obvious above the point of saturation, where further increments of PAR gave no increase in the rate of net photosynthesis. Here the additional input of solar energy to the leaf served only to increase the energy load, to be dissipated by transpiration, re-radiation and convection, and to increase the risk of thermal stress. From such data, therefore, it appears that higher rates of net photosynthesis of individual leaves are associated with lower levels of efficiency of use of intercepted radiation, and that the decline in efficiency becomes more serious above the irradiance corresponding to saturation.

Response curves such as Figures 4.6 and 4.10a can, with little loss of accuracy, be re-interpreted in terms of two straight lines intersecting at the irradiance corresponding to saturation (e.g. Figures 3.28, 4.20) (Monteith 1981a). Taking this approach, photosynthetic efficiency is constant at all levels of irradiance between the compensation point and saturation; above the irradiance corresponding to saturation, photosynthetic efficiency will decrease progressively. This pattern of response is consistent with the observation that the total dry-matter production of temperate crops is generally proportional to the total amount of PAR intercepted, irrespective of the irradiance at which the radiation is supplied (Sinclair and Muchow 1999; Figures 1.3, 4.2; see also the pioneering observations on soybean by Shibles and Weber 1966).

Furthermore, lowered efficiency at levels of irradiance above the saturation point for individual leaves need not hold for *stands* of crop plants. Consider a

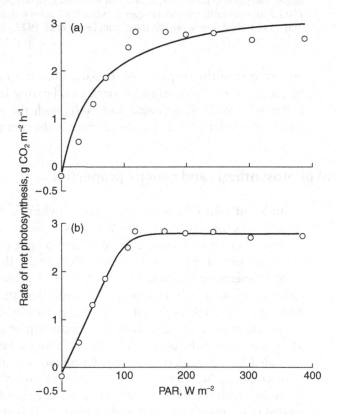

Figure 3.28 (a) The photosynthetic light responses of a fully-expanded wheat flag leaf, measured in the field in Central England, with a rectangular hyperbola fitted to the data points; (b) an alternative bilinear function, modified to include terms which account for respiration, fitted to the same data points (from Marshall and Biscoe 1980).

canopy that is intercepting 95% of the available solar energy at midday, at an irradiance of, say, 250 W m^{-2} PAR; if the leaves are large, thick and horizontal, then most of the interception will take place at the top of the canopy. The uppermost, light-saturated leaves will receive much more radiation than they can use for photosynthesis, whereas the poorly illuminated lower leaves will be unable to contribute much to dry matter production. The photosynthetic efficiency of the canopy will be low, the leaves will be exposed to stress, and potential dry matter production will be lost.

If, however, the leaves are more inclined to the vertical, then the efficiency of the canopy to utilise intercepted radiation, and crop growth rate, can increase for two reasons. First the angled leaf blades at the top of the canopy will intercept a smaller proportion of the incoming radiation at higher solar elevations (i.e. the projected leaf areas of the inclined leaves, which can be seen by an observer looking vertically downwards into the canopy, will be smaller than for horizontally disposed leaves of equal area; Monteith and Unsworth 1990). An increase in the angle of the leaves in relation to the horizontal will, therefore, lead to an improvement in photosynthetic efficiency at the top of the canopy, as the irradiance incident upon the leaf surface is reduced to the saturation point. If the solar radiation incident upon a leaf is reduced further by increasing leaf angle, then the rate of photosynthesis will fall without any significant gain or loss of efficiency.

Second, PAR that is not intercepted at the top of the canopy will be available for photosynthesis by the lower leaves, which can now contribute to the productivity of the crop stand, at the same efficiency as the more highly illuminated leaves at the top of the canopy, as long as the radiation supply exceeds the light compensation point. Overall, this analysis, which takes no account of the variation in photosynthetic potential with leaf age (Figures 4.20, 6.3; Biscoe *et al.* 1975a, b), predicts that, for temperate C$_3$ crops, the highest rates of canopy photosynthesis will be achieved when all of the incident radiation is intercepted by leaf blades that are disposed in space in such a way that no leaf is more than just saturated.

The relationship between the rate of net photosynthesis and irradiance is quite different for the leaves of C$_4$ crop species, most of which evolved under high irradiance in the dry subtropics (Fitter and Hay 2002; Section 4.4). Typical response curves for the predominant C$_4$ species, maize (Figure 4.16b), are steeper than for C$_3$ crop species, and virtually linear up to irradiances (300–350 W m^{-2} PAR) that are well above those corresponding to the saturation of C$_3$ leaves. At higher levels, the slope of the response curve does diminish progressively but without reaching an asymptote within the terrestrial range of irradiance (Chapter 1). Consequently, the efficiency of photosynthesis of individual maize leaves declines only at very high irradiance, and leaf angle has a smaller influence upon canopy photosynthesis in C$_4$ species than in C$_3$.

Leaf angle (α) varies with genotype, with ontogeny, within a leaf, and (within and between canopies) in response to the environment. In general, cereal and grass crops have more erect leaves than dicotyledonous crops, with typical values of greater than 60° (from the horizontal) for the later leaves of maize, wheat or perennial ryegrass stands (e.g. for maize, Lambert and Johnson 1978;

perennial ryegrass, Figure 3.31a). Nevertheless, the early leaves of some wheat varieties are very prostrate, and highly fertilised grass crops can have large more horizontally orientated leaves at the top of the canopy (e.g. Faurie *et al.* 1996). Leaves exposed to water stress can wilt, thus lowering leaf angle. To deal with such variation, the pattern of leaf angle displayed by a canopy can be described quantitatively using the cumulative frequency of α (e.g. for soybean, Blad and Baker 1972; Trenbath and Angus 1975). Although there have been relatively few thorough measurements of crop leaf angle, most species have been found to conform to one or other of the following idealised distributions:

- erectophile: α predominantly greater than 60° (many grasses and cereals),
- plagiophile: α predominantly between 30 and 60° (sugar beet, rape, soybean),
- planophile: α predominantly less than 30° (clover, bean, sunflower).

3.4.4   Canopy extinction coefficient

Leaf angle is only one of several factors governing the penetration of PAR into the crop canopy. Others include:

- leaf surface properties, affecting reflection,
- leaf properties, including thickness, affecting transmission;
- leaf size, shape and degree of dissection;
- characters affecting the three-dimensional arrangement of leaves within the canopy, including phyllotaxis and the potential for heliotropism.

The radiation environment within a canopy is also profoundly influenced by the elevation of the sun, and the proportions of direct and diffuse radiation.

In elementary treatments of canopy interception, plant and canopy characteristics are combined into a single composite property, the extinction coefficient (k), defined by the Monsi and Saeki equation:

$$I = I_0 \, e^{-kL} \tag{3.2}$$

where:

I_0 is the irradiance above the crop canopy,
I is the irradiance at a point in the canopy above which there is a leaf area index of L, both I_0 and I measured by horizontally disposed sensors, and
k and L are dimensionless (Monsi and Saeki 1953).
In some investigations, $s = e^{-k}$ is used in place of k.

This relationship is derived from the Bouguer–Lambert–Beer Law, which was originally formulated to describe the passage of monochromatic light through a homogeneous light-absorbing solution, and it is the basis of spectrophotometric methods of chemical analysis. Figure 3.29a presents model profiles of interception of direct solar radiation (from high solar elevation) for stands of highly-angled leaves ($k = 0.3$) and for more horizontally disposed leaves ($k = 0.8$), showing the substantial differences in the extent of penetration of

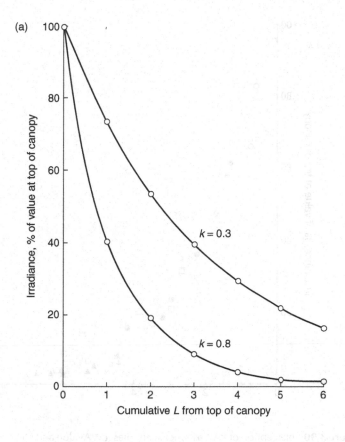

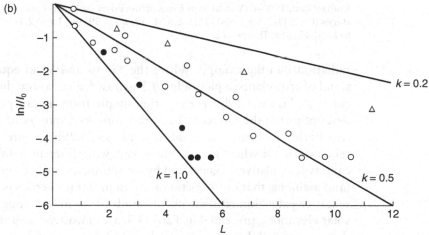

Figure 3.29 Solutions of the Monsi and Saeki equation: (a) profiles of radiation interception for model canopies with $k = 0.3$ or 0.8; (b) field measurements of interception by canopies of barley (△), pasture grasses (*Lolium perenne* and *Dactylis glomerata*, ○) and white clover (●) compared with values for model canopies with $k = 0.2$, 0.5 and 1.0 (adapted from Brown and Blaser 1968).

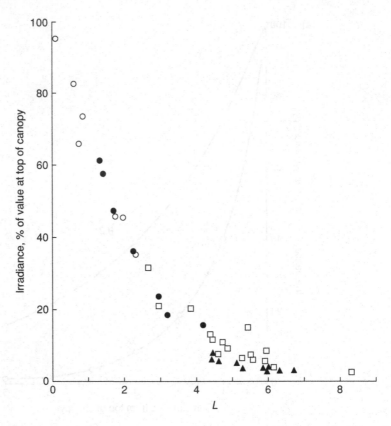

Figure 3.30 Interception of PAR by wheat canopies (cv. Avalon and Hustler) growing at Rothamsted, 1980–6. Variation in L was achieved by including canopies at different growth stages (○ GS11–30; ● GS30–31; □ GS 31–39; ▲ GS 59–69; Table 2.1). The curve corresponds to $k = 0.45$ (from Thorne *et al.* 1988).

radiation into the canopy. Where the Monsi and Saeki equation holds for a stand of crop plants, a plot of $\ln\{I/I_0\}$ against L gives a straight line whose gradient is k (Figure 3.29b); even in this simple form, it can provide satisfactory descriptions of the interception of radiation by canopies of a range of species (e.g. barley, pasture grasses and clover; Figure 3.29b). Figure 3.30 illustrates the relationship for wheat in a different way, with different pathlengths within the canopy (cumulative L) generated by measuring the same crop at different stages (and assuming that ontogenetic variation in leaf properties is small). The representative extinction coefficients for stands of the major crops, measured at high solar elevations, presented in Table 3.3, are consistent with the generalisations: that species with broad and more horizontal leaves intercept more radiation in the upper strata than do those with narrow more inclined leaves; but that there can be considerable variation within species, largely explained in terms of α.

The use of a single mean value of α in the interpretation of canopy extinction coefficients is clearly an oversimplification; for example, Figure 3.31a shows that there are large, and opposing, trends in leaf angle in stands of ryegrass and clover, and that α can vary by more than 45° within a clover canopy. Such variations are common in the canopies of many crop species (Monteith 1976). How, then, can

Table 3.3 Representative measurements of canopy extinction coefficient for a range of crop species.

Species	Extinction coefficient	Source
Wheat	0.38 (erectophile variety, before anthesis)	Yunusa *et al.* 1993
	0.59–0.60 (plagiophile) (Australia)	
	0.63–0.76 (same varieties, after anthesis)	
	0.46 (before flag leaf appearance, UK)	Thorne *et al.* 1988
	0.34–0.51 (at 3rd node visible) (UK)	Whaley *et al.* 2000
	0.33–0.50 (flag leaf appearance)	
	0.32–0.45 (anthesis) (19–338 plants m^{-2})	
Maize	0.66–1.0 (including a variety with	
	horizontally-disposed leaves; USA)	Hatfield and Carlson 1979
	0.47 (35 cm row spacing)	Flénet *et al.* 1996
	0.34 (100 cm, USA)	
	0.64–0.83 (simulated from field	
	data, France)	Maddonni *et al.* 2001
Soybean	0.56–0.68 (water stress)	Sivakumar and Shaw 1979
	0.58–1.0 (irrigated)	
	0.83 (25 cm row spacing,	
	peak *L*, USA)	Willcott *et al.* 1984
	0.62 (75 cm)	
	0.52 (35 cm) – 0.32 (100 cm)	Flénet *et al.* 1996
Sunflower	0.83 (USA)	Rachidi *et al.* 1993
	0.89 (Australia)	Bange *et al.* 1997
Pea	0.55–0.75 (leafless variety, UK)	Heath and Hebblethwaite 1985
	0.33–0.49 (leafed variety)	
Potato	0.5 (*L* = 1) – 0.7 (*L* = 3)	Khurana and McLaren 1982

straight-line relationships be obtained, for example in Figure 3.29b, when it would be predicted, from the distribution of α, that k would vary considerably between the top and base of the canopies measured? There are two important reasons for the relative constancy of k values within canopies. First, the interception of solar radiation tends to be dominated by the upper half of the canopy, where leaf angles tend to vary less than towards the base (e.g. Figure 3.31a). The dominance of the upper strata can be particularly marked in soybean canopies, where leaves attached to several nodes are concentrated (Willcott *et al.* 1984). Second, the uppermost leaves tend to receive radiation from a range of angles, especially under diffuse lighting, whereas, because of the filtering activity of these upper leaves, the lower strata tend to receive radiation from almost directly overhead. Thus, for example, the effect of the more horizontally disposed lower leaves of a ryegrass sward in increasing the value of k will tend to be counteracted by the fact that radiation from high elevations penetrates the canopy more easily.

The experimental verification of the Monsi and Saeki equation, and its application to crops in the field, are laborious procedures, involving the measurement of the irradiance at a series of levels above and within the canopy (Varlet-Grancher *et al.* 1989), and stratified measurements of leaf area. Many series of measurements, using appropriate sensors, must be replicated at different stations and continued over time, to allow for the variability of the crop canopy, and ontogenetic changes in leaf characteristics. Fortunately, indirect and slightly less arduous approaches, such as the use of inclined point quadrats, can be used successfully in the estimation of *k* (e.g. for grass–clover mixture; Lantinga *et al.* 1999), and there is considerable potential for photographic and three-dimensional digitising techniques (e.g. for clover canopies; Rakocevic *et al.* 2000).

With the continuing trend towards the modelling of crops and plants (Chapter 9), the deficiencies of the simple extinction coefficient approach to describing radiation interception by the canopy have become clear. For example, the Monsi and Saeki equation does not take into account the fraction of incident solar radiation that is lost from the crop stand by reflection. This is normally relatively small, even with horizontally disposed leaves (0–10%, depending upon wavelength; Fitter and Hay 2002), but, in the special case of flowering oilseed rape crops, losses can be much greater (e.g. an additional 7% of incident radiation lost over 20 days, compared with an apetalous variety; Fray *et al.* 1996). The approach does cover transmitted light, although the related changes in spectral composition can have wider implications for the development of crop plants and emerging weeds.

As it does not contain a height or distance term, the relationship does not distinguish among canopies in which the same *L* is distributed over different crop heights (e.g. owing to differences in internode length, as in dwarf and normal stature isolines of wheat and barley). Since the distance between layers of leaves can have an important influence upon the distribution of radiation reflected within the canopy, and upon the interception of radiation from low solar elevations, it may be important to describe the canopy in terms of leaf area density (i.e. the leaf area per unit volume of canopy; Monteith 1976). Another problem that is common to all measurements of *L* is how to classify senescent tissues, which intercept radiation but do not contribute to production.

Two contrasting examples of pasture canopy architecture (Figure 3.31) serve to emphasise the fact that crops can differ widely in overall height, in the vertical distribution of *L* and α, as well as in total *L*. In pure stands of ryegrasses, the combination of low *L* in the top strata and the highly angled disposition of the youngest leaves (*k* = 0.5 for perennial pasture grasses; Figure 3.29b) means that PAR can penetrate deeply into the crop stand, and photosynthesis can be spread over a large leaf area (e.g. up to 10 units of *L* in stands of *Lolium rigidum* in the classic experiment of Stern and Donald 1962). In contrast, in a pure stand of clover, the top layers of the rather shorter canopy contain a larger proportion of the total *L* than in the grass canopy, and the leaves are disposed in a more horizontal plane (*k* = 0.9 to 1.0 for white clover; Figure 3.29b). Consequently, a much higher fraction of incoming PAR is intercepted at the top of the clover canopy, and photosynthesis is distributed over a smaller leaf area (e.g. up to 6 units of *L* for *Trifolium subterraneum*; Stern and Donald 1962).

Depending upon a range of factors, discussed in Section 2.2.5, the balance of the grass and clover components in mixed swards can vary considerably. Thus, in Figure 3.31a, at a generally low level of productivity (total *L* = 3), the low proportion of white clover, whose leaf area was displayed half way into the

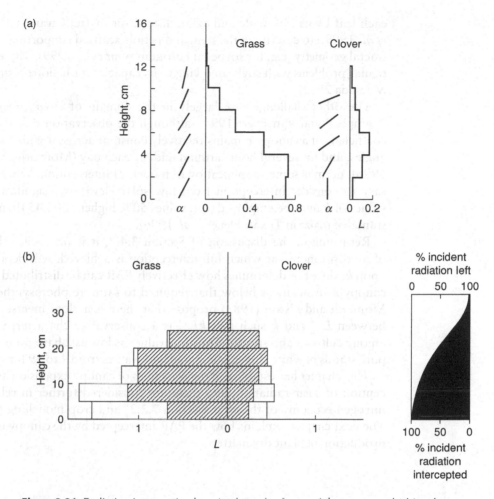

Figure 3.31 Radiation interception by mixed stands of perennial ryegrass and white clover in summer in England (equivalent to hay crops before harvesting). (a) Classic measurements of the stratification of L and α at relatively low soil fertility (Warren Wilson 1959); (b) stratification of L and the interception of radiation at higher fertility, total $L = 6.7$, grass component 5.1, clover 1.6; the hatched areas represent laminae, and the unshaded areas, sheaths or petioles (1984; J Woledge, personal communication).

canopy, would have intercepted a very small fraction of the incoming radiation. By contrast, in Figure 3.31b, at higher levels of productivity (total $L = 6.7$), and under conditions that favoured clover, the horizontally-disposed clover leaves at the top of the canopy, supported by extended petioles, would have captured a significant fraction of incoming radiation.

In addition to these issues, the principal set of interrelated problems faced by crop modellers involve: the non-random arrangement of leaves in the canopy (i.e. the canopy is not a homogenous solution); diurnal variation in the inclination of direct solar radiation; and variation in the proportion of diffuse radiation reaching the canopy at a range of inclinations. The non-randomness of leaf area, originating from variation in plant spacing, phyllotactic and ontogenetic traits, and branching pattern, has been tackled in a range of ways from: characterising each leaf by its leaf angle and azimuth (e.g. for maize; Girardin 1992; Drouet *et al.* 1999); to describing the stem and petiole scaffold supporting the leaves by fractal geometry (e.g. for soybean; Foroutan-pour *et al.* 1999). There can be particular problems with soybean owing to its capacity for heliotropism (Ikeda and Matsuda 2002).

The other challenges are largely in the domain of environmental physics (Campbell and Van Evert 1994), although the observation that the extinction coefficient of a canopy remains relatively constant across a wide range of solar angles, and for several hours around midday each day (Monteith and Unsworth 1990), permits some simplification of models of interception. Where it is necessary to consider interception from low solar elevations, significantly greater values of k must be employed (e.g. values 50% higher at 08.45 than at 11.45 in stands of maize in Texas; Flénet *et al.* 1996).

Returning to the discussion of Section 3.4.2, it is now clear that the L_{crit} of a crop canopy, at which full interception is achieved, will depend strongly upon k, since this determines how effectively PAR can be distributed through the canopy at irradiances below that required to saturate photosynthesis. Indeed, Monteith and Elston (1983) propose that there is a close inverse relationship between L_{crit} and k such that kL_{crit} is a conservative character, varying little among cultivars and species. Thus L_{crit} values as low as 2 have been recorded for pure stands of white clover, compared with values from 4 to >9 for wheat.

This chapter has established the principles of canopy expansion and the interception of solar radiation, and these are considered further in relation to the nitrogen economy of the crop (Section 7.2.2) and crop modelling (Chapter 9). The next chapter explains how the PAR intercepted by the canopy is used in the production of plant dry matter.

Chapter 4

Photosynthesis and photorespiration

Practically everything we see about us has involved photosynthesis at some stage or other. The gardener often talks about 'feeding' plants when he applies fertilizers and the notion that plants derive their nourishment from the soil is one that is commonly held. They do not. Plants take up minerals from the soil, they derive their nourishment from the air.

(Edwards and Walker 1983)

4.1 Introduction

The preceding chapter described how crops develop a leaf canopy and intercept solar radiation. This chapter describes how crops convert intercepted solar energy to chemical energy by photosynthesis, thereby generating simple carbohydrates from water and CO_2. Such carbohydrates are transformed into a wide range of more chemically reduced products, mainly proteins and lipids, which together with carbohydrates form the basis of food chains. The biochemical and physiological processes by which plants absorb and lose carbon are not only important in determining crop yields but also play a crucial role in regulating the composition and energy balance of the Earth and its atmosphere. Plants have an extraordinary capacity to concentrate atmospheric C from 0.036% in the atmosphere as CO_2 to about 40% ($C/CH_2O = 12/30 = 0.40$) as carbohydrate, the first constituent of dry matter; an increase in concentration of over 1000-fold. The principal enzyme responsible for the fixation of CO_2 in leaves and other green tissues (ribulose bisphosphate carboxylase-oxygenase or Rubisco) is the most abundant enzyme on Earth, and may form up to 30% of the protein present in leaves. As a rudimentary calculation, this would mean that, within the approximately 560 Gt of C in the terrestrial biosphere, there is about 0.8 to 1.1 G ($= 10^9$) t of Rubisco.

The approach taken in this chapter is that photosynthesis is a series of processes that occur at the three scales of cell, leaf and canopy and that it is necessary to understand CO_2 fixation at each scale. It is also possible to define potential upper rates for CO_2 fixation at each of the scales, and environmental conditions where water, nitrogen and temperature limit the potential rate. The chapter considers C_3 and C_4 photosynthesis in describing the steps by which the major C_3 (wheat, soybean, rice) and C_4 (maize, sorghum, millet) grain crops produce dry matter. It covers the responses of photosynthesis to water stress and nitrogen shortage and ends with a section on the effects of the gaseous pollutant gas ozone (O_3) on CO_2 uptake by field crops.

The rate of net or apparent photosynthesis (P_n), or dry matter production of a plant or crop is determined by the balance between three processes:

$$P_n = P_g - R_p - R_d \qquad (4.1)$$

where, P_g is the rate of gross photosynthesis, R_p is the rate of photorespiration and R_d is the rate of 'dark' respiration. Photorespiration, by definition, occurs only in the presence of light and it is intimately associated with P_g in C_3 plants, but its rate is greatly reduced at low levels of ambient O_2 and in plants using the C_4 photosynthetic pathway. Photorespiration occurs because Rubisco is able to bind with CO_2 or O_2 depending on their relative concentrations within the chloroplasts in the mesophyll cells (i.e. it is a carboxylase/oxygenase enzyme).

R_p and R_d are distinct processes but the term 'dark' respiration is both confusing and a misnomer because respiration, as a general metabolic process, is occurring all the time in plants and animals. Respiration increases with temperature but generally declines with leaf or plant age and long-term exposure to abiotic environmental factors such as water and nutrients (Lambers *et al.* 1998), but short-term exposure to water stress can increase respiration in proportion to photosynthesis (Fitter and Hay 2002). The main difference between photorespiration and 'dark' respiration is that the latter occurs in the light as well as in the dark but only in the mitochondria. In comparison, photorespiration also involves other leaf organelles such as chloroplasts and peroxisomes. It is thus less confusing to use the term mitochondrial respiration (R_m) for 'dark' respiration and this term will be used henceforth. Mitochondrial respiration is dealt with in Chapter 5. Thus, Equation 4.1 becomes:

$$P_n = P_g - R_p - R_m \qquad (4.2)$$

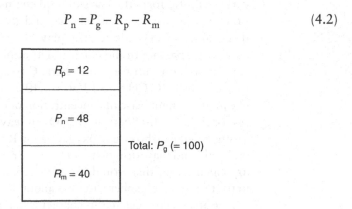

Figure 4.1 Schematic representation of the relationships between gross photosynthesis (P_g), photorespiration (R_p), net photosynthesis (P_n) and mitochondrial respiration (R_m), assuming that P_g has a value of 100 units of fixed CO_2.

The relation between the components of Equation 4.2 can also be shown schematically (Figure 4.1). For the sake of illustration it can be assumed that R_p is 12% of P_g, R_m is 45% of $(P_g - R_p)$ and P_g is set at an arbitrary 100 units of fixed CO_2. Accordingly, R_p will equal 12 units of CO_2, R_m will equal 40 units $(0.45(100 - 12))$, leaving $(100 - 12 - (0.45(100 - 12)))$ or 48 units as net CO_2 assimilation after all CO_2 efflux losses.

4.2   Photosynthetic efficiency

Photosynthetic efficiency can be expressed in a number of ways and at a number of scales. At the biochemical level, the number of C atoms fixed from the atmosphere per quantum of light energy can be calculated or its inverse – how many photons it takes to fix one molecule of CO_2. One quantum of light energy is called a photon. At the crop scale, the amount of dry matter produced per unit of radiation intercepted by a crop can be calculated – the radiation use efficiency (RUE). To determine the former, detailed measurements are made on isolated leaf cells or organelles under controlled laboratory conditions (e.g. Ehleringer and Björkman 1977; Lawlor 2001). For crops, the accumulated, usually aboveground, dry matter is expressed in terms of the amount of radiation either intercepted or absorbed by a crop (Section 3.4). As both total solar radiation and photosynthetically active radiation intercepted (PAR_{int}), with a wavelength between 400–700 nm, have been used in the calculation of RUE (Sinclair and Muchow 1999), it is necessary to specify the basis upon which RUE is calculated. In a sense, RUE is a form of canopy quantum efficiency and can be linked to Equation 4.2 as:

$$RUE = (P_g - R_p - R_m)/PAR_{int} \qquad (4.3)$$

Light has the property of behaving simultaneously as a continuous waveform and as a stream of discrete packets or quanta of energy, originally termed photons by Albert Einstein. As the energy content of a photon is inversely related to the wavelength of the light and not its intensity, a photon of blue light (wavelength 400 n (10^{-9}) m) contains 4.97×10^{-19} J, whereas a photon of red light (wavelength 680 nm) contains 2.92×10^{-19} J, about 60% of the energy content of the blue light photon. When absorbed by a chloroplast, a photon of red light has the necessary energy to raise an electron within the thylakoid membranes to a higher energy state, thus initiating the photochemical events that split water to release the protons and electrons that drive the chemical reactions of the Calvin cycle. This initiation of the very first steps in the photosynthetic process can of course be achieved by any light that has a wavelength between about 400 and 700 nm, termed the photosynthetically active radiation (PAR). These light-dependent processes result in the generation of the ATP (adenosine triphosphate) and NADPH (reduced nicotinamide adenine dinucleotide phosphate) necessary to provide the energy and reducing power for the light-independent biochemistry of the Calvin cycle.

As one photon of PAR contains an extremely small amount of energy, it is convenient to use, instead, Avogadro's number (6.02×10^{23}, the number of molecules per gram molecule), referred to as a mole of photons. Thus, for CO_2, which has a molecular weight of 44 ($12 + 16 + 16$), 44 g of CO_2 (one mol) contain Avogadro's number of CO_2 molecules. By analogy, one mol of 'red' photons contains $(6.02 \times 10^{23}) \times (2.92 \times 10^{-19})$ J $= 1.758 \times 10^5$ J $= 176$ kJ of energy and, assuming perfect efficiency, could initiate Avogadro's number of electron excitations. This is not the same as the number of photochemical reactions required to fix one mol of CO_2 because three mol of ATP and two mol of NADPH are needed to reduce one mol of CO_2 to carbohydrate (CH_2O). The advantage of using the mol as a basic unit is that it defines a quantity that is based on the same constant number of items, regardless of whether they are photons, molecules of CO_2 or any other.

The delivery of light energy is a flux or rate process, meaning that so much energy is delivered in a defined period over a particular leaf or ground spatial area. Common practice is to talk about a photosynthetic photon flux density (PPFD), defined as the number of mol of photons delivered per second and per m^2 of leaf area or ground with units of mol (or more conveniently µmol, i.e. 10^{-6}) m^{-2} s^{-1}. The rate of supply of PAR can also be expressed as radiant energy (with units of W m^{-2} = J m^{-2} s^{-1}) but, as this does not take account of the spectral distribution of the light as mentioned above, there is no single relationship between PPFD and PAR irradiance. As a guide, full sunlight on a bright summer day in the UK of 500 W m^{-2}, delivers a photon flux of about 2200 µmol m^{-2} s^{-1}. The conversion is approximately: PAR irradiance (J m^{-2} s^{-1}) $\times$ 4.4 is equivalent to PPFD (µmol m^{-2} s^{-1}), so that there are about $(4.4 \times 10^{-6}) \times (6.02 \times 10^{23}) = 26.5 \times 10^{17}$ photons per joule of PAR. Irradiance levels in the sunniest places on Earth can be up to twice as high as those found in north-western Europe.

The photochemical events of photosynthesis, the Hill reactions, provide the energy to reduce CO_2 to CH_2O by generating NADPH and ATP. NADPH donates protons for the reduction, and ATP provides chemical energy for the light independent Calvin cycle by the loss of one inorganic phosphate group to become ADP (adenosine diphosphate), which is then recycled to ATP. Three molecules of ATP and two of NADPH are needed to reduce one molecule of CO_2 to CH_2O and this requires 525 k ($= 10^3$) J per mol of CO_2 reduced. In a world illuminated only by red light (176 kJ/mol photons), three mol photons of 'red' light would be required, as a minimum, to reduce each mol of CO_2 (i.e. three photons per molecule of CO_2). Conversely, the absolute maximum quantum efficiency (or quantum yield) is 1/3 or 0.33 mol CO_2 per mol photons, assuming all photons are used for CO_2 reduction. Given that one mole of CH_2O contains 468 kJ, and three mol quanta of photons provide 525 kJ of energy, this means that the theoretical maximum biochemical efficiency of fixation of CO_2 is 468/525 or nearly 90%. However, the photochemical process does not operate at full efficiency; some photons provide the energy for 'wasteful' chlorophyll florescence and others are absorbed by non-photosynthetic pigments and are dissipated as heat. Such unproductive photochemical reactions raise the quantum requirement to about eight and lower the quantum yield to 0.125, giving an efficiency of about 33% (i.e. $468/(8 \times 176)$). However, the story does not end there. At the level of the single leaf, when simultaneous measurements are made in special chambers of leaf CO_2 fluxes and PPFDs, CO_2 losses by photorespiration and mitochondrial respiration (Chapter 5) mean that 15–20 mol photons are required per mol of fixed CO_2: a maximum quantum yield between 0.067 and 0.050. A range of values exists, principally because of differences between C_3 and C_4 species, and because of the influence of variation in temperature, with higher temperatures lowering quantum yield under ambient conditions for C_3 species (Farquhar *et al.* 1980).

A quantum yield of 0.067 mol CO_2 mol^{-1} photons (as PAR) represents an efficiency of 'conversion' of radiation of about 17% in energy terms. This is calculated as above, with one g mol of CH_2O, yielding 468 kJ on combustion, requiring 15 mol photons (each containing 176 kJ) for the reduction of one mol of CO_2. Thus, the efficiency = 468 kJ/(15 × 176 kJ) = 17.7%. Strictly speaking, it is correct to refer to the efficiency of energy use and not to the efficiency of energy conversion in photosynthesis because energy is not converted into dry matter directly during photosynthesis. Light energy is the initiator of metabolic processes that start by raising the energy levels of sensitive pigments, leading to the assimilation of CO_2 and the synthesis of plant dry matter. In moving from the biochemical reaction of CO_2 reduction at the level of the chloroplast to that of the single leaf in a C_3 species such as wheat, the efficiency of use of the absorbed photons has reduced from a theoretical maximum of nearly 90% to one of 33–17% but the efficiency measured in the field is much lower.

The next step is from the single leaf to the crop. Given that 17% is an optimistic figure for the efficiency of use of radiant energy in the production of CH_2O, about 15% of the light incident on a crop is lost due to incomplete interception by the leaf canopy, reflection, and transmission through the leaves without absorption (Section 3.4). This 'lost' radiation has to be accounted for in the radiation balance sheet before the 17% energy use is considered. For C_3 species, which include most of the major cereals except maize, there is a further 30% loss because the upper leaves in the canopy are not capable of utilising all the incident radiation. This is because the rate of photosynthesis in C_3 species under ambient conditions does not increase, or becomes saturated, above irradiances of about 180–200 W m^{-2}; levels that are less than half of full sunlight. However, the situation is complicated by the fact that planophile leaves in sparse canopies early in the growing season, but in full sunlight, can become saturated at higher irradiance levels. Additionally, as leaves age they photosynthesise less efficiently because of lowered stomatal conductance (Marshall 1978); thus Monteith's initial analysis (Monteith 1977) that a further 30% of incident radiation is 'lost' owing to this lack of its full utilisation has to be interpreted cautiously since the interaction between its radiation balance and photosynthesis depends on the state of the leaf canopy, its architecture and age-structure. There will also be periods of incomplete PAR interception and CO_2 capture during a growing season, caused by sub-optimal conditions such as low temperatures that limit photosynthesis and water deficits that do likewise. Finally, production of CH_2O and subsequent plant growth is a 24-h process and does not just occur only during sunlight but also in the dark. During the dark, CO_2 loss by mitochondrial respiration (R_m in Equation 4.2) may amount for 40% of the CO_2 fixed during

Table 4.1 The declining efficiency of energy use of incident PAR (as %) for the production of carbohydrate (CH_2O) in a C_3 crop under ambient conditions, expressed as a percentage of incident PAR.

Incident PAR	100
15% loss *via* unabsorbed PAR due to reflection and transmission	85
Maximum efficiency of use of PAR, 17%	14.5
30% 'loss' of PAR *via* light saturation of the upper leaves in the canopy	10.2
40% loss of CO_2 *via* mitochondrial respiration	6.1

the light, allowing for the fact that that the CO_2 fixed during the light is P_n (Equation 4.2) and thus net of photorespiration (R_p).

The complete efficiency balance sheet for a theoretical C_3 crop is shown in Table 4.1. The result is that the maximum overall efficiency of utilisation of radiation by a crop is about 6% – that is, about 84% lower than the process of photosynthesis at the chloroplast level.

An efficiency of about 6% is realistic for only a few crops where other environmental conditions such as nitrogen fertilisation, irrigation and weed and disease control are optimal (Table 4.2). Even the highest recorded season long values of PAR utilisation seldom reach 6%, with 4–5% being more typical. Measurements over periods shorter than a complete season commonly result in higher estimates of PAR utilisation than those presented in Table 4.2. It is also notable that the three C_4 species in Table 4.2 do not have markedly higher PAR utilisation percentages than the C_3 species, even though their rates of photosynthesis saturate at much higher levels of irradiance than C_3 species (Section 3.4.3).

Table 4.2 The highest recorded values of season-long radiation use efficiencies (RUE, g MJ^{-1}) for field grown C_3 and C_4 crops and the implied efficiency of use of intercepted PAR. Data calculated from Table 1 in Sinclair and Muchow (1999) and taken from Table 3.2 in Hay and Walker (1989) where references to the cited original sources for the estimates can be found. Percentage utilisation of PAR assumes that 1 g of carbohydrate plant dry matter yields 17.5 kJ of energy on combustion (Monteith 1977).

Crop	Location	RUE	Utilisation of PAR (%)
C_3 species			
Wheat	USA	3.24	5.7
Potato	UK	3.52	6.2
Barley	Australia	2.60	4.6
Rice	Japan	2.78	4.9
Soybean	Japan	2.20	7.7
Ryegrass	UK	2.78	4.9
Clover	Australia	1.65	2.9
C_4 species			
Maize	Australia	3.36	5.9
Sorghum	USA	3.02	5.3
Sugarcane	Australia	3.92	6.9

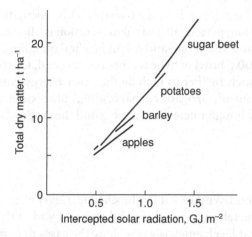

Figure 4.2 The relationship between total dry matter at harvest and the amount of solar radiation intercepted throughout a season (Monteith 1977).

On average, the RUE of temperate C_3 crops has been measured as 2.8 g MJ^{-1} intercepted PAR (Monteith 1977) and that of C_4 species as 3.4 g MJ^{-1}. The RUE relationship has been shown to be linear for a number of crops and to be fairly conservative across species (Figure 4.2). Thus, the RUE of the C_3 wheat examples presented in the data of Sinclair and Muchow (1999) varied by about 19%, whereas the C_4 maize examples by about 12%. Assuming an energy content of 17.5 kJ per gram of carbohydrate-based dry matter, 2.8 g MJ^{-1} gives an average efficiency of PAR use of just under 5%. While the value for RUE is reasonably conservative among crop types that have the same photosynthetic pathway, as in Figure 4.2, different crops can absorb different amounts of radiation and thus achieve higher dry matter totals. The total accumulated dry matter is affected by the length of time that a crop grows with an active leaf canopy and whether the crop is a perennial or an annual (Figure 4.2). Crops, such as soybean, that produce high energy primary products such as lipids and proteins rather than carbohydrates may have low RUE but a high level of PAR utilisation. This paradox is explained by the fact that RUE is solely based on dry weight gain per unit of intercepted radiation whereas PAR utilisation is a true energy ratio. Thus, soybean that has a much higher lipid and protein content than cereals can have a lower dry matter yield per unit intercepted radiation, and thus RUE, but a comparable or even high utilisation of solar energy. The usefulness of RUE as an index of growth depends on the chemical constitution of the dry matter being produced (Chapter 5).

Given the above information, it is now possible to calculate for how long the sun has to shine to deliver a bowl of breakfast cereal, assuming the cereal is the dry product of a wheat crop. One hundred grams of such a breakfast cereal as dry matter contains about 1500 kJ of energy. As seen above, this represents about 5% of the incident PAR that fell onto the wheat crop. The incident PAR was thus about 30 000 kJ. Given that on a bright summer day, of length 12 h, about 500 W (or 500 J s^{-1}, equivalent to 21 600 kJ per 12 h) of PAR falls onto 1 m^2 of crop, (30 000 kJ/21 600 kJ × 12 h) = 16.7 h are required to make the wheat for a breakfast cereal. This calculation assumes that during the grain filling period, all assimilated carbon is allocated to the grain. Thus, the sunshine from about one-and-a-half bright days falling on 1 m² of crop is needed to give a 100 g bowl of 'wheaty' breakfast cereal. Of course, this analysis can be extended much further to include the other energy costs involved, such as fertiliser, pest control, irrigation, harvesting, processing and transport but these issues, although interesting, are beyond the scope of the present analysis.

4.3   Photosynthetic processes

Photosynthesis at the biochemical level means measuring and modelling rates of the relevant processes as they occur within the mesophyll and other cells. At the leaf level, emphasis is on how the gases that affect photosynthesis and leaf water balance get into and out of leaves exposed to radiation and temperature. At the canopy level, interest is in how a group of leaves or plants with a particular spatial arrangement interact with the radiative environment to determine the rate of canopy photosynthesis. Clearly, it is conceptually possible to extend the spatial and temporal scales of photosynthesis to encompass the productivity of biomes at one end, and the nanosecond effects of individual photons at the other end, of the spectrum. Each scale of observation has its own methods and techniques of measurement and has its own conceptual models that attempt to describe and predict the processes at the scale in question (Curran *et al.* 1997). Another guiding principle in the discussion that follows is that, at each scale, it is possible to define a potential rate of fixation that is determined by the availability of radiation, the temperature level and the concentration of CO_2. Where conditions are suboptimal (water shortage or excess, imbalance in mineral nutrition – Chapter 7) the rate will fall below this potential. As an example, the effect of the pollutant gas, ozone (O_3) on leaf and canopy photosynthesis will be examined *via* its effects on photosynthesis at the biochemical level.

4.3.1   Photosynthesis as a cellular biochemical process

At the biochemical level, CO_2 makes its way from the air *via* stomata into the intercellular spaces within the leaf, dissolves in the water overlying the mesophyll cells, moves across the mesophyll membrane and ends up in the thylakoids within the chloroplasts. In what follows, the so-called light-dependent reaction of photosynthesis is described briefly, where photons are used to split water, releasing electrons, protons and O_2. These electrons are used to generate, *via* the sequence of photosystem II and photosystem I, the reducing agent (NADPH) and the biochemical energy (ATP) to convert CO_2 to simple carbohydrates and other plant products. However, the efficiency of photosynthesis will be described in terms of the number of mols of CO_2 fixed per mol of photons received.

The dimensions of the cells and subcellular organelles that perform these functions are important. Lawlor (2001) provided a semiquantitative analysis of the photosynthetic system of an 'average' C_3 leaf that has been scaled down here to conform to the area (0.025 m^2) of a typical wheat leaf. The fresh mass of such a leaf is about 4.25 g with a dry weight of 1 g. It will contain 1.8×10^8 mesophyll cells, each of which has a volume of 1.4×10^{-14} m^3. These mesophyll cells contain on average 50 chloroplasts, giving 9×10^9 chloroplasts per leaf. The volume of each chloroplast is about 3.3×10^{-17} m^3, so that they occupy in total about 8% of cell volume. Such a leaf would contain about 15 μg of chlorophyll (1.5×10^{-5} moles), corresponding to 7×10^8 molecules of chlorophyll per chloroplast and an individual chloroplast weight of about 20×10^{-20} g.

The stoichiometry, or mass balance, of the light-independent or Calvin cycle is summarised in Figure 4.3. To make one molecule of carbohydrate ($C_6H_{12}O_6$), the Calvin cycle has to operate six times, each time capturing one C atom from CO_2. Over such a six-course cycle, six molecules of ribulose 1,5-bisphosphate (RUBP) join with the six C atoms from CO_2 to make 12 molecules of the 3C phosphoglyceric acid (PGA). Using the 12 NAPDH and 12 ATP generated from the light-dependent reaction, the 12 PGA molecules are reduced to 12 3C

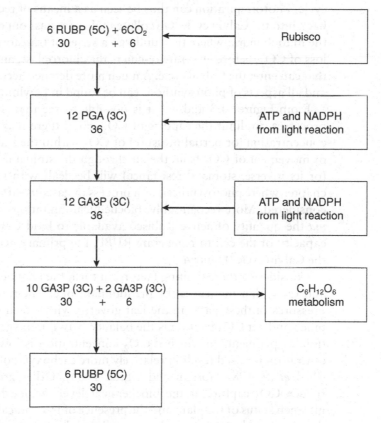

Figure 4.3 Diagram of the principal steps in the Calvin cycle showing the major transformations. Numbers in brackets are the numbers of C atoms per molecule. 36 C atoms are conserved at each step in the cycle to produce one molecule of glucose ($C_6H_{12}O_6$). The 6 RUBP molecules left at the end of the cycle form the receptors for the next six molecules of CO_2, catalysed by Rubisco. Abbreviations: RUBP, ribulose 1,5 bis-phosphate; PGA, 3-phosphoglycerate; GA3P, glyceraldehyde 3-phosphate; ATP, adenosine 3-phosphate; NADPH, reduced nicotinamide adenine dinucleotide phosphate.

glyceraldehyde 3-phosphate (GA3P) molecules, also known as triose phosphate. Six molecules of ATP are needed to regenerate the six CO_2 acceptor 5C RUBP molecules with which the process started, from ten of these 12 3C GA3P molecules. The remaining two 3C GA3P molecules enter the plant metabolic pathways from which many valuable food and fibre products eventually emerge. The important point is that the number of C atoms entering and leaving the Calvin cycle is conserved (Figure 4.3).

Although Rubisco has a lower affinity for O_2 than for CO_2, as shown by their respective Michaelis-Menten constants (e.g. $K_{CO_2} = 441$ µmol mol^{-1}, $K_{O_2} = 248\,000$ µmol mol^{-1} at 25°C for soybean), at low internal CO_2 levels RUBP binds to O_2 in preference to CO_2 and thus starts the sequence of reactions known as photorespiration of which the major transformations are shown in Figure 4.4. Low internal CO_2 levels can occur when stomata close, for example, under water stress or in response to aerial pollutants. Six RUBP molecules bind with six O_2 molecules to form six 2C glycollate molecules (photorespiration is sometimes called the 2C cycle), releasing six 3C PGA molecules, which re-enter the Calvin cycle. Photorespiration can thus be seen as a means of recovering 'wasted' PGA back into the Calvin cycle. Glycollate molecules pass out of the chloroplasts into the mitochondria where they undergo a series of transformations, including the loss of CO_2, before glycerate re-enters the chloroplast, and is changed into PGA that can enter the Calvin cycle. A much more detailed account of the Calvin cycle and all aspects of photosynthesis can be found in Lawlor (2001).

From Figures 4.3 and 4.4 it is possible to see that, within a C_3 leaf, three processes can limit the capture of CO_2. First, there may be a limitation to the concentration (or partial pressure) of CO_2 within the leaf caused by a low rate of movement of CO_2 from the air through the stomata. Stomatal conductance (or its inverse, stomatal resistance) will be dealt with more fully later in this chapter where photosynthesis as a process of gaseous diffusion at the leaf level is considered. More recognisably, biochemical limitations to CO_2 capture can be *via* the quantity of active Rubisco available to bind CO_2 to RUBP and by the capacity of the cell to regenerate RUBP, the primary acceptor and initiator of the Calvin cycle (Figure 4.3).

Considering these limiting factors, in turn, the relative concentration of CO_2 and O_2 within the leaf is the predominant one. The ratio between the partial pressures of these gases in the leaf governs with which of them RUBP mostly binds and net CO_2 fixation is the balance between oxygenation and carboxylation. Experiments in which the O_2 concentration is lowered or the CO_2 concentration is raised result in relatively more carboxylation and less oxygenation (Bird *et al.* 1982; Jordan and Ogren 1984). Other growing conditions that reduce CO_2 capture at the biochemical level include high temperature, low nitrogen status of the plant and the presence of free radicals that damage and can destroy the chloroplast membranes and thylakoids, the active sites of the Calvin cycle. A key to understanding the biochemistry of photosynthesis is the competition between O_2 and CO_2 for reaction sites on the Rubisco enzyme and thus leading RUBP down either the oxygenation route to CO_2 loss or the carboxylation route to CO_2 fixation. The balance of this competition is generally weighted in favour of CO_2 fixation at the current ambient level of CO_2 and Long (1994)

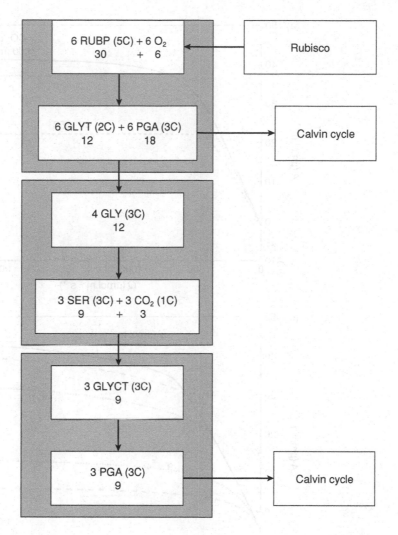

Figure 4.4 Diagram of the principal steps in photorespiration giving the major transformations. Numbers in brackets are the numbers of C atoms per molecule. Abbreviations: RUBP, ribulose 1,5 bis-phosphate; PGA, 3-phosphoglycerate; GLYT, glycollate; GLY, glycine; SER, serine; GLYCT, glycerate. Processes in the upper and lower box occur in the chloroplast and processes in the middle box occur in the mitochondria.

has argued that evolution has set a high value on the efficiency with which Rubisco binds CO_2 to RUBP in C_3 plants. This is shown by constancy in the high affinity of Rubisco for CO_2 and its selectivity *vis-à-vis* O_2 in C_3 plants, a less consistent feature amongst C_4 plants.

Figure 4.5 presents generalised responses of the net assimilation of CO_2 (A, equivalent to P_n) by leaves of a C_3 plant in response to variation in incident light level (Q) and internal leaf CO_2 concentration (c_i) (Long 1994). Figure 4.5a shows that, at low light levels, A increases linearly with Q up to approximately 500 μmol m^{-2} s^{-1}. In this part of the curve, A is limited by the amount of active CO_2 acceptor (RUBP) because the formation of RUBP is dependent on the light-dependent generation of ATP and NADPH.

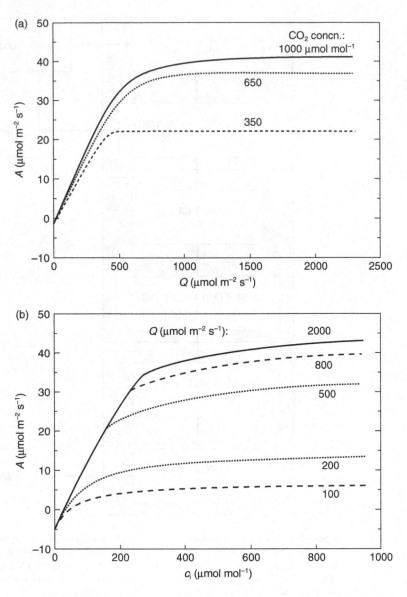

Figure 4.5 (a) The response of the photosynthetic rate of CO_2 uptake per unit leaf area for a C_3 leaf (*A*) against incident photon flux (*Q*) at three atmospheric CO_2 concentrations; (b) the response of *A* to the intercellular CO_2 concentration within a leaf (c_i) at five values of *Q* (after Long 1994).

At higher levels of *Q*, the control of *A* shifts either to the concentration of CO_2 or the concentration of Rubisco. In either case, raising the level of CO_2 in the atmosphere will raise *A*, as is seen in Figure 4.5a, but does not alter the basic asymptotic shape of the curve. Conversely, raising *Q* at different c_i concentrations (Figure 4.5b) will progressively reduce limitations to *A* caused by lack of RUBP and the increasing levels of c_i will also reduce limitations from stomatal resistance and the level of Rubisco but with other processes, such as later

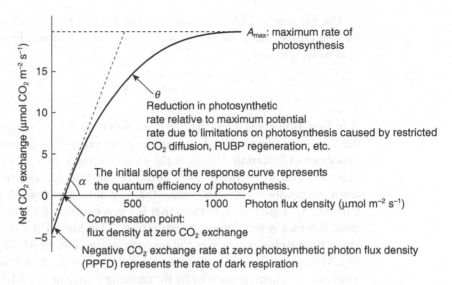

Figure 4.6 Cardinal points of a CO_2 assimilation – light response curve (after Fitter and Hay 2002).

enzymatic reactions in the Calvin cycle, becoming limiting in their turn. For example at high light and high c_i levels, the rates of removal of glucose and subsequent steps of metabolic pathways may inhibit fixation rates via negative feedback controls, although another reason may be photoinhibition, caused by high light damage to the chloroplasts. Combined high light and c_i effects can be seen in Figure 4.5b: at a c_i of about 300 µmol mol^{-1}, increasing Q from 800 to 2000 µmol m^{-2} s^{-1} increases A only slightly. It is also notable that, whereas a plateau is reached in the A/Q curve (Figure 4.5a), the asymptote is approached more gradually in the A/c_i curve in Figure 4.5b, meaning that the controls on A at high light and high c_i are imposed more gradually, as a result of more interacting processes.

As summary, the cardinal points of the photosynthetic light response curve are illustrated diagrammatically in Figure 4.6 (Fitter and Hay 2002). The initial slope of the curve is a measure of the photon or quantum efficiency of the light reaction with slope, α; the light compensation point is that at which irradiance gives zero P_n. The value of respiration at zero irradiance (Figure 4.6) gives the mitochondrial respiration and θ measures the rate of the transition from a light to a CO_2 carboxylation limited rate of CO_2 assimilation. The curve in Figure 4.6 is a rectangular hyperbola that can be described by the equations (Equations 9.6 and 9.7) used in the AFRC2 wheat model and described in Chapter 9.

Farquhar *et al.* (1980) developed a mechanistic model of photosynthesis at the biochemical level using the notion of a steady-state equilibrium between the carboxylation and the oxygenation processes referred to above. The main features of the model are outlined below. Further details are given in the original paper by Farquhar *et al.* (1980) and to further papers by Evans and Farquhar (1992), Long (1994) and Ewert (2004).

The net rate of CO_2 assimilation (A or P_n) can be represented as the balance between three rates, one of which (V_c) absorbs CO_2, the other two (V_o and R_m) lead to CO_2 losses:

$$A = V_c - 0.5 V_o - R_m \qquad (4.4)$$

A in Equation 4.4 is equivalent to P_n in Equation 4.1, V_c is equivalent to P_g and $0.5 V_o$ is equivalent to R_p, but with the rates expressed in terms of biochemical processes in Equation 4.4. V_c is the rate of carboxylation, V_o the rate of oxygenation and R_m is the mitochondrial respiration representing losses of CO_2 from respiratory activities other than photorespiration (Equation 4.2 and Chapter 5). Equation 4.4 states that for every mol of CO_2 fixed by carboxylation, 0.5 mol is lost by photorespiration. This loss of 0.5 mol of CO_2 occurs every time Rubisco catalyses the reaction of one mol of RUBP with one mol of O_2. As seen above (Figure 4.6) for ambient conditions, V_c is limited either by the amount of Rubisco present or by the capacity to regenerate RUBP. The Farquhar model hypothesises that it is the greater of these two potential limitations on V_c that determines the carboxylation component of A (Equation 4.4). In other words, V_c is the lower of the RUBP-regeneration rate or the Rubisco limited rate and is calculated from a combination of three Michaelis–Menten terms, describing the carboxylation reaction, the oxygenation reaction and the RUBP reaction with CO_2; terms 1, 2 and 3, respectively in Equation 4.5. Thus,

$$V_c = V_{cmax} \; (c_c/(c_c + K_c))(o_c/(o_c + K_o)^{-1}) \; (R/(R + K_r)) \qquad (4.5)$$
$$\quad\quad\quad\quad\quad\;\; 1 \quad\quad\quad\quad\quad\; 2 \quad\quad\quad\quad\; 3$$

where, V_{cmax} is the maximum rate of carboxylation; c_c is the concentration or partial pressure of the substrate CO_2 at the site of carboxylation in the mesophyll cells; K_c is the Michaelis–Menten constant for carboxylation; o_c is the O_2 concentration in the mesophyll; K_o is the Michaelis–Menten constant for oxygenation; R is the concentration of available RUBP, i.e. that RUBP which is not already bound to CO_2 and K_r is the Michaelis–Menten constant of the regeneration reaction. If this last reaction is not limited by the availability of RUBP then the last term in Equation 4.5 equals 1. The part of Equation 4.5 involving the O_2 term is set as an inverse, because this is operating against carboxylation and in the direction of C oxygenation or photorespiration. Rearranging Equation 4.5 and setting the last term to equal one gives the maximum carboxylation rate or velocity when the availability of RUBP is not a limitation (W_c):

$$W_c = V_{cmax} \, c_c/(c_c + K_c(1 + o_c K_o)) \qquad (4.6)$$

W_c increases when c_c increases and will reduce when o_c increases relative to c_c, although it should be noted that K_c is much smaller than K_o. Increase in o_c relative to c_c will increase the denominator in Equation 4.6 and reduce W_c. The C and O Michaelis–Menten constants (K_c and K_o) vary between photosynthetic organisms but have similar values between higher plants (Bainbridge *et al.* 1995).

When the regeneration of RUBP is limited at, for example low light levels, the maximum carboxylation rate is given by:

$$J' = J/(2(2 + 2\phi))$$ (4.7)

J' is the maximum rate of carboxylation allowed by electron transport and J is its potential rate when one photon is absorbed by each of the electron transport pathways of photosystems I and II, leading to the transfer of one electron from H_2O to $NADP^+$ and the initiation of the photochemical reactions. However, two electrons are required per NADPH molecule and this is the reason for the first 2 in the denominator of Equation 4.7. ϕ is the ratio of the number of oxygenation reactions to carboxylation reactions that occur and is the ratio of V_o to V_c, where the equation for V_o has the same general form as that for V_c (Equation 4.5) with the o_c and c_c functions inverted, i.e.:

$$V_o = V_{omax} (o_c/(o_c + K_o))(c_c/(c_c + K_c)^{-1})(R/(R + K_r))$$ (4.8)

J is related to the photosynthetic photon flux density (PPFD) according to:

$$J = 0.5(1 - f)Q_a$$ (4.9)

The 0.5 is included because a photon has to be absorbed by both of the photosystems, f is the fraction of light lost by absorption by non-chloroplast structures and Q_a is the leaf-absorbed PPFD.

In summary, the essential reactions in the absorption of CO_2 at the biochemical level are the respective rates of carboxylation and photorespiration and the rates of the reactions that lead to the regeneration of RUBP. The overall result is that assimilation proceeds at the lower of either the RUBP regeneration or the Rubisco-mediated carboxylation rate. In the Farquhar model, it is proposed that the *actual* V_c value is the lower of the two possible maximum carboxylation rates derived above, i.e. that allowed by the light reaction or by the Rubisco meditated reactions, that is:

$$V_c = \min(W_c, J')$$ (4.10)

There are many elegant nuances to the Farquhar model that are not covered in this basic description and readers are referred to Lawlor (2001, Chapter 11) and references therein for further consideration. One of the main reasons that the model has proved attractive to researchers is that it provides a description of the biochemical mechanism of photosynthesis and describes processes that are measurable from the organelle level to the leaf canopy. One of the advantages of the model has been its ability to represent the responses of photosynthesis to temperature. This was a necessity since the Farquhar model was developed for leaves growing at 25°C, whereas leaves can function over a temperature range from about 0°C to between 35°C and 40°C. Temperature is an important determinant of the behaviour of biological systems in general and photosynthesis in particular. Membranes change their fluidity and permeability with temperature and the activity rates of enzymes, such as Rubisco, are affected. High temperatures favour photorespiration in C_3 plants and membrane functions are impaired by both high and low temperatures. As well as reducing K_o and the solubility of CO_2 relative to O_2, thereby increasing the affinity of RUBP for O_2, high temperatures reduce the activity of Rubisco from an optimum at 25°C to zero activity between 40°C and 45°C (Law and Crafts-Brandner 1999).

All of the main variables in the Farquhar model change their value with temperature and further models have been developed to describe such responses (Brooks and Farquhar 1985; Long 1991; Bernacchi *et al.* 2001). Attempts to use such biochemical models to model canopy photosynthesis in particular are complicated (e.g. de Pury and Farquhar 1997), difficult to parameterise and rely on measurements of the nitrogen distribution and leaf area index of the modelled crop in the calculations. Others (Ewert 2004) have argued that progress in modelling the responses of crops to environmental change is unlikely without a change of emphasis from modelling the details of photosynthesis to modelling canopy development and structure. However, the Farquhar model is very useful when trying to predict the combined effects of pollutant gases, such as ozone, and CO_2 on leaf and crop photosynthesis (Ewert and Porter 2000), as will be seen later.

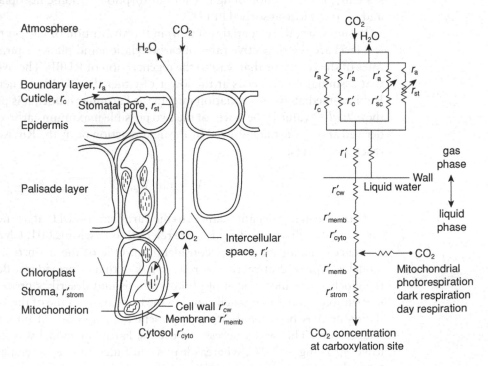

Figure 4.7 Pathway for CO_2 entry into a leaf in the light shown as a diagram of the structures involved and an electrical analogue with resistances denoting sites of restricted diffusion of CO_2 gas (after Lawlor 2001).

4.3.2   Photosynthesis as a leaf diffusive process

The CO_2 concentration within the leaf (c_i) plays a central role in the biochemical processes of photosynthesis, in particular. CO_2 assimilation can be measured and modelled as a function of c_i and can be interpreted in terms of the competitive rates of oxygenation and carboxylation reactions. However, CO_2 must move from air, with a concentration of c_a, external to the leaf, into the leaf and then to the sites of carboxylation in the mesophyll cells. This movement is largely a diffusive process, initially gaseous and then in liquid water. CO_2 molecules move from the bulk air to the site of fixation down a concentration gradient, created by the consumption of CO_2 by photosynthesis. The concentration of CO_2 at the sites of carboxylation in the chloroplasts (c_c) is lower than both c_a and c_i since CO_2 is continually being removed *via* carboxylation and incorporation into the Calvin cycle. Thus c_a close to the leaf surface is higher than c_i by a factor of about 1.6 under ambient conditions and c_c is about 30% lower than c_i when leaves are photosynthesising at maximal rate (Loreto *et al.* 1992, 1994; Evans and von Caemmerer 1996). The CO_2 diffusion pathway and the resistances within it from the boundary layer over the leaf, *via* the stomatal resistance to CO_2 influx to the leaf internal resistances (Figure 4.7) are important since Rubisco has a low affinity for CO_2. This means that the difference between c_a and c_c, i.e. the size of the complete CO_2 gradient, affects the efficiency of photosynthesis in C_3 species more than would be the case if Rubisco had a higher affinity for CO_2. Thus, the level of c_i is a result of the level of CO_2 conductances (or resistances) into the leaf chamber and the rate at which CO_2 is removed from it by photosynthesis. Both conductances into the leaf chamber and removal of CO_2 from it have units of mol m^{-2} (leaf plan area) s^{-1}. The physical process of CO_2 diffusion into a leaf can be described by Fick's Law, which relates generally the rate of diffusion to the difference in concentration between two points separated by a distance:

$$F = -D((c_1 - c_2)/\Delta z) \qquad (4.11)$$

F is the rate of movement of a substance in a period of time; D is the specific diffusion coefficient for CO_2, which is considered to be constant in air under standard meteorological conditions; and the concentrations are given by c_1 and c_2, separated by the distance, Δz. It is conventional to indicate a flux from a higher to a lower concentration by a negative sign; thus in Equation 4.11, c_1 is larger than c_2.

$\Delta z/D$ in Equation 4.11 has the general units of time per unit distance and can be expressed, for example, as s m^{-1}, i.e. that of a resistance. Fick's Law can be seen as a particular example of the more general physical relationship linking a rate of flux to a resistance and a potential concentration gradient or difference. In applying Fick's Law to leaves, Equation 4.11 can be rewritten as:

$$F_{CO_2} = (c_a - c_i)/R = \Delta CO_2/R \qquad (4.12)$$

where R is the total resistance of the leaf to the inward movement of CO_2 and ΔCO_2 is the difference in CO_2 concentration between the bulk air and the intercellular space. If $\Delta z/D$ is a resistance, R, then its inverse $D/\Delta z$ is equal to $1/R$ or a conductance with units of m s^{-1}, which is also a rate or speed. Conductances can thus be thought of as the speed with which CO_2 molecules move from the bulk air external to the leaf to the sites of carboxylation in the chloroplasts, passing through a series of 'road-blocks' or resistances on the way. Expressed thus, Fick's Law is analogous to the familiar Ohm's Law that links the flow, i.e. electron flux or current, of electricity to a potential difference, i.e. a concentration gradient, and a resistance in an electrical circuit. Conductances vary in proportion to the flux but in a system with more than one resistance, the resistances can be summed in series or parallel to give a total resistance for the pathway. Strictly, this is under the condition that a pathway is unidirectional (Parkhurst 1994), which is often not the case in plants where, for example, leakage of CO_2 out of bundle-sheath cells in C_4 photosynthesis complicates the measurement of leaf internal CO_2 transfer conductances in crops such as maize and sorghum.

Figure 4.7 shows the many resistances to CO_2 diffusion, but four resistances dominate the flow of CO_2 from the air outside the leaf to the sites of carboxylation in the chloroplasts. These are the boundary layer resistance r_a, the stomatal resistance r_s, the cuticular resistance r_c and the mesophyll resistance r_m. The boundary layer resistance operates in series with the other resistances but the stomatal and cuticular resistances operate in parallel with each other (Figure 4.7). The domains of these resistances are outside the leaf surface (r_a), the leaf surface (r_c), the stomata (r_s), and the internal leaf intercellular spaces and carboxylation sites within the chloroplasts (r_m). Until CO_2 crosses into the mesophyll cell wall, it is in a gaseous form; thereafter it is dissolved in liquid water (Figure 4.7). As a result, environmental conditions, such as humidity, wind speed and radiation level and plant phenotypic characters such as leaf form and stomatal density affect the size of the resistances in the gaseous phase more than in the liquid phase. Conductances in the liquid phase are more influenced by factors such as leaf age, leaf internal anatomy and the efficiency of the photosynthetic processes described above. Following Figure 4.7, Equation 4.12 can be expanded to give the net CO_2 flux or rate of photosynthesis (P_n) over the whole pathway as:

$$P_n = (c_a - c_c)/(r_a + (1/r_s + 1/r_c) + r_m) \tag{4.13}$$

The still air that surrounds a leaf and separates it from the turbulent air of the atmosphere is the source of the boundary layer resistance (r_a). Even in very still conditions there is turbulent air above a leaf as a result of the convective exchange of heat between plant and atmosphere. Turbulent air consists of small revolving packets of air, known as eddies, which are established as a result of the frictional drag on air from its contact with a surface, such as a leaf or the ground. Mass (e.g. CO_2 and water) and energy can be exchanged between a crop and the atmosphere *via* the revolving eddies. As would be expected, the turbulent mixing of air increases with wind speed so that the CO_2 concentration above a crop is more constant on a windy day than on a calm one (Figure 4.8).

Equation 4.14 is an empirical relation (after Nobel 1974) linking boundary layer thickness (δ) to wind speed and the downwind dimension of a leaf with parallel sides, such as wheat or maize;

$$\delta = Kd^{0.5}/u^{0.5} \tag{4.14}$$

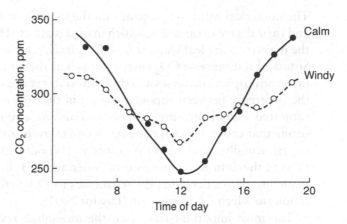

Figure 4.8 The effect of wind conditions on the CO_2 concentration above a maize canopy during the day (after Uchijima 1970).

where, K is a constant, d is the downwind dimension and u is the wind speed. Thus, boundary layers increase for long parallel-sided leaves in still conditions (Nobel 1974) and the thickness of the leaf boundary layer determines the laminar resistance to gas flow and the convective heat flux of plants. An empirical relationship (Hay and Walker 1989) linking r_a (s m^{-1}) to δ (cm) for a cereal leaf is:

$$r_a = 633.2\delta - 6.3 \quad r^2 = 0.99 \tag{4.15}$$

How r_a changes generally with δ is complicated by such factors as leaf hairiness and little is known of the genetic control of the factors controlling leaf boundary resistance, even though it represents an important link between plants and the atmosphere. For cereal leaves such as wheat and maize, which are long relative to their breadth, r_a values range from 100 s m^{-1} for wind speeds of the order of 3–4 m s^{-1} to 900 s m^{-1} for very calm meteorological conditions. Since wind speeds can be much lower within a canopy than at the top, boundary layer resistance can be an important barrier to CO_2 influx into leaves low in the canopy. However, low levels of PAR deep in the canopy are more likely to be limiting for photosynthesis than boundary layer limitations on their own.

In comparison with the resistance offered by the boundary layer to CO_2 movement, that offered by the stomata (r_s) is generally larger, but in still conditions r_a can dominate the total external (i.e. r_a plus r_s) leaf resistance. At the maximum values for both, the ratio r_s/r_a is about 10–20 from which it is clear that in most cases the major diffusive resistance to CO_2 is at its point of entry into a leaf. This ignores the resistance to CO_2 provided by the waxy leaf cuticle that is many orders of magnitude larger than either r_a or even r_s and offers, in effect, total blockage to CO_2 influx. From Equation 4.16 it can be seen that as r_c is very much larger than r_s and r_a, then r_l is basically proportional to ($r_s + r_a$), and it is generally the case that r_s offers the larger resistance. If stomata are closed then r_s equals r_c.

$$r_l = (1/r_c + 1/r_s) + r_a \tag{4.16}$$

The numerical value of r_s depends on the number of stomata per unit leaf area and their degree of openness, which in turn is affected by many factors, of which the main ones are leaf water status, CO_2 level, air humidity and radiation level. Stomatal resistance to CO_2 entry increases as the air dries and/or the CO_2 level in the atmosphere increases and the plant water status decreases. It reduces when the difference between vapour pressure in the air and that inside the leaf, the saturated vapour pressure, reduces and also as a result of toxic gases such as ozone that cause stomatal opening. Stomatal resistance is commonly measured experimentally by diffusion porometers (Woodward and Sheehy 1983). The ratio of the diffusion coefficients of water and CO_2 in air ($CO_2 = 1.6 \times H_2O$) is made use of in calculating the resistances to CO_2 influx into a leaf under conditions in which c_a/c_i is constant (Lawlor 2001).

The most important feature of the mesophyll resistance, in contrast to r_a and r_s, is that it contains diffusional components in line with Fick's Law but biochemical events and structures, such as cell walls and membranes, occurring within the liquid phase and within the mesophyll cells and chloroplasts, contribute to r_m (Figure 4.7). Thus factors that influence r_m include the biochemical processes of photosynthesis outlined above. Conceptually, r_m can be disaggregated into a more physically defined transport resistance component that covers the diffusion of CO_2 in solution from the walls of the mesophyll cells to the chloroplasts and which can be understood, if not measured, as a diffusion process. The remainder is thought of as a biochemical resistance (r_x) that is related to the efficiency of CO_2 carboxylation. This comprises both the photochemical and Rubisco limited processes described above. It is the latter that is of most interest as a mesophyll resistance, because the transport component contributes only about 0.2 s m^{-1} of the total r_m of about 200–1000 s m^{-1} for C_3 plants and 200 s m^{-1} of C_4 plants. The question is how to arrive at a value of the biochemical component.

An early method of estimating r_m was based on the idea that the value of c_c was the same as the CO_2 compensation point (Γ) at which P_g exactly balances R_m plus R_p (Equation 4.2), making P_n zero. Equation 4.13 describes the diffusion of CO_2 through all of the four identified resistances, and can be rewritten to apply solely to r_m as:

$$P_n = (c_i - c_c)/r_m \tag{4.17}$$

Thus,

$$P_n = (c_i - \Gamma)/r_m \tag{4.18}$$

or

$$r_m = (c_i - \Gamma)/P_n \tag{4.19}$$

If P_n is measured at high light levels but with different intercellular CO_2 concentrations, the photochemical limitation is removed and the 'true' mesophyll

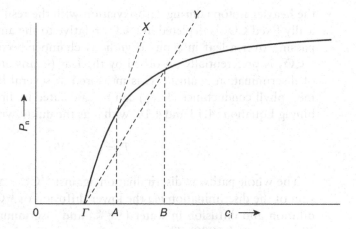

Figure 4.9 The response of net photosynthesis (P_n) to intercellular CO_2 concentration c_i and the relation between the initial slope of the relationship to the mesophyll resistance, r_m. The slope (Γ to X) of P_n for values of c_i between Γ, the CO_2 compensation point, and A is the carboxylation efficiency (CE in Equation 4.20) and r_m equals 1/CE. The reciprocal of the lower slope (Γ to Y) in P_n seen for c_i between Γ and B, is the sum of r_m and CO_2 intercellular diffusion resistance and structural resistances (r_{id}). Thus, the difference between the reciprocals of the two slopes gives r_{id} (after Jones 1983).

resistance is the result. If the carboxylation efficiency (CE) is defined as the amount of P_n per unit of c_i, then CE can be written as:

$$CE = P_n/(c_i - \Gamma) \qquad (4.20)$$

As Equation 4.20 is the reciprocal of 4.19, r_m equals 1/CE. Thus, the relationship between P_n and c_i, in its linear phase in the region of the CO_2 compensation point, provides an estimate of CE, as shown in Figure 4.9.

In Figure 4.9, at higher values of c_i, P_n approaches an asymptote, where P_n does not increase much with c_i. Other non-biochemical resistances, such as mesophyll cell wall and membrane barriers and intercellular diffusion are thus playing a role at this point. Thus at the CO_2 concentration, marked point B in Figure 4.9, CE is affected by a 'true' carboxylation resistance plus the diffusion and structural resistances in the gaseous and liquid internal leaf phases (Figure 4.7). The difference between the reciprocals of the 'true' mesophyll CE and the total CE gives the residual resistance of the gaseous and liquid phases. An important assumption in the above reasoning is the equivalence between Γ and c_c, or that the CO_2 concentration at the sites of carboxylation equals the CO_2 compensation point. Lawlor (2001) has argued that these cannot be synonymous since, by definition, P_n is zero at Γ but c_c continues to result in a positive P_n. Thus there has been a need to measure c_c and this led to the development of two techniques to measure the resistance to CO_2 movement within a leaf.

The first technique relies on the fact that about 1.1% of c_a is composed of the heavy C isotope, ^{13}C, and making [13]CO_2 heavier than the more ubiquitous [12]CO_2 and thus slower to diffuse in air and water. Rubisco discriminates against

the heavier isotope during carboxylation with the result that any photosynthetically fixed CO_2 is depleted in $^{13}CO_2$ relative to the ambient air. Therefore, air
passing over a leaf in a photosynthesis chamber becomes richer in $^{13}CO_2$ as
$^{12}CO_2$ is preferentially absorbed by the leaf (Evans *et al.* 1986). If the degree
of discrimination against ^{13}C is measured at several levels of irradiance, then
mesophyll conductance ($1/R_m$) can be calculated by first calculating c_i by combining Equations 4.13 and 4.16, with c_i as the unknown:

$$P_n = (c_a - c_i)/r_l \qquad (4.21)$$

The whole pathway discrimination against $^{13}CO_2$ (Δ, ‰) is calculated as the
sum of the discrimination *via* the lower diffusion of $^{13}CO_2$ in air (4.4‰), lower
dilution and diffusion in water (1.8‰) and discrimination by Rubisco (29‰)
(Farquhar *et al.* 1989). Thus:

$$\Delta = 4.4(c_a - c_i)/c_a + 1.8(c_i - c_c)/c_a + 29c_c/c_a \qquad (4.22)$$

At low irradiances c_i almost equals c_c and the middle term in Equation 4.22
approaches zero, but as irradiance increases so do P_n, the difference between c_i
and $c_c(c_i - c_c)$ and Δ. By analogy with Equation 4.21:

$$P_n = (c_i - c_c)/r_m \qquad (4.23)$$

where P_n, in this case, is the rate of the carboxylation reaction. A plot of P_n for
different irradiances against Δ gives a line with slope r_m. Thus, with simultaneous
measurements of c_a, P_n and Δ it is possible to calculate the mesophyll resistance.
Further details of these calculations are given in Evans and von Caemmerer
(1996) as is a description of how measurement of chlorophyll florescence and
calculation using the equations linking the rate of electron transport to NADPH
in photosystem II (Gentry *et al.* 1989) also can be used to calculate r_m (Harley
et al. 1992). Where these two methods of calculating r_m have been compared
(Loreto *et al.* 1992) they have returned similar values.

Taking an overview of the size of the various resistances in the pathway of CO_2
from ambient air to the sites of carboxylation, r_a is about 300–400 s m^{-1} for
wheat leaves in low wind speeds but drops to about 40 s m^{-1} with increasing wind
speed; r_s ranges from about 100 s m^{-1} when stomata are fully open to 5000 s m^{-1}
when stomata are closed and r_m varies between 200 and 1000 s m^{-1} in C$_3$ plants
but is under 200 s m^{-1} in C$_4$ plants. From these approximate but indicative values,
it can be seen that CO_2 diffuses more slowly in that part of the leaf where it
encounters water – such as the films of water surrounding the mesophyll cells
and at 15°C, the diffusion coefficient for CO_2 in water is 10^{-4} that of its value in
air. A strong inverse correlation has been found between P_n and r_m for a wide
range of leaves of different species (Loreto *et al.* 1992, 1994) and it is has been
well established that c_c changes from about 70% of the value of c_i when leaves
are photosynthesising in high irradiances to almost equal c_i at low irradiances.

In summary, as CO_2 moves from the ambient atmosphere to the sites of
carboxylation in the chloroplasts it encounters a series of resistances. Under

reasonably turbulent conditions, the resistances generally increase as CO_2 moves from external air into the water-saturated body of the leaf. Some of these resistances are more easily measured and estimated than others and they also vary by orders of magnitude in their absolute values. Finally, the resistances or conductances can be affected by factors such as leaf nitrogen and water stress, topics that will be considered later.

4.3.3   Photosynthesis as a crop canopy process

As assemblages of leaves and shoots, canopies possess different micro-meteorological and physiological properties from leaves or cells. One of the most striking ways in which canopies differ from these lower-scale structures is in their photosynthetic responses (Figure 4.10).

At the leaf level (Figure 4.10a and see also Figure 3.28) the light response curve has its characteristic asymptotic shape of declining photosynthetic response with increasing irradiance. However, at the canopy level and over the same range of irradiance, canopy net photosynthesis has a much more linear response (Figure 4.10b) and an absence of saturation at higher irradiance levels (Section 3.4.2). The impetus for this insight was the realisation in the early part of the twentieth century that light interception was critical in plant growth coupled to the fact that light is not evenly distributed down a leaf canopy, with many leaves in the shade. Erectophile canopies, in which leaves are held more vertically, have a more even distribution of radiation among leaves than more planophile, or horizontal, leaf postures. This topic has been thoroughly reviewed by Sinclair and Muchow (1999) to whom the reader is referred for further details.

Theoretical and experimental studies have shown that RUE has a fairly constant value across a wide range of latitude and sites and is insensitive to solar

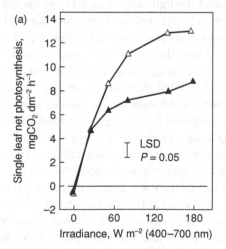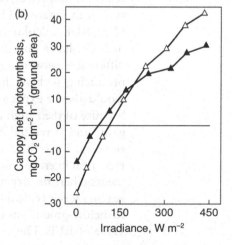

Figure 4.10 Light response curves of (a) leaf net photosynthesis and (b) canopy net photosynthesis at high (open triangles) and low (filled triangles) nitrogen levels from communities of perennial ryegrass (*Lolium perenne*) (after Robson and Parsons 1978).

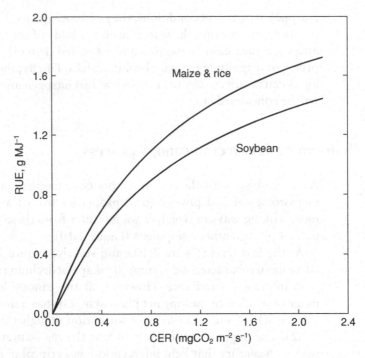

Figure 4.11 Calculated RUE (radiation use efficiency) (total radiation basis) as a function of light-saturated leaf photosynthetic rate (CER, CO_2 exchange rate) (Sinclair and Horie 1989).

altitude and leaf angle for leaf area indices (m^2 leaf m^{-2} ground) of less than about six. Besides varying among species, RUE is mainly sensitive to changes in the maximum rate of leaf photosynthesis, the proportion of the radiation that is diffuse or direct, and the nitrogen status of the canopy. The theoretical relationship between RUE and the light-saturated CO_2 exchange rate (CER) is shown in Figure 4.11 for maize, rice and soybean (Sinclair and Horie 1989). Here, RUE, calculated in terms of total intercepted radiation, which is twice that of PAR, has a curvilinear relationship with CER, and there are the expected differences between crop species: that, for the same CER, the protein- and lipid-producing soybean has a lower RUE than maize or rice. Similarly low RUE would also be expected in perennial forage crops, such as lucerne (alfalfa) both because of their high protein content but also because RUE involves dry matter gain, and perennial forage crops grow initially on the basis of stored root reserves. Thus, their apparent dry matter increases will not keep pace with their radiation interception, with low apparent RUE the result. Conversely, in experiments where the dry matter of lucerne roots and shoots are both measured, RUE reaches levels typical for C_3 species (Table 4.2). Lucerne devotes a lot of its DM to below-ground storage so that measuring the shoots alone will give a low value of RUE. The essential point is that the value of RUE for a crop depends not only on whether it uses the C_3 or C_4 pathway (Section 4.3) but also on the form of its growth cycle and the subsequent partitioning of dry matter among plant parts.

Sinclair and Muchow (1999) argued that the nature of the response of RUE to saturated CER was important in explaining the stability of RUE. Thus, in Figure 4.11, halving CER from the extremely high level of 2 $mgCO_2$ m^{-2} s^{-1} (equivalent to *c.* 45 μmol m^{-2} s^{-1}) only reduces RUE (total radiation) from 1.6 g MJ^{-1} to about 1.2 g MJ^{-1}. The commonly observed total radiation RUE for non-legume C_3 crops of 1.4 g MJ^{-1} implies a maximum CER of *c.* 27 μmol CO_2 m^{-2} s^{-1}, again a high value compared with many measured values.

One possible cause of variation in RUE could arise because of differences in nitrogen content per unit leaf area (g m^{-2}). The reasoning is that Rubisco levels are directly related to leaf nitrogen level (Section 7.2.2) and, as CER is a light-saturated rate, then this should be sensitive to Rubisco. When this has been examined it has been found that RUE again varies little over a range of leaf nitrogen values commonly found in crops (Figure 4.12) but that differences are found among crop species. However, for crops in which nitrogen is limiting there is a sharp reduction in RUE with declining leaf nitrogen content (Figure 4.12). This can be expected to be seen more commonly in natural plant communities. The evidence also seems to be that the distribution of nitrogen within the leaf canopy (Section 7.2.2) has little effect on the RUE and that it is the average level of leaf N per unit leaf area that is important. This may be because most studied crops are well supplied with N and thus even leaves with the lowest N contents are in the asymptotic part of the curves shown in Figure 4.12. By the same

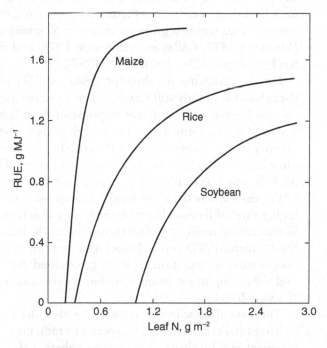

Figure 4.12 Calculated RUE (total radiation basis) as a function of mean leaf N content per unit area for crop canopies of maize, rice and soybean (CER, CO_2 exchange rate) (Sinclair and Horie 1989).

reasoning, leaf N distribution may be more important in determining RUE for crops that are short of N.

The factor that has been found to influence RUE the most is the composition of radiation in terms of the proportion of direct to diffuse radiation received by the crop. Diffuse radiation reaches the crop canopy from a range of angles owing to scattering by clouds and atmospheric particles. In terms of providing radiation to leaves within a canopy, diffuse radiation is more effective for two reasons. The first is that the sum of direct radiation plus scattered diffuse radiation decays more slowly down a canopy than simple direct radiation, and thus penetrates the canopy more deeply; second, scattered diffuse radiation is available to all leaves regardless of whether they are directly sunlit or in the shade and thus 'evens out' the non-linearity of the photosynthetic light response curve. Hammer and Wright (1994) explored these considerations in detail and concluded that increasing the fraction of the diffuse component on a cloudy day accounted for an increase in RUE of 0.15 g MJ^{-1}. Counterbalancing this statement is the fact that although the efficiency of diffuse radiation use by a canopy may be higher than that for direct radiation, a very cloudy day will not deliver as much total radiation as a sunny day and thus the dry matter produced will always be higher on a bright sunny day than on a cloudy day, all other things being equal.

RUE has been criticised as a concept because it embodies circularity in that one cannot have dry matter growth without radiation interception and vice versa: RUE is a spurious correlation because it is based on a relationship between two cumulative variables. However, in spite of strict statistical disentitlement it has proved to be a durable and unifying concept in crop physiology, increased understanding of crop growth, and has acted as a mechanistic index of growth since the idea was first given precision by Monteith and colleagues in the 1970s (Monteith 1977; Gallagher and Biscoe 1978) and the mechanism behind RUE has been explored by Dewar *et al.* (1997).

Finally, reviewing the three processes of CO_2 accumulation in plants from the cell to the canopy still leaves one very interesting question: how is it that the seemingly universal non-linear response of P_n to light and/or CO_2 at the cell or leaf level is transformed into a linear relationship between intercepted radiation and dry matter accumulation at the crop level (Figure 4.10)? The answer to this question is mostly to be found in the arrangement of leaves in a canopy (Chapter 3). It is only leaves at the top of a canopy that are likely to be in the range of PPFD levels where the photosynthetic response is saturated and the limitation to higher rates of fixation lies with the level and activity of Rubisco. Leaves deeper in the canopy are likely to find themselves in the linear, RUBP limited, portion of the P_n versus PPFD curve (Figure 4.6) and thus have a linear response of CO_2 assimilation to radiation level. A generalised representation of the radiation and other important profiles within a crop canopy is shown in Figure 4.13 (Lawlor 2001).

The overall conclusion seems to be that the non-linearity of the response of leaves and cells of mainly C_3 species to radiation is the result of mainly photochemical and biochemical processes, whereas the linearity of response at the crop level is the result of the physical placement of leaves and shoots and their physical interaction with incident radiation.

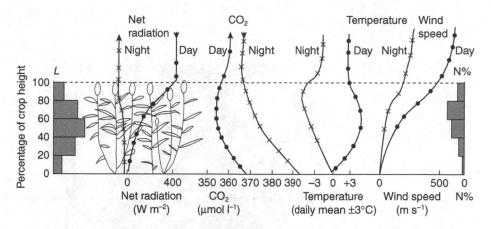

Figure 4.13 Generalised profiles of leaf area index (*L*) and % of above-ground nitrogen (%N) in different leaf strata of a wheat crop at maximum *L*. Interception of incident radiation during the day in relation to *L* is indicated. The profiles of CO_2, temperature and wind speed are shown for day and night situations and the direction of the CO_2 and radiation fluxes are indicated (after Lawlor 2001).

4.4 The C_4 photosynthesis mechanism

C_4 plants form an important category of crop plants, originating from and still mainly grown in warm dry areas. The C_4 method of photosynthesis can be interpreted as a set of adaptations to a combination of high temperature and irradiance, although C_4 species do occur naturally in a range of other environments (Fitter and Hay 2002). C_4 photosynthesis is characterised by a large reduction in photorespiration when compared with C_3 plants, and lower mesophyll resistances to CO_2 movement, leading to significantly lower water use per unit of dry matter produced. The most important C_4 crops are maize, sugar cane, sorghum and millet. Whereas, at one time, there was optimism about the opportunities given by C_4 photosynthesis to increase crop yields, now it is generally understood that C_4 plants will not, in all circumstances, give higher dry matter production than C_3 plants. This is because the optimal climatic and growing conditions for photosynthesis in the two plant types are not identical.

C_4 plants minimise photorespiration by concentrating CO_2 using an alternative CO_2 acceptor, with a higher affinity for CO_2, and a series of carriers that delivers CO_2 at high concentrations into the Calvin cycle. As indicated earlier (Section 4.3), the functioning of Rubisco as a carboxylating or an oxygenating enzyme in the Calvin cycle depends on the relative concentrations of CO_2 and O_2. The action of concentrating CO_2 by the combination of a specialised biochemical pathway and a particular arrangement of the so-called Kranz (German for wreath) cells around the vascular bundles within a leaf (Figure 4.14), largely suppresses the oxygenase activity of Rubisco.

The mesophyll cells that surround the Kranz cells are rich in chloroplasts but these are not the sites of the initial CO_2 fixation; Rubisco is not the primary enzyme involved in fixation, as in C_3 plants, and sugars are not synthesised at this point. The primary carboxylation in all C_4 plants is the conversion of

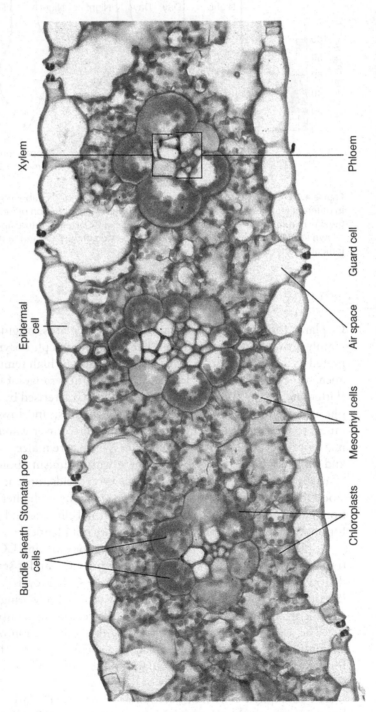

Figure 4.14 A portion of the cross-section of a leaf of maize (*Zea mays*), which is a C$_4$ plant, showing the Kranz bundle sheath cells surrounding the xylem and phloem. Mesophyll cells surround the Kranz cells and show large numbers of well-developed chloroplasts (from Stern *et al.* 2003).

phosphoenolpyruvate (PEP) to the 4-carbon acid oxaloacetate by the binding of CO_2 with the 3-carbon PEP, catalysed by the enzyme PEP carboxylase. This enzyme is restricted to the cytosol of the mesophyll cells. Oxaloacetate is short lived and is rapidly reduced to malate in maize, for example. It is the translocation of the 4-carbon acids from the many mesophyll cells to the fewer Kranz cells and their decarboxylation therein to inject CO_2 at high concentrations into the Calvin cycle that epitomises the C_4 process. In maize and sugar cane, the major pathway of decarboxylation involves malate surrendering its CO_2 and regenerating pyruvate that is phosphorylated to PEP as the acceptor, as shown diagrammatically in Figure 4.15.

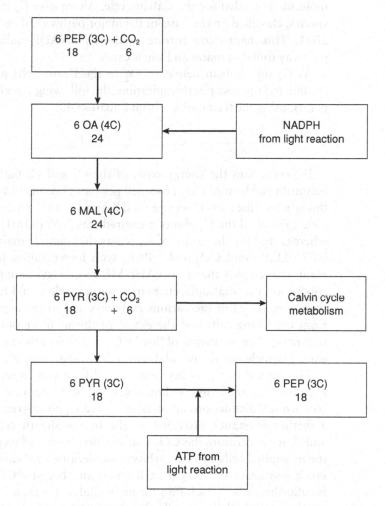

Figure 4.15 Diagram of the principal steps in one form of the C_4 cycle showing the major transformations. The number in brackets is the number of C atoms per molecule. 18 C atoms are conserved at each step in the cycle to insert CO_2 into the Calvin cycle. The 6 PEP molecules left at the end of the cycle form the receptors for the next six molecules of CO_2, catalysed by the enzyme PEP carboxylase. Abbreviations: PEP, phosphoenolpyruvate; OA, oxalacetate; MAL, malate; PYR, pyruvate; ATP, adenosine 3-phosphate; NADPH, reduced nicotinamide adenine dinucleotide phosphate.

As with the C_3 cycle, light generates the reducing agent (NADPH) and the energy carrier ATP but more of these are needed to drive the C_4 cycle than the C_3 cycle and thus there is an extra energy cost to C_4 photosynthesis. In addition to the three ATP and two NADPH molecules needed to fix one molecule of CO_2 in the Calvin cycle, one NADPH and two ATP molecules are needed to generate the 4-carbon acids and decarboxylate CO_2 in the mesophyll and Kranz cells. There is a net saving of one NADPH in the C_4 cycle since one molecule is generated in the conversion of malate to pyruvate. The net result is that, of a total of five ATP molecules required to fix CO_2 into carbohydrate using the C_4 cycle, three are required for the Calvin cycle and two for the PEP-oxalacetate-malate-pyruvate-PEP pathway. One NADPH molecule is needed for the C_4 pathway and a net one molecule is needed for the Calvin cycle. Alternative C_4 pathways exist in other species, classified on the basis of the major pathway of decarboxylation (Lawlor 2001). This chapter concentrates on the main NADP-malic enzyme (NADP-ME) pathway found in maize and sugar cane.

As C_4 metabolism delivers CO_2 to the Kranz cells at concentrations high enough to suppress photorespiration, the following equation describes P_n in C_4 plants, using the terminology from Equation 4.2:

$$P_n = P_g - R_m \qquad (4.24)$$

Differences in the energy costs of the C_3 and C_4 pathways infer that their quantum yields (mol CO_2 absorbed per mol photons of PAR) should differ and this is indeed the case. Ehleringer and Pearcy (1983) found that the mean quantum yield (SE) of all the C_3 plants measured was 0.05 (0.001) mol CO_2 mol^{-1} PAR, whereas that for the group of C_4 plants that include maize and sugar cane was 0.07 (0.001) mol CO_2 mol^{-1} PAR, with lower values for the dicotyledenous plants that employ the same NADP-ME decarboxylation pathway. Their overall conclusion was that differences in quantum yield could be linked to the respective energy costs of the various pathways and to the degree of leakage of CO_2 from the Kranz cells and the extent of the small amount of photorespiration occurring. Any refixation of 'lost' CO_2 by the 4-carbon pathway would impose extra photochemical costs and thus reduce quantum yields.

The main differences between C_3 and C_4 leaves in terms of pathway resistances to CO_2 are found within leaves and occur because of the anatomical and biochemical distinctions involved in C_3 and C_4 photosynthesis. C_4 leaves possess a further resistance provided by the bundle-sheath cells that decarboxylate and thus concentrate the CO_2 that has first been carboxylated into C_4 acids in the mesophyll cells by PEP carboxylase. Because PEP carboxylase discriminates much less against $^{13}CO_2$ than Rubisco and because CO_2 tends to leak from bundle-sheath cells back into the intercellular spaces, it is not possible to measure the mesophyll plus bundle-sheath resistance to CO_2 diffusion as is possible in C_3 plants. However, it is possible to calculate a maximum mesophyll transfer resistance from c_i to c_c in C_4 plants, if it is assumed that c_c equals Γ. Evans and von Caemmerer (1996) show that at 25°C, under high irradiance and at ambient CO_2, c_i in C_4 species is about 10 Pa, whereas in C_3 species it is about 25 Pa. Γ in C_4 species is close to 0 Pa and about 5 Pa in C_3 species. Thus following Equation

4.19 with a common value (x) of P_n for C_3 and C_4, shows that the minimum diffusion resistance in a C_4 species, $x/(10 - 0)$, needs to be about half that in a C_3 species, $x/(25 - 5)$, to maintain the same rate of net photosynthesis. Conductance across the bundle-sheath cells is 10–25 times lower than that in the mesophyll but this is thought to be essential for the C_4 process as the high resistance prevents the leakage of the decarboxylated CO_2 out of the bundle-sheath. C_4 plants have evolved a photosynthetic metabolism that can operate at higher rates than C_3 plants, especially at temperatures above 30°C, but which is built up of a resistance pathway that has a longer overall path length and in parts has higher barriers to CO_2 diffusion than C_3 plants. The nature of leaf resistance to CO_2 in C_4 plants seems as much to keep CO_2 within the plant as to prevent its influx and incorporation.

The significance of the C_4 pathway for crop production is seen mainly in the temperature and irradiance responses of the photosynthesis of crops that have either C_3 or C_4 metabolism. Photorespiration increases with temperature in C_3 species, because O_2 is more soluble than CO_2 at higher temperatures and the carboxylase affinity of Rubisco for CO_2 declines with temperature. The quantum yield of C_3 plants has been measured to halve when leaf temperature is increased from 16°C to 36°C (Hay and Walker 1989; Lawlor 2001). Over the same range, C_4 species show no change in quantum yield but they do exhibit lower quantum yields than the C_3 plants at temperatures in the range 15–20°C. Thus, there is a temperature crossover point for the quantum yields of C_3 and C_4 plants at which the loss of quantum efficiency of C_3 plants with temperature is balanced by the higher intrinsic energy costs of C_4 photosynthesis (Figure 4.16a). At leaf temperatures lower than about 16°C, C_3 plants have higher quantum yields but from 16°C to 28°C, C_4 species are increasingly more efficient.

The implications of these temperature responses are that, at low temperatures and radiation levels, rates of C_3 photosynthesis will be higher than those of C_4 photosynthesis because of the combined effects of temperature on the RUBP- and Rubisco-limited portions of the P_n versus PPFD curve. Even at saturating irradiance

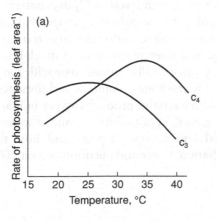
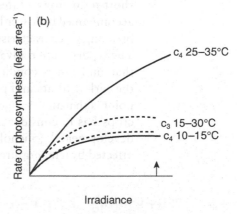

Figure 4.16 (a) Typical temperature response curves for photosynthesis in C_3 and C_4 plants. (b) Typical photosynthetic light response curves of C_3 and C_4 plants at different temperatures. The arrow indicates full sunlight (after Edwards and Walker 1983).

but low temperatures, C_3 plants can outperform C_4 species (Figure 4.16). However, at high temperatures and high irradiances the combination of the CO_2 concentrating mechanism and the absence of a reducing effect of temperature on quantum yield means that even at full sunlight C_4 photosynthesis may not be saturated. For these reasons, it is predictable that C_4 plants will be less responsive to increases in ambient CO_2 concentration than C_3 plants at high temperatures and irradiances.

The temperature and radiation responses of C_3 and C_4 species go some way to explain the natural distribution of C_3 and C_4 grasses that are the close relatives of many of the current staple human food crops; humankind largely has members of the Gramineae at the base of its food chain. Thus, maize is restricted in its range from low latitudes to about 50° from the equator; wheat extends up to perhaps 60° where oats, rye and temperate forage grasses take over. The most important legume crop, soybean, ranges from low latitudes to about 50°. These are only broad ranges and there are exceptions – for example sunflower and rice, both C_3 plants, are adapted to high irradiance environments and approach photosynthetic saturation at comparable irradiances to those of C_4 maize. Conversely, maize is better adapted to lower temperatures and irradiances and has a much wider distributional range than other C_4 plants such as sorghum and millet. It is not possible to discuss the differences between C_3 and C_4 plants without mentioning differences in their rate of use of water and this is considered in Section 7.

In summary, C_4 crops are consistently able to produce higher levels of dry matter in those mainly tropical and subtropical dry zones, characterised by high temperature and radiation, to which they are adapted. Where the productivity of C_3 and C_4 crops has been compared at similar levels of irradiation, maximum short-term growth rates for C_4 species are in the range 50–54 g dry matter m^{-2} d^{-1} whereas C_3 species have rates of 35–40 g dry matter m^{-2} d^{-1}. (Monteith 1978; Sinclair and Muchow 1999). However, as both C_3 and C_4 cereals are usually determinate in growth habit (Chapter 3), the higher temperatures that C_4 crops are likely to experience more frequently than C_3 crops will tend to lengthen the duration of C_3 relative to C_4 canopies. Overall productivities for C_3 and C_4 crops are not as large as would be predicted by solely considering short-term growth rates. As noted, high levels of C_4 dry matter production are accompanied by lower levels of water use than in C_3 crops, making the notion of breeding C_4 characteristics into C_3 crops an attractive if extremely difficult goal. The C_4 dry matter advantage in temperate situations is much less and it is likely that, under these conditions, C_3 crops will continue to provide the greater part of the carbohydrate and protein in the diets of humans and livestock. The general point is that any evaluation of the relative productivity per unit area of different crops has to consider both growth (photosynthetic and respiratory) and crop developmental (phenological and canopy) aspects and how these are both affected by temperature, radiation, water and nutrition (Ewert 2004).

4.5   Water shortage and photosynthesis

It is important to understand the effects of lack of water on photosynthesis in terms of stomatal and non-stomatal influences. Two general points need to be made. The first is that, in addition to its photosynthetic effects, drought affects both the expansion and senescence of the leaf canopy (Sections 3.2, 3.3). Such canopy responses normally take effect earlier than those on CO_2 exchange and can influence dry matter accumulation more drastically than those on photosynthesis. Second, there is little scientific consensus about which components of the dry matter balance of plants are affected by drought and even less about any mechanisms. In his in-depth treatise on photosynthesis, Lawlor (2001) concluded that 'Despite many studies of the effects of water stress on basic photosynthetic metabolism, there is no generally accepted model of the processes occurring in dehydrating leaves'. Water stress has two interacting causes; it can be the result of high atmospheric demand for water that tends to be short-lived or it can be a more gradual and longer-term process associated with soil drying. The situation in the field for crop plants is likely to be a combination of the two, thus adding a further layer of complexity.

The majority of crop plants are classified as mesophytes, meaning that they are not especially adapted to survival in dry conditions. In contrast to xerophytes that can survive leaf water potentials (ψ) as low as -10 MPa, most crop plants start to exhibit signs of physiological damage and leaf wilting at about -1.5 MPa (Hsiao *et al.* 1976; Fitter and Hay 2002). Leaf and cell water potential are used as indices of drought or water stress in plant physiology, with the convention that ψ is said to drop and drought increase as water potential becomes more negative. Water flows from regions of higher to lower water potential (less to more negative); thus water flows from a soil at field capacity where ψ_{soil} is zero to the leaves where ψ_{leaf} can be -0.5 MPa to -1.0 MPa on a day when leaves are freely transpiring. Water stress for a typical mesophyte has been characterised into three classes (Hsiao 1973): mild stress, ψ between 0 and -0.5 MPa; moderate stress, ψ between -0.5 MPa and -1.2 to -1.5 MPa; severe stress, ψ below -1.5 MPa. Further details regarding the calculation and use of ψ in plant ecophysiology can be found in Fitter and Hay (2002, Chapter 4). The temporal development of a crop drought, in terms of the changes in water potentials and crop transpiration, is described in Section 7.1.1 (Slatyer 1967).

Mesophytes exhibit a gradation in the effects of water stress on physiological processes such that 'sink' related processes such as cell and leaf expansion and synthesis show reduced rates at water potentials of -0.3 to -0.4 MPa (Hsiao *et al.* 1976; Connor and Jones 1985). Increased abscisic acid synthesis, stomatal closure, reduced CO_2 fixation and effects on respiration are seen from -0.25 MPa to -1.5 MPa and extensive disruption of xylem and phloem function occurs between -1.0 MPa and -2.0 MPa at which point permanent wilting is well established (Fitter and Hay 2002). Thus growing tissues are more sensitive to drought than growth processes that are in turn more affected than transport within the plant. Complications in all considerations of the effects of water stress are first, that as a soil becomes drier, it transports fewer nutrients to roots thus confounding nutrient and water stress; second, stomatal closure reduces evaporative cooling and thereby increases plant temperature with a multitude of growth and developmental consequences.

The initial and main effects of water stress on net photosynthesis come about through closure of stomata, with more severe drought leading to increases in

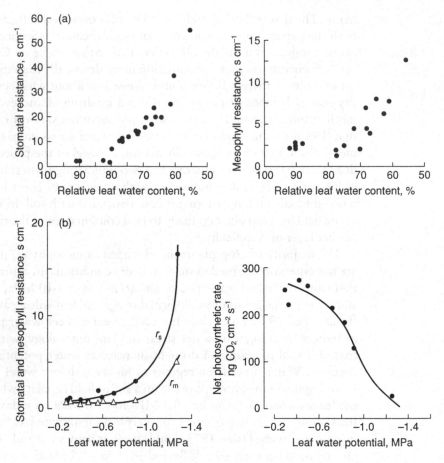

Figure 4.17 Effect of lowering leaf water potential on the stomatal and mesophyll resistances in (a) cotton and (b) millet, with associated effects on P_n (after Troughton 1969; Ludlow and Ng 1976).

biochemical resistances within the mesophyll cells. The responses of the various resistances to variation in leaf water status (expressed as leaf water potential or relative leaf water content) are illustrated in Figure 4.17.

As relative water content or leaf water potential falls, stomatal and mesophyll resistances increase, with the suggestion in both the C_3 cotton and the C_4 millet that the increase in stomatal resistance precedes that of the intracellular mesophyll resistance. The r_m increases from a baseline of about 250–300 s m^{-1} to about 1100 s m^{-1} over an approximately 50% decrease in leaf water content in the C_3 cotton; r_s rises from about 200–300 s m^{-1} to about 5000 s m^{-1} for the same change in leaf water content. In the C_4 millet, r_m rises from a characteristic-ally very low level to about 500 s m^{-1} whereas r_s rises to about 1700 s m^{-1} at −1.4 MPa, a value close to the wilting point of −1.5 MPa. The net effect of this level of water stress is almost to curtail photosynthesis completely. Although the data in Figure 4.17 describe the general responses to drought of the components of P_n, care must be taken in interpreting them since the level of the CO_2 compensation point (Γ) increases with drought. Ignoring this effect or assuming that

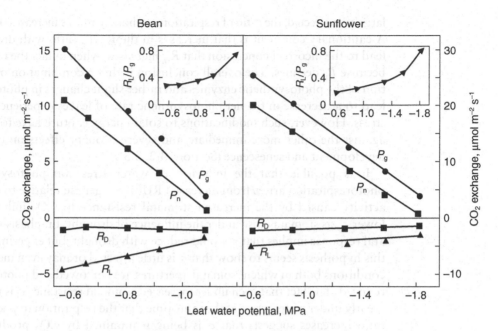

Figure 4.18 Rates of gross (P_g), net (P_n) CO_2 uptake and CO_2 evolution in the light ($R_L = R_p$) and dark ($R_D = R_m$) of bean (*Phaseolus*) and sunflower leaves of various water potentials. The insets show R_L/P_g ratios at the different water potentials (after Krampitz *et al.* 1984).

c_i is close to zero will result in an overestimation of r_m and it is important to emphasise that an increased Γ both indicates and determines reduced net photosynthetic capacity.

A full assessment of the effect of drought on P_n must involve not only an examination of the responses of the CO_2 pathway resistances, but also of the four components of net photosynthesis, P_g, P_n, R_p and R_m, while recognising the near-impossibility of isolating their individual roles. Where this has been attempted (e.g. Krampitz *et al.* 1984; Figure 4.18) there appeared to be little effect of lowered water potential on respiration, except for a reduction in R_p in bean. Conversely, both P_g and P_n fell linearly with declining water potential with the result that the ratio R_p/P_g rose from about 0.3 for mild water stress.

Additionally, there is evidence that drought reduces the activity of Rubisco, although the sensitivity is rather low, such that Rubisco activity drops by only about 50% in wheat and barley leaves even when leaf water potential is as low as −3.3 MPa, a value that indicates very serious drought (Johnson *et al.* 1974; Hay and Walker 1989).

The relatively few data on the effects of drought on photorespiration and mitochondrial respiration show that they decline slightly but to a much lower degree than does either P_g or P_n, a generalisation that seems to apply equally to C_3 and C_4 crops (Krampitz *et al.* 1984). The general pattern seems to be that both P_n and P_g decline linearly as leaf water potential becomes more negative at levels corresponding to low to moderate water stress, whereas R_m is largely unaffected and R_p is slightly reduced, meaning that less CO_2 is gained but also less CO_2 is lost as drought develops. As P_g declines with drought but respiration is largely unaffected, the ratio of respiration to photosynthesis increases (Figure 4.18). A cautionary comment is that increases in the R_m/P_g ratio with drought might lead to the incorrect conclusion that R_m increases, when in fact the ratio changes because P_n declines. The overall conclusion is that a combination of the inhibition of the photosynthetic enzymes and rather slight changes in photorespiration lead to a decrease in the net photosynthetic rate of leaves experiencing drought stress. However, such modifications to solar energy capture have to be weighed against the other more immediate and severe drought effects on crop canopy development and senescence (Sections 3.2, 3.3).

It is possible that the influence of water stress on photosynthesis and photorespiration arises from increased RUBP oxygenase relative to carboxylase activity caused by the increased stomatal resistance to CO_2 influx, and thus lower c_i, as stomata close under the influence of drought. Emphasising the stomatal response implies that c_i would reduce with drought, but experimental test of this hypothesis seems to show that c_i is little affected or may even increase under conditions both in which stomatal apertures are narrowed and photosynthesis is reduced. The fact that even under increased stomatal resistance c_i is not changed greatly under water stress at the same time that the respiration to photosynthesis ratio increases suggests that c_i is being maintained by CO_2 produced during photorespiration. For a leaf water potential lowering of about −1 MPa one could expect a five-fold increase in r_s and a decrease in P_n by a factor of about 3 or 4 compared with the absence of drought conditions. The conclusion must therefore be that the intracellular resistance r_m is directly affected by water stress and that the parallel increases in r_s and r_m seen in Figure 4.17 have independent causes. This interpretation is strengthened by studies that show how little P_n changes as a function of c_i for plants grown under water stress conditions (Figure 4.19). However, it is clear from Figure 4.17 that although P_n declines with increasingly negative leaf water potentials, there is a large range of c_i for which P_n does not change. Figure 4.19 also shows that the carboxylation efficiency is reduced with drought as the initial slope of the P_n response curves declines, further indicating non-stomatal control of the drought response.

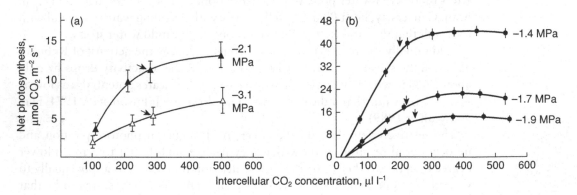

Figure 4.19 The relation between leaf water potential and the response of net photosynthesis to intercellular CO_2 concentration for (a) cotton (C_3) and (b) sorghum (C_4) made over a range of c_i levels. Note the P_n levels of the C_3 and C_4 species and the arrows indicate P_n for c_a levels of about 340 μmol mol^{-1} CO_2 (after Hutmacher and Kreig 1983; Kreig and Hutmacher 1986).

A systematic view would be that the causal chain that links the development of drought to the response of the plant is an example of a negative feedback loop that serves to maintain a fairly constant c_i, thereby offering the best trade-off between water loss and carbon gain under dry conditions. The possible feedback is that an initial increase in stomatal resistance reduces c_i but then reducing and independent effects of drought on P_n reduces the rate of carboxylation of CO_2 and also closes stomata further. The level of c_i may be maintained by increased respiration but the main observation is that the respiration to photosynthesis ratio increases. Lowering of leaf water potential is mainly *via* the connected network of hydraulic resistances from the soil to the leaf with the soil controlling the rate and extent of leaf rehydration. Stomatal closure is now thought to involve the plant hormone abscisic acid produced in the leaves and roots (Davies *et al.* 1994; Davies and Gowing 1999), the level of which affects stomatal aperture. Such a hypothesis challenges the control that above-ground meteorological conditions have on transpiration but is challenged by the question of the ability of roots to sense the dryness of the soil and then respond with increased hormone production. Roots and areas of soil moisture are not uniformly distributed in soils, thus raising the question of which soil water potential is affecting which roots to generate a hormonal response.

4.6 Nitrogen effects on photosynthesis

Nitrogen (N) deficiency causes significant reductions in crop growth rate and it would be reasonable to assume, since Rubisco forms the dominant protein in leaves (Section 7.2.2), that the main effect of lack of N would be through reduced rates of photosynthesis. This certainly happens but the main effect of N on crop photosynthesis is through its effects on radiation capture (Sections 3.2, 3.3) rather than the radiation-powered reduction of CO_2 and water into carbohydrates and other compounds. The interception of radiation in relation to crop N status has been discussed in Chapter 3 and this section concentrates on the direct effects of N on rates of photosynthesis per unit leaf area and the general effects of reduction in the amount of N per unit leaf area. In addition, it must be remembered that as well as affecting the level of Rubisco in leaves, nitrogen nutrition is also important for the size and development of the chloroplasts essential for the initial light harvesting processes of photosynthesis. Typical responses of P_n to variation in N supply are shown for field-grown wheat plants in Figure 4.20.

Flag leaves from plants given no N had lower light-saturated P_n levels (P_{max}) than plants given a moderate to high amount of fertiliser N per hectare. Unlike the effects of water stress (Figure 4.19), the carboxylation efficiency, as measured by the initial slope of the P_n curve, is largely unaffected by fertilisation level or leaf age. This statement has to be qualified by the fact that low leaf N levels reduce the incident radiation level (Figure 4.5a) at which the switch from RUBP-limited to Rubisco-limited photosynthesis occurs. The importance of this can be seen in C_4 plants, such as maize and sorghum, which have lower N per unit leaf area and Rubisco levels than C_3 plants, but are able to maintain higher levels of P_{max}. P_{max} in C_3 rice doubled from 10 μmol m^{-2} s^{-1} for a doubling of leaf N from

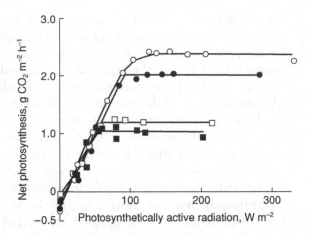

Figure 4.20 Photosynthetic light response curves for flag leaves of winter wheat given either no nitrogen fertiliser (closed symbols) or 151 kg N ha^{-1} (open symbols) and measured at anthesis (circles) and three weeks later (squares) (after Gregory *et al.* 1981).

1 to 2 g N m^{-2}, whereas, over a range from 0.5 g N m^{-2} to about 1.5 g N m^{-2}, P_{max} in sorghum rose from 10 to 35 µmol m^{-2} s^{-1} (Sinclair and Horie 1989).

It is also clear from Figure 4.20 that wheat flag leaves measured at anthesis had higher P_{max} levels than those measured three weeks later, regardless of N fertilisation. Thus, P_n levels for the N-starved younger leaves were higher than those for older leaves even when these were given reasonable levels of N. It is also suggested by Figure 4.20 that the difference in light-saturated P_n for the nitrogen treatments was less for the older leaves than for the younger ones. The implication is that flag leaves are drained of nitrogen during grain filling and thus the effects of leaf age and nitrogen status on CO_2 exchange become confounded. Results such as those in Figure 4.20 point to the other important function of Rubisco in leaves: that of a reserve of N for grain filling. This will be dealt with in more detail later (Chapter 8) when the notion of crop quality is discussed. It is sufficient now to appreciate the role of Rubisco and leaf nitrogen as a protein reserve in crop plants that is translocated to the harvested fraction of the crop during the programmed senescence of the leaf canopy, be it for grain or tuber crops. In pasture crops, since leaves and shoots act as sources of nitrogen and protein for livestock, the reservoir function of leaf nitrogen and Rubisco is similarly important.

A full summary of the relationship between leaf nitrogen level and aspects of photosynthesis (Figure 4.21) has been given by Lawlor (2001). As already seen, P_{max} (or A_{max}) declines linearly with declining leaf N as do protein levels but, importantly, these latter components decline at different rates from each other, which could have consequences for the biochemistry of photosynthesis but is unlikely to be important in the field. Of more field importance is the relationship between the *a* and *b* light-capturing chlorophyll pigments and leaf N. The total chlorophyll level drops with declining leaf N, but the chlorophyll *a:b* ratio is remarkably constant even for extremely low leaf N per unit area levels. It is also important to compare the relative rates of decline in P_{max} and total chlorophyll with leaf N per unit area because this has consequences for attempts to

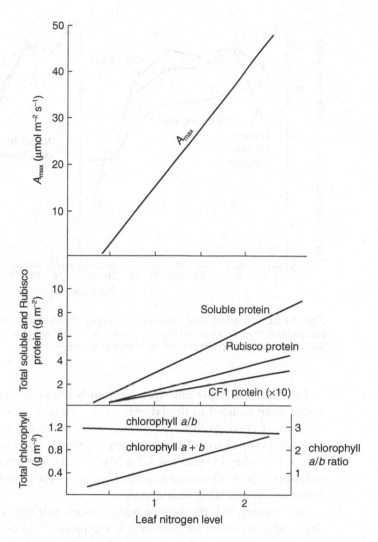

Figure 4.21 The dependence of maximum assimilation rate, A_{max} (or P_{max}), and various protein and chlorophyll fractions on leaf nitrogen level (g N m^{-2} leaf) (from Lawlor 2001).

sense leaf nitrogen levels remotely as a means of monitoring declining plant growth. The question is how to discriminate between reflectance from loss of N and reflectance from changes in the pigment leaf chlorophyll content such that the pigment reflectance does not swamp the N reflectance, which is highly correlated with the Rubisco level and thus photosynthesis. Where patterns of decline in chlorophyll and photosynthesis with leaf age have been simultaneously measured (Figure 4.22), it seems that, in senescing leaves, photosynthesis declines before chlorophyll. The leaf is almost functionally dead as a fixer of CO_2 before any changes in chlorophyll level are apparent. Fortunately, there is a reasonably strong correlation between leaf N per unit area, Rubisco level and chlorophyll level in leaves, meaning that crop reflectance aimed at measuring plant nitrogen status has to be corrected to account for the fact that it is chlorophyll that is being measured as a surrogate for plant N status.

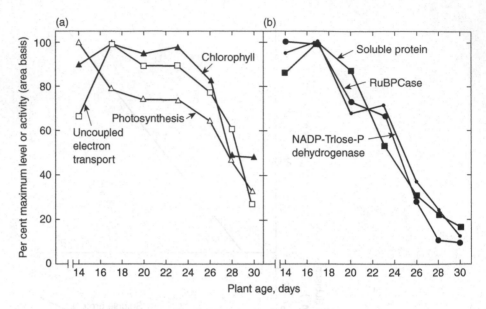

Figure 4.22 Changes in total chlorophyll, photosynthesis and electron transport (a) and protein and enzymes involved in photosynthesis, including Rubisco (b) during senescence of the second leaf of a wheat plant (from Camp *et al.* 1982).

Levels of leaf N per unit area commonly found in C_3 crop plants in the field rather than induced in the laboratory are unlikely to offer serious limitations to rates of photosynthesis, especially when it is remembered that leaves near the radiation-rich top of a leaf canopy are young. They will have leaf N per unit area values often in excess of 1.5 g N m^{-2} and any shortages of nitrogen will be manifested in the dynamics and morphology of the canopy *via* smaller leaf sizes, fewer shoots and faster senescence.

In summary, N is the most important plant nutrient as far as accumulation of dry matter is concerned because it is the basis of the enzymes that regulate the rate of photosynthesis. However, the first effects of shortage of nitrogen are seen in the crop canopy, as lower shoot numbers, smaller leaves and faster leaf and shoot death. It is questionable whether N availability in the field commonly approaches the limits that affect photosynthesis, although this may well be true in poor parts of the world where crops are chronically starved of N, because N is mobile in soil moisture and plants have roots that can progressively explore the soil, locate and then capture nitrogen (Section 7.2.1). There are open questions as to whether water or nitrogen has relatively the more severe effect on photosynthesis and also whether they act independently or interact synergistically when both are limiting factors.

4.7   Ozone effects on photosynthesis and crop productivity

Besides the ubiquitous and relatively common gases found in the atmosphere, such as N_2, O_2 and CO_2, there are trace gases in very low concentrations that can nevertheless exert influences on photosynthetic and growth processes. For example, the trace gas ozone (O_3) is found at concentrations measured in parts per billion of air or 10^{-9} mol (nano or nmol), that is a thousand times less than the concentration of CO_2. Ozone in the lower atmosphere is formed by the photochemical action of sunlight on the hydrocarbon emissions of motor vehicles, as well as some natural processes involving lightning and electrical storms. This tropospheric ozone, which can be harmful to plants and animals, is distinguished from the protective layer of O_3 that occurs in the higher stratosphere and reduces the transmission of ultra-violet radiation to the Earth's surface. O_2 is a stable molecule but O_3, a very strong oxidising agent, is intrinsically unstable, readily releasing free radicals ('active' oxygen) that can react with, and damage, living tissues. Cell and organelle membranes, including thylakoid membranes, with unsaturated lipids are particular targets. The effect is to disrupt the structure and thus the functioning of many biological processes. The mode of action of some contact herbicides such as the bipyridylium compounds diquat and paraquat is based on the generation of free radicals, and studying the mode of action of O_3 on plants can give some indications of how such herbicides have their effect.

The general effect of tropospheric O_3 on a range of crop plants exposed for the whole growing season to a range of O_3 concentrations (Figure 4.23) is for yields to decline even at low O_3 levels and to fall progressively with increased exposure. Some species are more susceptible to O_3 exposure than others, with soybean in the middle range in terms of the loss of crop yield, whereas lettuce is more resistant.

Detailed analysis of the effects of O_3 on soybean indicates that O_3 affects growth in a number of ways, consequently affecting the components of yield.

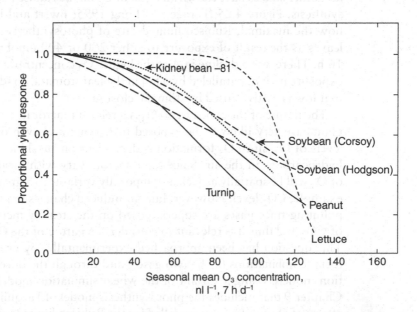

Figure 4.23 The relation between seasonal mean ozone concentration and the proportional yield response of a number of crops (after Unsworth *et al.* 1994).

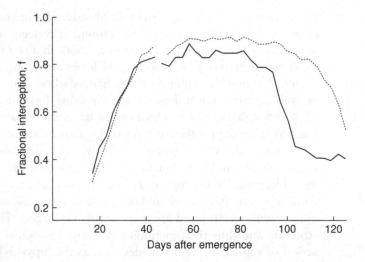

Figure 4.24 The effect of ozone exposure on the fractional interception of PAR for a soybean crop during a season (after Unsworth *et al.* 1994). The upper line is from open-top chambers ventilated with charcoal-filtered air and the lower line with air containing 60 nmol mol^{-1} O$_3$.

Ozone reduced radiation interception by the canopy from about the middle of the growing season (Figure 4.24), and this effect became more evident later in the growing season. The grain-filling period was shortened leading to a reduced number of pods per plant and seed number per pod. Although these results seem to indicate that most of the O$_3$ effect can be understood *via* the action of the gas on radiation capture (through enhanced leaf senescence) rather than radiation use, other studies have shown an effect of O$_3$ on the Rubisco limited rate of photosynthesis. Figure 4.25 (Farage and Long 1995; Ewert and Porter 2000) shows how the maximal, Rubisco limited rate of photosynthesis is reduced in wheat leaves as the result of exposure to either 200 or 400 nmol mol^{-1} O$_3$ for 4 h or 16 h. There was a decline in *A* at the higher O$_3$ concentration and for the longer exposure period, paralleled by a fall in the leaf stomatal conductance, suggesting that it was partly caused by stomatal closure.

The nature of the exposure to O$_3$ is a relevant factor in crop response because plants are very likely to be exposed to varying and even diurnally fluctuating levels of O$_3$, since O$_3$ formation is dependent on sunshine and the presence of hydrocarbons in the air. Since these factors vary with locality, concentrations of O$_3$ will be spatially as well as temporally variable to a greater degree than, for example, CO$_2$ level. However, human-induced changes in the concentrations of polluting trace gases are superimposed on the steadily increasing global levels of CO$_2$ and thus it is relevant to consider the nature of the O$_3$–CO$_2$ interaction. This question has been approached experimentally, by growing plants in factorial combinations of the two gases, and through the development of simulation models. Using a variant of the wheat simulation model to be presented in Chapter 9 that includes the photosynthesis model of Farquhar *et al.* (1980), the effects of O$_3$ uptake were modelled for the Rubisco limited rate of photosynthesis and leaf senescence (Ewert and Porter 2000).

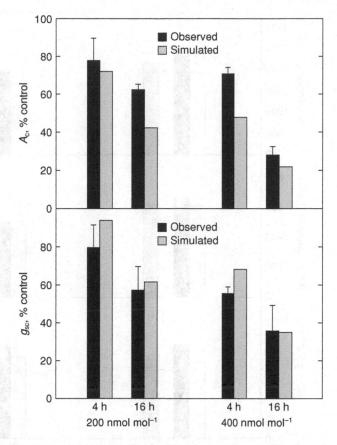

Figure 4.25 Observed and simulated effects of ozone exposure on the Rubisco limited rate of photosynthesis (A_c) and stomatal conductance (g_{sc}) after 4 h and 16 h exposure to 200 and 400 nmol mol^{-1} ozone for wheat leaves (after Farage and Long 1995; Ewert and Porter 2000).

In wheat, a reduction in carboxylation capacity appears to be the main cause of O_3-induced decline in the light-saturated rate of CO_2 uptake. Elevated CO_2 has been shown to protect against the ozone-induced reduction in photosynthesis in wheat primarily through reduction in stomatal conductance (Figure 4.25).

In terms of the principles developed in this book, it is important to understand to what extent the change in final crop biomass of crops growing under elevated ozone or CO_2 conditions, or where both gases are simultaneously elevated, can be explained by changes in the interception of PAR or its efficiency of use in making dry matter, defined by the RUE. Where these processes have been investigated in crops fumigated in the field (Figure 4.26), the answer seems to be that final biomass increases with elevated CO_2, partly because of increased interception of PAR probably through an increase in tiller production, but more by the stimulation of RUE. O_3 on its own has a directly opposite effect; the reduction in total biomass is caused by a slightly larger reduction in the interception of PAR than in RUE. When fumigated with both gases the overall effect is to leave biomass production slightly raised compared with the control, with the elevation of RUE being responsible for the net effect (Figure 4.26). Figure 4.26 also shows

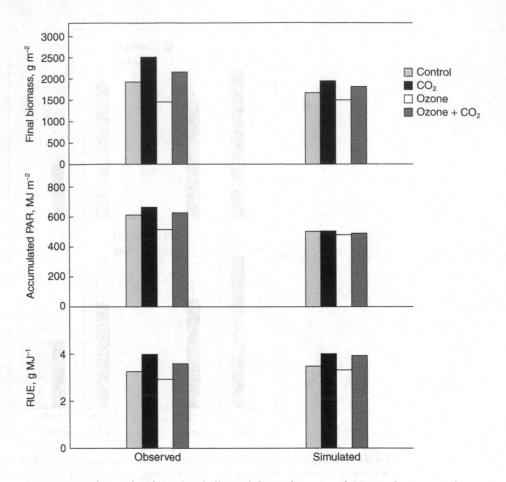

Figure 4.26 Observed and simulated effects of elevated ozone and CO_2 on above-ground biomass, accumulated PAR and RUE of spring wheat grown in open-top chambers (after Ewert and Porter 2000; Mulholland *et al.* 1998).

that it has been possible to simulate these crop reactions using a crop simulation model.

In summary, trace gases that are by definition present at low concentrations in relation to other gases in the atmosphere can have serious effects on the production of crops of many kinds. Some crops are more sensitive to these pollutants than others. There are noticeable reductions in the amount of PAR intercepted in crops exposed to ozone, interpreted in terms of the enhancement of canopy senescence. However, the presence of O_3 has also been seen to reduce the efficiency with which intercepted radiation is utilised to produce dry matter, probably by reducing the effectiveness of Rubisco and directly damaging thylakoid membranes in the chloroplast. Finally, there is interplay between the effects of CO_2 and O_3 in terms of their opposing effects on stomatal conductance, a finding that goes some way to explaining how a high level of CO_2 is able to mitigate the yield-reducing effects of O_3. The final chapter in this book will consider the future role that modifications to the above processes could have for the growth and development of crops in a world with 9 billion instead of 6 billion people.

Chapter 5
The loss of CO$_2$: respiration

To the physiologist or biochemist, respiration is a process by which materials a, b and c are converted into materials x, y and z. To the systems analyst it is a leak in a 'black box'. It is time we attempted to reconcile these two views.

(McCree 1970)

5.1   Introduction

Attempts to estimate the productivity of the 1.4×10^9 ha of cultivated arable land in the world and the development of crop growth models in the 1980s and 1990s, have challenged agricultural scientists to measure the gains but also the respiratory losses of CO$_2$ from crops and to understand the factors that control non-photosynthetic respiration. Many of the general principles of respiration as a plant and crop process were formulated in the 1970s and 1980s and have not fundamentally altered since then. What has changed has been the level of sophistication in the understanding and calculation of respiration costs of different plant products and processes and how these are affected by the abiotic environment, particularly CO$_2$ level, drought, nutrition and temperature. Crop simulation modellers and physiologists, such as C T de Wit and R S Loomis, drove the need for a conceptualisation of respiration into its growth and maintenance components in order to have better algorithms of respiration with which to calculate whole-crop carbon balances (Amthor 2000). At the biochemical level, the loss of CO$_2$ by plants is an unavoidable cost of the energy use needed to transform the simple molecules of water, CO$_2$ and nitrate and other nutrients into initially simple and then complex sugars and the other structural and functional building blocks of lignins, lipids, proteins and nucleic acids and thereafter to maintain them. The biochemical steps involved in the respiratory glycolytic and the tri-carboxylic acid (TCA or Krebs) cycles, mitochondrial electron transport, pentose phosphate and oxidative phosphorylation pathways are well understood and details can be found in biochemistry textbooks (e.g. Lambers *et al.* 1998).

Understanding the size, mechanisms and importance of respiration in crops and other plants started with the question of how, and to what degree, respiration is linked to photosynthesis and growth. This is an important question because it is natural to assume that photosynthesis, growth and respiration are in some way 'in opposition' to each other and that growth is a 'zero-sum' trade-off between assimilate gains and losses, which it is not. High growth rates depend on maximising the difference between the gain, *via* photosynthesis, and the loss, by respiration, of carbon and the assimilation of nutrients and water. However, research has shown that fast-growing plants, including crops, have both high rates of resource assimilation and respiration. Thus, respiration is an unavoidable accompaniment to high rates of plant growth; and in almost all cases levels of environmental factors that increase or decrease growth rate, such as drought and nutrition, affect respiration in the same manner. A common but not universal exception to this generality is the lowering effect on respiration of elevated CO_2 that at the same time raises photosynthetic rates (Amthor 1991; Drake *et al.* 1997), at least over short time-intervals. Understanding of respiration has also progressed *via* the development of the growth and maintenance respiration paradigm (McCree 1970, 1974) and has became more refined *via* investigations into the respiratory costs of the biochemically distinct groups of plant compounds such as carbohydrates, lipids and proteins (Penning de Vries 1974, 1975; Penning de Vries *et al.* 1974, 1983; de Visser *et al.* 1992; Poorter 1994). As crop species, for example legumes and cereals, contain different amounts of each of the major substrates, they will have different respiration costs since substrates such as lipids are more 'expensive' to produce than less chemically reduced carbohydrates. More recent insights into plant respiration have elucidated the respiratory costs of different plant processes (Amthor 1994, 2000; Cannell and Thornley 2000; Thornley and Cannell 2000). Such processes include the structure and function of alternative respiratory pathways that have unclear roles in plant metabolism; the effects of abiotic environmental factors, including elevated CO_2 and temperature, for the loss of CO_2 (Drake *et al.* 1999); and measurements of CO_2 fluxes from cropping systems. As the global population moves towards nine or ten billion people in the next 30–40 years (Evans 1998), estimates of crop production and global food supply again rely on understanding the plant and crop processes that govern the acquisition and loss of carbon and other elements by crops. Interest in carbon gains and losses from cropping systems has now extended to include measurements and modelling of CO_2 losses from heterotrophic soil components – detritivorous bacteria and fungi and herbivorous soil fauna – in addition to the autotrophic plant component, in order to measure and model net agro-ecosystem C exchange and its role in greenhouse gas fluxes (e.g. Robertson *et al.* 2000).

In common with photosynthesis (Chapter 4), respiration is a plant and crop process that occurs and can be measured and modelled at scales from the cell to the crop stand to the complete cropping system. As with photosynthesis, respiration measurements may be made on isolated cells or their components, individual organs, whole plants or populations of crops. At the larger scales, such biomass may be actively growing or be largely mature. The influx of O_2 or the efflux of CO_2 per unit mass (usually dry mass and known as the specific respiration rate) is higher in younger tissues and organs than when they are mature. The ratio of CO_2 efflux to O_2 influx, known as the respiratory quotient, is also much influenced by the composition of the organ or material that is respiring. When carbohydrates are the respiring substrate, such as in cereal grains, the CO_2/O_2 ratio is close to unity. In seeds in which lipids and proteins form a significant substrate fraction, such as in many legumes, the respiratory quotient is less than unity, the reason being that lipids and proteins are more chemically reduced than carbohydrates and therefore require more O_2 to achieve complete oxidation to CO_2 (Lambers *et al.* 1998). The importance of substrate type in the energetics of respiration will be discussed later in the chapter.

It is generally the case that CO_2 efflux is used by crop physiologists as an index of respiration than measurements of O_2 influx, which is a more biochemically relevant measure. Crop physiologists have an overriding interest in the overall carbon balance of crops, their economic yields and hence the magnitude and controls over fluxes of CO_2. Differences in O_2 influx enable comparisons between plants in terms of their respiratory efficiency (defined in terms of ATP production), the functioning of respiratory pathways and the effect of environmental conditions on respiratory processes. It is possible to measure CO_2 respiration on an area basis, when this can be the area of a single leaf or the ground area of a community of plants (Amthor 1991). The contemporaneous processes of respiration and photosynthesis make estimates of sole CO_2 efflux difficult to achieve and respiration, of course, occurs over a whole 24-h daily cycle. For crops over relatively short timescales, the general methodology is to capture a measured volume of emitted CO_2 over a time interval for a known leaf or ground area in the dark but other measures of mitochondrial respiration are O_2 absorption or CO_2 evolution in the dark, since photorespiration is thereby excluded. Since temperature and other environmental factors differ between light and dark periods thus affecting respiration rate, it is thought best to make respiration measurements in the dark shortly after a series of photosynthetic measurements have been made and to use the same leaf or plant material (e.g. Griffin *et al.* 1999). In addition to the difficulties of measuring the two components of the autotrophic CO_2 balance, it has also proved difficult to measure whole-plant or whole crop CO_2 losses over extended periods since autotrophic CO_2 losses are difficult to distinguish from heterotrophic ones, such as from soil animals. Therefore, respiration is somewhat of a 'black box' and the main concern of crop physiologists has been to determine the amount of CO_2 lost from the autotrophic components of a cropping system and the factors that regulate the level of loss. It is fair to say that respiration has received less scientific attention as a crop process than photosynthesis, mainly because it is difficult to measure and it lacks immediate apparency. Except for the study of 'alternative', non-cytochrome respiration (Lambers *et al.* 1998) that generates less ATP per unit O_2 absorbed than the above cytochrome-based pathways, understanding crop respiration has not progressed conceptually much beyond that of the 1970s and 1980s and the insights given by McCree in the USA, de Wit and Penning de Vries in the Netherlands and Thornley in the UK (Amthor 2000). However, the amount, form and controls over respiration remain crucial in determining the quantity and quality of yields of crop plants. Thus, the central issues in this chapter are:

the size and nature of crop non-photosynthetic respiratory losses, the growth and maintenance paradigm of respiration, the respiratory costs of biochemically distinct crop products, the effects of abiotic factors on respiration and the CO_2 fluxes of cropping systems.

5.2   The basis of crop respiration

In Chapter 4, it was seen that especially C_3 plants but also C_4 crops such as maize lose CO_2 to the atmosphere during photosynthesis *via* photorespiration. Non-photosynthetic respiration, however, produces the ATP energy needed for the synthesis of the carbon frameworks of new molecules, cells, tissues and organs (measured as biomass increase or crop growth); and the energy needed to acquire nutrient ions and drive internal plant transport processes. Besides making growth possible, respiration also maintains existing crop structure and function *via* processes such as the turnover of proteins and lipids and intracellular ion transport that require metabolic energy (Amthor 2000). To distinguish it from photorespiration, respiration associated with growth and maintenance, is known collectively as mitochondrial respiration (R_m) and is so, unless otherwise indicated, in this chapter. The division of R_m between that promoting growth and that supporting maintenance is more of an analytical concept rather than a physiological reality, since their shared expression is the efflux of CO_2 or the influx of O_2 into the plant. However, the growth and maintenance components of R_m respond differently to environmental factors such as temperature, drought and nutrient levels. Thus, temperature markedly increases the level of maintenance but has little effect on growth respiration, whereas growth respiration tends to increase relative to maintenance with a limiting nutrient supply but decrease with drought. There is also an ontological change throughout a crop's lifetime, during which the ratio of maintenance to growth respiration increases as a crop ages. Growing roots generally respire relatively more than growing leaves and shoots (Table 5.1, van der Werf *et al.* 1994; Lambers *et al.* 1998) and the level of respiration, in terms of the CO_2 emitted or O_2 absorbed, depends heavily on the level of chemical reduction of the synthesised plant products. Thus, lipids are more 'costly' to produce in terms of CO_2 lost than celluloses that, in turn, are more respiratory costly than carbohydrates. Estimates of the percentage of the C fixed and used by different organs and processes in conditions where plants have free access to nutrients are shown in Table 5.1.

Thus, in conditions of free nutrient availability and for fast-growing plants such as annual crops, shoots consume more than half of the total fixed C for growth and about 20% for their respiration. Roots use a much higher ratio of utilised C for respiration relative to growth than do shoots and use about 5% of the total fixed C for ion acquisition. In conditions of low nutrient supply, the fraction of fixed C devoted to all aspects of root function, including N_2 fixation in legumes, can increase dramatically.

Over seasonal timescales, data have shown that respiration (R_m)/photosynthesis (P_n) ratios have values in the range 0.35–0.80 (Amthor 2000). Maize, wheat and rice have R_m/P_n values of between 0.30 and 0.60 which are close to

Table 5.1 Estimates of the percentage of the C fixed and used by different organs and processes in conditions where plants have free access to nutrients (after van der Werf *et al.* 1994; Lambers *et al.* 1998).

Item	Utilised photosynthate as % of C fixed
Shoot growth	57
Root growth	17
Shoot respiration	17
Root respiration	8
Growth	3
Maintenance	0.6
Ion acquisition	4
Exudation	<5
N$_2$ fixation	Minimal

the calculated minimum for the ratio, an observation that means it is unlikely that biomass production in crops can be raised by reducing 'wasteful' respiration. Selected R_m/P_n values for an indicative range of vegetation types are shown in Table 5.2 (adapted from Amthor 1989, 2000), in which references to the original sources can be found.

Estimates of crop R_m/P_n ratios are typically lower than those for plants from natural vegetation. The relatively low values of R_m/P_n in crops can be related to the fact that much of the biomass in crops is, by definition, in storage and not growing organs such as seeds and tubers; each mol C of structure added to a tuber or a grain has a conversion efficiency of about 85% in terms of mol C of substrate used (Amthor 2000) and thus growth respiration is small as is maintenance also likely to be. An unintended consequence of selecting crops for high harvest indices may thus indirectly to have been to select them for a low R_m/P_n ratio. Other evidence from experiments where plants were grown under favourable conditions have shown that growth conversion efficiencies approaching 90% are close to the theoretical maximum, meaning that opportunities for reducing growth respiration in the hope of making larger plants are small. Some (e.g. Penning de Vries 1974) have argued that an attack on wasteful maintenance respiration processes would be more propitious.

Table 5.2 Estimates of annual or seasonal respiration (R_m) as a fraction of annual or seasonal photosynthesis (P_n) in intact ecosystems (adapted from Amthor 1989, 2000).

Ecosystem	Respiration/photosynthesis
Cropping (maize, rice, wheat)	0.30–0.60
Tallgrass prarie	0.61–0.65
Tropical moist forest	0.88
Evergreen temperate forest	0.72
Coniferous forest	0.72
Coastal sea marsh	0.77
Arctic tundra	0.50

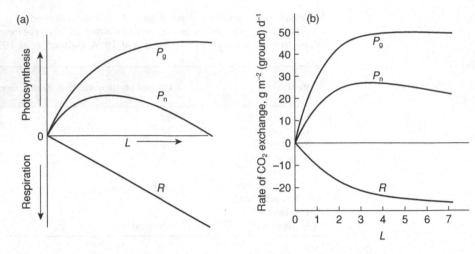

Figure 5.1 Relationship between the rate of CO_2 exchange of crop communities and leaf area index (L). (a) A theoretical linear response of respiration (R) to increasing L; (b) results from communities of white clover plants of differing L showing a non-linear relationship of respiration to gross and net photosynthesis (after McCree and Troughton 1966).

Three fundamental hypotheses concerning crop respiration were that it was a constant proportion of synthesised biomass; or of photosynthesis; or that respiration varied with photosynthesis and environmental conditions. Early work favoured the first hypothesis but later work has shown that it is necessary to consider the relationship between photosynthesis and respiration to explain the temporal and spatial variations in the rate of respiration. The early hypothesis of the relation between crop biomass and respiration was that total respiration (including photorespiration) was simply a constant proportion of biomass, mostly leaf dry matter. Thus, as leaf area increased, so proportionally would respiration. With a hypothesised linear relationship between respiration (R, Figure 5.1) and amount of biomass or L, P_n would initially rise but then fall with increasing L, giving an optimum L value that maximises P_n (Figure 5.1a). This could be because as L increases, the canopy becomes light saturated and Rubisco limited (Chapter 4) but respiration continues to increase with L with the result that P_n starts to fall. Elegant early experimental studies (McCree and Troughton 1966) showed that as crop growth increases, leaves lower in the canopy reduced their respiration rates to give the type of result shown in Figure 5.1b (McCree and Troughton 1966). Here, the relationship between P_n and leaf area has the same shape as that for P_g and leaf area. As McCree and Troughton (1966) saw this response, they concluded that the rate of P_g regulates the rate of respiration (Figure 5.1b) and thus the rate of respiration must proportionally fall at higher L values and be non-linear.

The hypothesis of linearity between respiration and L can be criticised as unrealistic on the basis that there would be no expansion of L if the rate of net canopy photosynthesis were light saturated. At this point, the optimum L would be the maximum L and there could be no further L expansion. A similar conclusion to this is also inherent in the notion of RUE, introduced in Chapter 4, since there can be no radiation interception without leaf area growth and *vice versa*. All three curves in Figure 5.1b start by increasing as L increases to a value of about 3 or 4. Thereafter they *all* become flatter as L increases. This means that neither the values of P_g minus R *nor* P_n minus R change very much for increasing values of L above these lower values. This insight permits an understanding of the reason for the relative constancy of RUE (Chapter 4) as the slope between the dry matter produced and radiation intercepted. For L values less than 3 or 4, the difference between P_g or P_n and R (i.e. dry matter production) increases, as does the interception of light. After L has passed a value of 4 neither is more radiation intercepted by the canopy nor is the production of dry matter proportionally increased, that is P_g or P_n minus R is almost constant. In other words, the point at which canopies stop intercepting more light is also the point at which the rate of change of dry matter production becomes zero, with the result that RUE retains a constant value over the whole range of L but for different reasons depending on the L value.

The conclusion from the foregoing experiments and analysis is that indirect, theoretical and direct analyses of the causes of variation in crop respiration discount the notion that it is related linearly to crop biomass but it is linked positively and non-linearly to photosynthesis and to growth – at least in the short term. These insights were able to explain the response of P_n or dry matter production to increasing L, and thereby the linearity of the RUE relationship seen in Chapter 4. The next stage in the analysis of crop respiration was to divide the mitochondrial respiration into components that support the growth of new structure on the one hand and the maintenance of existing biomass on the other.

5.3   Growth and maintenance respiration

The preceding discussion linked the rate of respiration to the production of respiratory substrate by photosynthesis. However, it is also possible to link respiration to the use that plants make of the carbohydrate they produce to grow and to maintain existing structure and function. McCree (1970) provided formal expression of this concept by showing that a simple linear equation (Equation 5.1) could describe the rate of respiration of small canopies of white clover grown in a controlled environment:

$$R_m = g_R G + m_R W \qquad (5.1)$$

where R_m and G are the rates of mitochondrial respiration and crop absolute growth rate (g new biomass s⁻¹) and W is crop dry matter (g), respectively; g_R is the growth respiration coefficient; and m_R is the maintenance respiration coefficient. Equation 5.1 says that respiration (R_m; mol CO_2 s⁻¹) is composed of respired CO_2 per unit of produced new biomass (mol CO_2 g⁻¹) per unit growth (g s⁻¹) or gross photosynthesis (g CO_2 s⁻¹; Amthor 2000), plus CO_2 lost per unit of existing biomass (mol CO_2 g existing biomass⁻¹ s⁻¹). Growth can be defined in many ways, but, in this context, it is the conversion of the non-structural products of photosynthesis into new structure such as structural carbohydrates, lipids, proteins, lignins and organic acids. Intra-plant transport of ions and sugars, is in some definitions (Lambers *et al.* 1998) identified as a separate respiratory cost but elsewhere (Amthor 2000) it is divided between growth and maintenance. Maintenance respiration comprises the CO_2 losses associated with the integrity of cellular processes and the dominating processes are protein turnover, the maintenance of ion gradients and intracellular transport processes. An important distinction is that g_R represents a simple proportion of newly fixed CO_2 lost in biosynthesis, whereas m_R comprises a series of biochemical processes whose rates are affected by temperature. The temperature response of m_R has been encapsulated in the use of a Q_{10} value of between 1.8–2.0, meaning that the losses of CO_2 *via* the maintenance component about doubles for a 10°C increase in temperature. The effect of temperature on m_R can be represented by Equation 5.2:

$$m_R = m_R' Q_{10}^{T_{mean}/10}$$
(5.2)

where m_R is the maintenance respiration coefficient, m_R' is the basal maintenance respiration coefficient, Q_{10} is the factor by which m_R' increases for a 10°C increase in temperature and T_{mean} is the mean temperature over a period that can range from an hour to a day. The use of the Q_{10} concept in maintenance respiration infers that linear increases in temperature produce non-linear increases in respiration rate, a response that has been demonstrated for potatoes growing in either sunny or shady conditions between 5°C and 35°C (Sale 1974; Figure 5.2).

The relationships inherent in the growth and maintenance paradigm between growth rate and the two components of respiration can be visualised (Figure 5.3) based on the work of Lambers *et al.* (1996). Figure 5.3 allows visualisation of

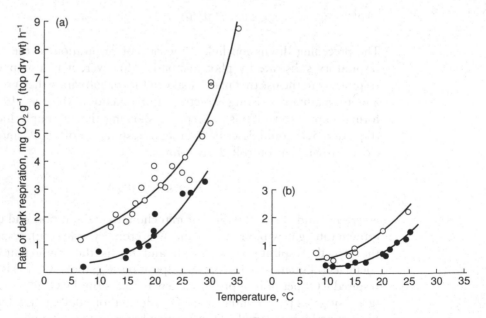

Figure 5.2 Relationship between temperature and the rate of dark respiration of the above-ground parts of potato plants during the day (open symbols) or during the following night (closed symbols) on (a) sunny days or (b) cloudy days (after Sale 1974).

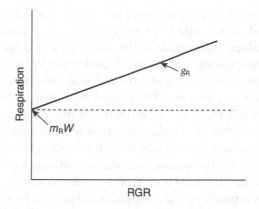

Figure 5.3 Schematic drawing of the relation between crop relative growth rate (RGR) and maintenance (m_R) and growth respiration (g_R) (after Lambers *et al.* 1996). *W* is crop weight.

what it could mean to design more efficient plants in terms of allowing high RGR levels but reducing the CO$_2$ losses of respiration. Lower bounds on the ratio between respiration and photosynthesis (R_m/P_n; Table 5.2) for major cereal crops have values from 0.3 to 0.6 (Amthor 1989) and as these are generally lower than R_m/P_n ratios of other forms of vegetation, they must represent values close to the ratio's physiological minimum value. The R_m/P_n ratio is synonymous with the slope of the g_R line in Figure 5.3. Three possibilities suggest themselves in any attempt to reduce CO$_2$ loss while preserving RGR. First, would be to lower the level of m_R and the level of the intercept on the respiration axis, perhaps by selecting plants with a lower level of protein turnover or ion transport, since these are the major components of maintenance respiration (Penning de Vries 1974). The slope of the g_R line seems to indicate that growth in cereals proceeds with close to maximal growth efficiency and this is probably because of the large fraction of biomass in crops in storage organs such as seeds and tubers, which have both low growth and maintenance costs. It may be the case that selection for high harvest indices in modern crop varieties has also meant selection for low whole-plant R_m/P_n (Amthor 2000). Whether the same conclusion applies to legumes, including grain legumes such as soybean, needs to be resolved. If it is unlikely that the slope of the g_R line can be reduced further then a third possibility might be to combine a reduction in the maintenance respiration intercept with an increase in the slope of the g_R function. This 'trade-off' between increased growth but reduced maintenance costs could mean that higher RGR values could be achieved for lower CO$_2$ emissions until the existing g_R line was crossed.

McCree found the equation constants g_R and m_R in Equation 5.1 to have values of 0.25 and 0.015 d^{-1}. They can be understood as saying that a total of 25% of daily gross photosynthate (g_R, growth respiration) plus 1.5% of the existing dry weight (m_R, maintenance respiration) gives the total respiration from this community of plants on a daily basis. Furthermore, considerable support for the concept came from the theoretical work of Penning de Vries (1972), who calculated the minimum amounts of carbon lost through respiration during the conversion of photosynthate into biomass of given chemical composition. These calculated values for growth respiration corresponded closely with measured values. Unfortunately, the rate of maintenance respiration cannot yet be derived theoretically with as much confidence because of inadequate knowledge about the materials being replaced and the rates of their degradation. However, Penning de Vries (1975) used measured rates of protein and lipid turnover in plants, together with estimates of the energy required for the maintenance of ion gradients, to calculate that maintenance requires the consumption of 15–25 mg glucose g^{-1} dry weight d^{-1}. This figure agrees with at least some of the measured rates of maintenance respiration, although rates of up to 80 mg glucose g^{-1} dry weight d^{-1} have been reported. This discrepancy is not surprising: first, because of differences in the conditions of growth between the plants used for measuring protein turnover rates and those used for the measurements of maintenance respiration; and second, because of uncertainties about some of the assumptions that have to be made in the calculation of maintenance requirements. Nevertheless, this 'quantitative biochemistry' has provided a valuable theoretical basis for the interpretation of CO_2 exchange data from whole-plant and crop experiments in terms of growth and maintenance processes.

Over days to months, the rate of mitochondrial respiration of plants declines as they age. This occurs in crop plants as the result of the translocation of nitrogen in the form of protein from leaves to the reproductive organs (Drake *et al.* 1997). Thus, the earlier conclusion that different amounts of metabolic energy are needed to build plant compounds of different chemical structure is reversed with pleasing symmetry when these high-energy compounds degrade during senescence. Respiration can thus be thought of as an index of a plant's entropy or level of disorder; as structure and order are being assembled and plants concentrate their environmentally acquired energy and mass thus reducing their level of entropy, energy is expended with respiration as the result. Death and senescence represent increases in disorder and thus lowered energy consumption and respiration. If growth is the maintenance of a disequilibrium between the gain and loss of environmental resources requiring respiration, then senescence is its thermodynamic and physiological symmetrically opposite twin.

5.4   The respiration of different plant substrates

The main crops being considered in this book are wheat, maize and soybean, which differ in their chemical composition, when measured as the proportion of carbohydrates, proteins, lipids, lignins and organic acids they contain. Such differences in chemical composition have consequences for the total respiratory costs of these major crop plants and the balance between their growth and maintenance respiration components. Pioneering work in this area was done by Penning de Vries and colleagues in the 1970s where they reviewed and identified the biochemical production costs of 61 plant compounds that were then categorised into the five major groups mentioned above. Taken together with data on the composition of crop plant storage organs (Penning de Vries *et al.* 1983; Amthor 2000) it is possible to calculate the costs of biosynthesis of wheat and

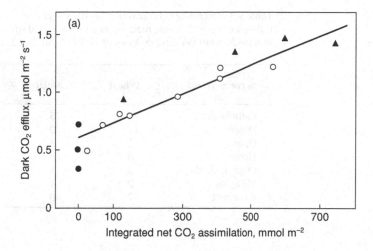

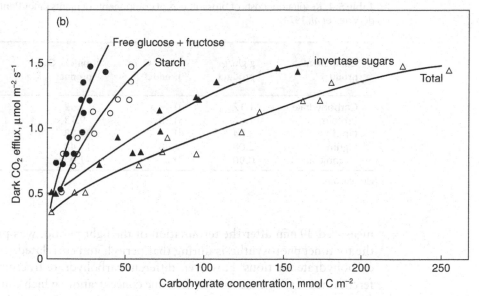

Figure 5.4 (a) Relationship between the rate of dark CO$_2$ efflux and integrated net CO$_2$ assimilation during the preceding light period in mature leaves of wheat for leaves photosynthesising at ambient CO$_2$ levels or at 800 µmol mol^{-1} CO$_2$ or for leaves selected at the end of the night period. (b) Relationship between dark CO$_2$ efflux and several carbohydrate fractions in mature leaves of wheat. The invertase sugars fraction is largely sucrose, but includes some low molecular weight fructosans (after Azcón-Bieto and Osmond 1983).

maize grains and ears, and the pod and seeds of soybean. These are illustrated in Tables 5.3–5.5. Experimental evidence for the respiratory cost of different substrates showed, in wheat (Azcón-Bieto and Osmond 1983), a linear relationship between the rate of CO$_2$ efflux in the dark and P_n (Figure 5.4a) and that the increase in dark CO$_2$ efflux was proportional to the concentration of several carbohydrate fractions that form the respiratory substrate (Figure 5.4b). Thus, variations in amount of soluble carbohydrate, the respiratory substrate, are responsible for this effect. The rate of respiration of mature leaves of wheat,

Table 5.3 Percentage composition (as dry weight) of wheat inflorescence and grains, maize cob and grains and soybean pod and seeds (from Penning de Vries *et al.* 1983; Amthor 2000).

% composition	Wheat	Maize	Soybean
Carbohydrate	76	75	29
Protein	12	8	37
Lipid	2	4	18
Lignin	6	11	6
Organic acids	2	1	5
Minerals	2	1	5
C content	47	48	53

Table 5.4 Respiratory costs of biosynthesis of components of plants (after Penning de Vries *et al.* 1974).

Component product	g glu/g product	mmol O_2/g product	mmol CO_2/g product	mmol ATP/g product
Carbohydrate	1.17	0	1.52	12.25
Protein	1.61	0.28	9.13	77.52
Lipid	2.84	0	32.28	50.80
Lignin	2.08	2.88	11.48	18.75
Organic acids	0.90	0	−1.04	−4.60

glu, glucose.

measured 30 min after the termination of the light period, was proportional to the total net photosynthesis during that period, and correlated with several leaf carbohydrate fractions. However, different carbohydrate fractions showed different rates of CO_2 emission for the same concentration, which can be interpreted as meaning that crops or crop organs with higher levels of reduced carbohydrates, such as seeds, would respire less than other organs, such as younger leaves, that contain high levels of simple carbohydrates such as glucose and fructose.

Making a calculation of the respiratory costs of plants with different chemical compositions starts with measurement of the percentage composition of the plant or organ that is respiring. Table 5.3 shows that wheat and maize yield components have similar carbohydrate levels but with differences in protein content (wheat higher than maize), lipids (both low) and lignin (maize higher than wheat). Their carbon contents are very similar. The legume soybean has an almost 60% lower carbohydrate content but a three-fold higher protein content and a many-fold higher lipid content than the graminaeceous cereals. Because the fractional composition of soybean includes more chemically reduced compounds, their total carbon content is higher in percentage terms than wheat or maize.

In wheat and maize the total production values of glucose for carbohydrates are clearly much higher than those for nucleic acids, amino acids and proteins,

Table 5.5 Calculated respiratory costs of production of wheat ear plus grain, or maize cob plus grain or soybean pod plus seed (based on data in Penning de Vries *et al.* 1974; Amthor 2000).

Crop	g glu	mmol O$_2$	mmol CO$_2$	mmol ATP
Wheat				
Carbohydrate	88.9	0	115.3	931.0
Protein	19.3	3.4	109.6	930.2
Lipid	5.7	0	64.6	101.6
Lignin	12.5	17.3	68.9	112.5
Organic acids	1.8	0	−2.1	−9.2
Total for 100 g	128.2	20.7	354.5	2066.1
Maize				
Carbohydrate	87.8	0	113.7	918.8
Protein	12.9	2.3	73.0	620.2
Lipid	11.4	0	129.1	203.2
Lignin	22.9	31.6	126.2	206.3
Organic acids	0.9	0	−1.0	−4.6
Total for 100 g	135.9	33.9	441.0	1943.9
Soybean				
Carbohydrate	33.9	0	44.0	355.3
Protein	59.6	10.4	337.8	2868.2
Lipid	51.1	0	581.0	914.4
Lignin	12.5	17.3	68.9	112.5
Organic acids	4.5	0	−5.2	−23.0
Total for 100 g	161.6	27.7	1026.5	4227.4

which in turn are higher than the production values for lipids. Conversely, the relative CO$_2$ emitted per unit of biomass product is lower than for soybean. These differences are reflected in the efficiencies with which substrate is converted into plant biomass of differing chemical composition (Tables 5.3 and 5.4). Thus, it takes 1.17 g of glucose to produce 1 g of carbohydrate, but 2.8 g of glucose per g of lipid and so on (Table 5.4). The high metabolic cost of synthesising high protein or lipid seeds such as soybean therefore imposes a limitation on yield that is independent of photosynthetic rates. Combining the information from Tables 5.3 and 5.4 enables calculation of the glucose requirement, O$_2$ uptake, CO$_2$ production and ATP requirement for seeds of a high protein legume crop such as soybean and high carbohydrate and lower protein crops such as wheat and maize. Note that the production of organic acids involves a small uptake of CO$_2$, because these exist in a more oxidised form than glucose and have thus a net production of ATP.

Table 5.5 illustrates that only small differences exist between the metabolic costs of production of maize and wheat and that the composition of the yield component makes some costs more expensive than others, but not all costs are similarly affected. For example, O$_2$ uptake for 100 g of a maize cob is more than 30% higher than for 100 g of wheat grain, but the total ATP needed is less because of the ratio between protein and lignin content in the two yield products of these crops. The ATP required for 100 g of soybean is more than twice that for the cereals with high relative costs of production for protein and lipids. Despite these ranges in energy costs, the overall cost in terms of C substrate of maize, wheat and soybean varies much less from about 1.3–1.6 g glu g^{-1} biomass. This conservatism in construction is because of the reciprocal variation between the classes of constituents in plant biomass, particularly between carbohydrates and lipids. It is generally the case that plant tissues have either high levels of protein allowing high metabolic activity or high levels of storage compounds such as carbohydrates and lipids – a clear example would be leaves (high activity) versus stems (high storage). In an assembly of data, Poorter (1994) showed that the average construction costs of leaves, stems, roots and seeds for over 100 examples ranged from 1.3 to 1.6 g glu g^{-1} biomass. Thus, the calculated values for maize, wheat and soybean fall centrally within those found for both slow- and fast-growing non-crop plants.

Respiration costs of different chemical components of seeds and other plant parts can be calculated with reasonable precision, given their chemical composition, and are clearly linked to plant quality, whether defined in terms of human nutrition or food processing. Plant quality is an expression of the fractional chemical composition of yield components, but can be broken down into the types of, for example, proteins and their molecular weights, sulphur content and degree of polymerisation. Such sophistication requires more refined estimates of the costs of component production, including the metabolic costs of important processes. The respiratory costs of different crop processes are discussed later in the chapter.

5.5   Growth and maintenance respiration in the field

Controlled environment and theoretical studies have indicated that young plants use 25–35% of their daily assimilate to support growth and 1.5–3.0% of their dry weight (CO_2 equivalents) for maintenance processes. These results imply a system of physiological control over the magnitude of plant respiration. Thus, McCree's formula should be applicable to crops growing in the field, although limitations to its use might be envisaged. For example, the steady-state conditions of controlled environments used to develop the concepts and measure the parameters of Equation 5.1 do not exist in the field. Over the life of a crop there will be considerable changes in the plants' or crop's chemical composition. There is also likely to be more variation in local temperature within a crop stand in the field than in a growth chamber.

In an attempt to validate McCree's approach in the field, Biscoe *et al.* (1975a) compared respiration rates derived from measured CO_2 fluxes above a barley crop with estimates based on McCree's formula (Figure 5.5). At night, the measured flux of CO_2 from the soil–crop system (R_a) is the sum of respiration by the canopy (R_c), by the roots (R_r) and by soil organisms (R_s). The net loss of CO_2 by plant respiration in the dark (R_d) can therefore be calculated as:

$$R_d = (R_a - R_s) \qquad (5.3)$$

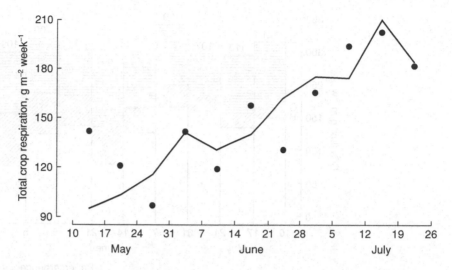

Figure 5.5 Comparison of the total respiration ($R_l + R_d$) of a barley crop calculated from CO$_2$ fluxes (●) with that estimated from McCree's formula (—) (from Biscoe *et al.* 1975a). The amount of CO$_2$ respired during the day (R_l) was estimated as the night-time loss (R_d) corrected for temperature differences by a Q$_{10}$ value.

Total loss of CO$_2$ from the soil ($R_r + R_s$) was estimated by cutting the foliage at ground level within an area of 500 cm^2, placing a weighed dish of soda-lime just above the soil surface, and covering the area with a tin box. Temperature differences between the air and the soil inside and outside the box were minimised by covering the box with aluminium foil. The increase in the weight of the soda-lime over a period of 72–100 h was attributed to the absorption of CO$_2$ released from the soil surface. Root respiration (R_r) over the same period was then calculated on the basis of measured changes in root dry weight and assumptions about the chemical composition of the roots, allowing R_s to be estimated by difference.

The amount of CO$_2$ respired during the day (R_l) was estimated as the night-time loss corrected for temperature differences by a Q$_{10}$ value:

$$R_l = R_d \, Q_{10}{}^{(T_l - T_d)/10} \qquad (5.4)$$

where T_l and T_d are the mean air temperatures in the canopy at the height of maximum foliage density during the day and night, respectively. A Q$_{10}$ of 2 and a 14-h day were assumed, allowing respiration during the light to be estimated as:

$$R_l = (14/10) \, R_d \, 2^{(T_l - T_d)/10} \qquad (5.5)$$

Weekly amounts of total crop respiration ($R_l + R_d$), and corresponding estimates of respiration in terms of the growth and maintenance requirements of the crop, are shown in Figure 5.6. The estimates were made using McCree's formula (Equation 5.1) and the values of a and b found for white clover. Thus, growth respiration was calculated as 0.25P_g and assumed to be independent of temperature, and the weekly maintenance respiration was calculated as:

$$7 \times 0.015 \times 2^{(\bar{T}-20)/10} \times \hat{W} \qquad (5.6)$$

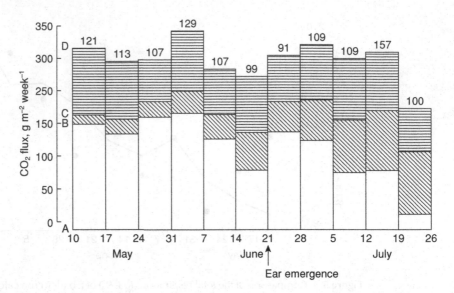

Figure 5.6 Weekly amounts of gross photosynthesis, P_g (AD), maintenance respiration (BC), growth respiration (CD), and net photosynthesis, P_n (AB), of a crop of barley sown on 18 March 1972 and harvested on 21 August. See text for method of deriving P_g, P_n and total respiration (BD = $R_l + R_d$). The total measured respiration was divided into growth (R_G) and maintenance (R_M) components in the proportions derived from estimates of R_G as $0.34P_g$ and R_M as $7 \times 0.015 \times 2^{(\bar{T}-20)/10} \times \hat{W}$ (see text). The numbers above the histograms are the weekly totals of incident solar radiation (MJ m^{-2}) (adapted from Biscoe *et al.* 1975a).

where $\bar{T}$ is the mean air temperature during the week and $\hat{W}$ the standing crop dry weight (CO_2 equivalents), taken to be the mean value of successive weekly harvests. Weekly amounts of P_g were calculated as the sum of net photosynthesis, derived from CO_2 fluxes above the canopy, and corresponding totals of $(R_l + R_d)$.

There was good agreement between the respiratory losses calculated from CO_2 fluxes and those derived from McCree's formula. The actual values of a and b for the barley crop were estimated by correcting the measured rates of respiration to a standard temperature of 20°C, and dividing Equation 5.1 through by P_g to give $R/P_g = a + bW/P_g$. The values of a and b were then given as the intercept and slope when R/P_g was plotted against W/P_g. The values obtained, $a = 0.34$ and $b = 0.012$, are consistent with those found in controlled environment experiments, and very close to those found for field-grown cotton ($a = 0.37$, $b = 0.013$) (Baker *et al.* 1972).

The application of Equation 5.1 over an extended period to crops growing in the field therefore allows the course of respiration to be considered in terms of its growth and maintenance components. This is important because it has helped to relate variations in respiration rate to the prevailing weather, and hence has aided our understanding of the effects of weather on dry matter production. Respiration should not be considered in isolation, however, because other factors, notably the size and activity of the photosynthetic system and the receipt of solar radiation, operate simultaneously to determine dry-matter production. Biscoe *et al.* (1975a) have shown how the seasonal pattern of net CO_2 uptake by the barley crop results from the interplay of these factors (Figure 5.6).

During the first 5 weeks, total respiration changed little. Net photosynthesis therefore followed changes in gross photosynthesis, which depended on the amount of radiation intercepted. L was still increasing during this period, so that the proportion of incident radiation intercepted by the canopy increased from 70% to 95%. Increased maintenance respiration was associated with the development of the ear during the week before emergence, and resulted in a reduction in net CO_2 uptake from 57% of gross photosynthesis to 39%. By this time, maintenance respiration had increased from 11% of total respiration to 40%, in response to a four-fold increase in standing crop dry weight and an increase in mean daily temperature from 9.1°C to 12.4°C. In the week after ear emergence, increased gross photosynthesis, attributed to assimilation by the ears and peduncles, reinstated net CO_2 uptake to its former level. However, over the next 4 weeks an increasing respiratory load superimposed upon declining amounts of gross photosynthesis caused a dramatic decline in net assimilation, from 150 to 27 g CO_2 m^{-2} week^{-1}. This reduction in gross photosynthesis reflected the decreasing size and activity of the photosynthetic system associated with the senescence of the canopy (Figure 5.7) and, to some extent, with water stress. The increased respiratory losses were due to the continuing high growth respiration (30–35% of gross photosynthesis throughout the first half of July) and the steadily increasing maintenance component: in the final week, respiration accounted for 88% of gross photosynthesis, and 60% of this respiration was required for maintenance. The overall picture is thus one of a relatively constant absolute level of growth respiration through time, but of a steadily increasing

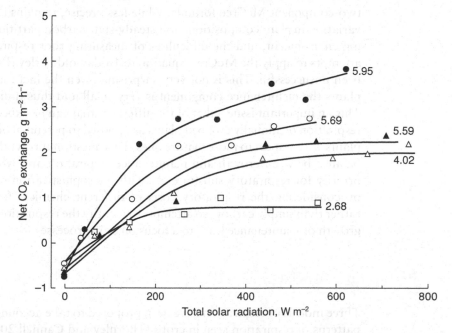

Figure 5.7 Relationship between the net CO_2 fixation of a barley crop and irradiance at weekly intervals after anthesis. The green leaf area index on each occasion is shown. The curves were derived from hourly averages of solar radiation and the corresponding CO_2 fluxes measured throughout the day (after Biscoe *et al.* 1975a).

maintenance component, such that the proportion of total respiration that is devoted to the support of existing structure increases as the crop ages and senesces. The decline in gross photosynthesis towards the end of growth also plays a role in the changing pattern of respiration. In very large perennial plants such as trees, maintenance respiration is responsible for almost all their CO_2 efflux.

Net photosynthesis, and therefore dry-matter production, during the early part of the growing season is related directly to intercepted radiation almost until ear emergence (Figure 5.6). Thereafter, solar radiation becomes a less important factor, as L declines (perhaps exacerbated by water stress) and respiration increases (Figure 5.6). The extent of the respiratory losses will be increasingly dependent on temperature as the maintenance component becomes more important. Thus, a 3°C rise in temperature during the week 17–24 May, would reduce dry-matter production by only 3%, but 2 months later the same rise in temperature would reduce dry-matter production by 27%. Biscoe and Gallagher (1977) suggest that the increase in maintenance respiration and the consequent reduction in crop growth rate with increasing temperature may explain an association between cool summers and improved crop growth. These studies of Biscoe and co-workers in the 1970s and later have provided the clearest description of the changes with time in the balance between growth and maintenance respiration costs to cultivated crops. Although the goals and ideas behind these classical studies with wheat have been used elsewhere, they have not been bettered.

The field-scale experiments discussed above show that measurements of the respiration of crops growing in the field could be analysed in the same way as those of plants in controlled environments. The description of respiration by the two-component McCree formula, while less precise, remains adequate, despite variation in plant composition, non-steady-state carbon partitioning in a changing environment, and the difficulties of measuring root respiration. However, attempts to apply the McCree equation to 18-day-old barley (Farrar 1980) have been unsuccessful. This is not very surprising given the fact that in such young plants the maintenance component is very small and thus difficult to measure. This an important issue – that of the difference that can be expected between the respiration of annually cropped plants as opposed to perennial ones and of young plants with respect to more mature ones. This question throws the focus onto the balance between growth and maintenance respiration and which of these has priority for respiratory substrate. As with the sophistication introduced above by considering the respiratory costs of different chemical fractions of crops rather than simply carbon, so attempts to identify the respiration associated with growth or maintenance leads to a focus on crop processes.

5.6   Respiration associated with crop processes

Three models of respiration have been proposed to take account of the observed patterns of respiration seen in crops (Thornley and Cannell 2000). The models differ in whether they give priority to maintenance or growth respiration and whether 'new' or 'old' biomass can be used as a respiratory substrate. Figure 5.8 is a simplified schematic description of the possible routes involved in the

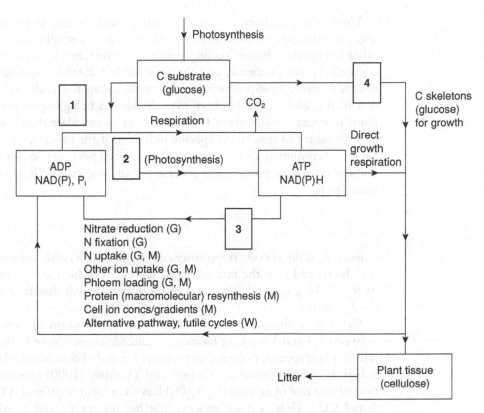

Figure 5.8 A simplified scheme of the possible routes for C substrate in growth and respiration. Arrows indicate fluxes of substrate and CO_2; G, growth process; M, maintenance process; W, waste process; P_i, inorganic phosphate. Numbers 1–4 refer to the pathways described in the text (adapted from Cannell and Thornley 2000).

consumption of C substrate (glucose) for growth and respiration as it moves from a product of photosynthesis. Consumption of C substrate results in the production of energy, as ATP, and reducing power, as NADPH, for the energy-requiring processes in plants and the skeleton of structures used for growth.

Route 1 in Figure 5.8 portrays the use of C substrate and the loss of CO_2 as the energy source for the production of ATP and NADPH from their precursors, ADP, NAPD and P_i (inorganic phosphate). Note that the production of metabolic energy and reducing power are required for both maintenance and growth respiration. Route 2 indicates that photosynthesis (Chapter 4) can also directly produce ATP and NADPH during periods of high PAR input and can thus directly support the energy costs of metabolism without the loss of CO_2 (Lawlor 2001). Route 3, in which ATP and NADPH are used in metabolism represent the main processes by which plants lose CO_2. The processes do not fall neatly between growth and maintenance aspects of respiration with several processes, for example N uptake, being involved in the production of new and the turnover and maintenance of existing biomass. The fourth route is where C substrate provides the structural building blocks of C skeletons for growth, for example as cellulose in new cell wall assembly. Note that the CO_2 loss involved in the production of new structure, principally cellulose, occurs *via* the processes associated with route 3.

Thus, nine processes can be identified as contributing to respiration. These are: growth, nitrate reduction, N_2 fixation, root N-uptake (as NO_3 or NH_4), other ion uptake, phloem loading, protein turnover, maintenance of cell ion concentrations and gradients, and the apparently wasteful respiration associated with alternative heat-producing and cyanide-insensitive pathways, the function of which is unclear but has been hypothesised as being important in regulating plant temperature (Lambers 1997). In order to calculate their respiratory cost it is necessary to specify the specific unit cost of the process, as the amount of substrate consumed, CO_2 emitted or O_2 absorbed per unit product of the process and the rate of the process. For process (n), the respiration (R_{m_n}) rate associated with this is:

$$R_{m_n} = sc_n \times r_n \qquad (5.7)$$

where, sc_n is the specific respiratory cost of process n (with units of g (CO_2) g^{-1} (product)) and r_n is the rate of the process (g (product) s^{-1}), making the units of R_{m_n} to be g CO_2 s^{-1}. The total respiratory cost will therefore be the sum of the R_{m_n} terms.

Growth respiration costs in terms of the production of new biomass are between 1.3 and 1.6 g C (g biomass)$^{-1}$ and are conservative both between and within plant species, organs, and tissues, as noted above. In calculation of the C substrate costs of processes, Cannell and Thornley (2000) assumed that oxidation of one mol of glucose ($C_6H_{12}O_6$) has a net yield of 30 mol ATP, 24 H^+ and 6 mol CO_2. Thus, a plant process that has an energy and a reducing power requirement of (a mol ATP + n mol H^+) would require:

$$(a/30 + n/24) \text{ mol of glucose} \qquad (5.8)$$

This requirement could also be expressed in g C given that the molar mass of C is 12 and multiplication by (44/12) would convert any C loss to that of CO_2. This assumption has been followed below, but more accurate estimates of the net yields of ATP and H^+ from the consumption of glucose would modify calculations of the respiratory costs of processes. The essence of the calculation of the respiratory cost of different processes is first to determine the amount of ATP and/or NADH required in the reaction and then to equate this to an amount of glucose, and thus C and thus CO_2, based on Equation 5.8. However, the calculation is based on the stoichiometry of one mol of glucose yielding 24 protons and 30 ATP molecules. Further details of the metabolism of glucose can be found in basic biochemistry and plant molecular biology textbooks (i.e. Heldt 1997).

Nitrate reduction uses 8 mol H^+ mol N^{-1} (Cannell and Thornley 2000) and thus is takes (8$H^+ \times 6C \times 12/24$ (expressing H^+ in terms of C)) g C per mol of N with molecular mass 14, or $(8 \times 6 \times 12/24)/14 = 1.7$ g C (g N reduced from NO_3 to NH_3)$^{-1}$ or 6.2 g (i.e. $44/12 \times 1.7$) CO_2. Symbiotic N_2 fixation is a high-energy demanding process, requiring 16 ATP and 4 NADPH (mol N_2 reduced to NH_3)$^{-1}$. Although this gives a theoretical minimum C requirement of 2 g (g N_2 reduced to NH_3)$^{-1}$, values cited in Cannell and Thornley (2000) are up to three times the minimum, yielding 15–22 g CO_2 (g N_2 reduced to NH_3)$^{-1}$. The specific cost of N

Table 5.6 Summary of the specific unit costs of processes with readily quantifiable respiratory costs in terms of g CO_2 (g delivered product)$^{-1}$ (based on data in Cannell and Thornley 2000).

Process	g CO_2 g^{-1} product
Growth	5.4
NO_3 reduction to NH_4	6.2
Symbiotic N_2 fixation	15–22
NO_3 uptake	1.2
NH_4 uptake	0.6
Uptake of other ions	0.2
Phloem loading	0.2

uptake depends on whether NH_4 or NO_3 is the ion involved, with uptake cost of NH_4 about half that for NO_3. The figure arrived at in Table 5.6 for NO_3 uptake assumes that each mol of NO_3-N (i.e. 14 g) requires 2 mol H$^+$ or 2 mol ATP or 0.5 mol O_2, since consumption of one mol of O_2 produces *c*. 5 mol ATP. In terms of C this translates to 2/5 mol C or (2/5 × 12) g C. In terms of mass of NO_3-N, the C cost is $((2/5) \times (12/14)) = 0.34$ g C respired (g NO_3-N taken up)$^{-1}$, or 1.2 g g^{-1} when converted to the CO_2 equivalent. The costs of uptake of other ions are thought to be less than that of NO_3-N and to require 1 mol H$^+$ (mol ion)$^{-1}$ or 1 ATP per ion. Using the same calculation logic as for NO_3-N leads to the estimate of 0.06 g C (g mineral ion taken up)$^{-1}$ or 0.2 g CO_2 (g mineral ion taken up)$^{-1}$, assuming that the average mineral relative molecular mass is 40 (*cf*. 14 for N). Clearly, a more detailed analysis could be performed for each of the non-N macro- and micro-elements but weighted by their concentration in tissues. The respiratory costs of phloem loading, starting in the chloroplasts and ending with apoplastic loading into the phloem, take their starting-point in energy requirement of 2.5–4.0 mol ATP (mol sucrose loaded)$^{-1}$ or 0.5–0.8 mol CO_2 (mol sucrose loaded)$^{-1}$ given that the respiration of 6 mol C *via* oxidative phosphorylation provides 30 mol ATP. Converting to C mass gives 0.06 g C (mol sucrose loaded)$^{-1}$ or 0.2 g CO_2 (g sucrose loaded)$^{-1}$. Phloem unloading is assumed to occur passively through the symplast and be without respiratory costs. A summary of the costs of the above processes is given in Table 5.6.

Respiratory processes associated primarily with maintenance are more difficult to quantify than those of growth, because they inevitably involve breakdown and synthesis of compounds and bi-directional fluxes rather than the more uni-directional flow of material associated with growth. Cannell and Thornley (2000) identified protein turnover, the maintenance of cell ion concentrations and non-oxidative 'wasteful' respiration as falling under this cloud of uncertainty.

Protein turnover is catalysed by proteases and is thus under metabolic regulation that requires energy. The use of protein during leaf development, in cellular membranes and, most importantly, as Rubisco (Chapter 4) means that a major part of maintenance respiration is associated with protein turnover. Penning de Vries (1975) calculated that 50–60% of maintenance respiration is associated

Table 5.7 Estmiated specific costs of the component processes involved in protein turnover (based on Amthor 2000).

Process	Metabolic cost (no. ATP per amino acid)
Total protein breakdown cost (to amino acids)	0.13–2.0
Protein synthesis	
Amino acid activation	2.0
Editing of tRNAs	0.2
Polypeptide initiation and elongation	$2 + 1/n$[a]
Editing non-cognate tRNA	0.01
Methylation, acetylation etc.	0.1
Phosphorylation	0.1–0.3
mRNA turnover	0.2–0.4
Signal synthesis	0.2–1.0
Total synthesis cost	4.8–6.0
Total cost (breakdown and synthesis)	4.9–8.0

[a] n is the number of amino acids in a protein.

with protein turnover in C_3 plants, although others (e.g. Barneix *et al.* 1988) have given values of 27–36% of maintenance respiration as being for protein turnover. Up to 20% of proteins may be recycled on a daily basis, although values of 2–7% are more usual (de Visser *et al.* 1992; Bouma *et al.* 1994; Zerihun *et al.* 1998). The overall costs of protein turnover are high and may consume 3–5% of the dry matter produced per day, but details of the metabolic costs of individual processes in protein turnover are rare. However, Amthor (2000) made a bold attempt to estimate the specific ATP costs of component processes of protein turnover, calculated as ATP per amino acid. His conclusions as to the metabolic costs of the separate processes in protein turnover are shown in Table 5.7.

The total cost of 5–8 mol ATP (mol amino acid)$^{-1}$ infers a C cost of 0.2–0.3 mol or 0.9–1.6 g CO_2 (g amino acid)$^{-1}$. However, there is a high level of uncertainty associated with these estimates. The larger costs are associated with synthesis rather than breakdown of proteins and within the synthesis component, the production of the amino acids and their assembly into polypeptides can be up to ten times more costly than the finer-scale RNA translation processes.

It was seen in Equation 5.7 that the total respiration is the combined result of the rate of respiration of a process and its specific cost. Little has been said about the rates of respiratory processes, mainly because it is difficult to find cases where the specific costs and the rate are disentangled. Amthor (2000) attempted to calculate the rate for protein turnover using data from Zerihun *et al.* (1998), who presented data showing that between 6.5% and 21% of total protein is turned over daily. Thus, plants cycling 10% of their biomass at a rate of 0.15 (i.e. 1/6.5) d^{-1} would cycle 0.015 mol protein (mol biomass)$^{-1}$ d^{-1} as a specific cost. Given that the mean molecular mass of amino acids is 0.120 kg mol^{-1}, this means that 0.015/0.120 or about 125 mmol (or 5.5 g) CO_2 (kg biomass)$^{-1}$ d^{-1} is the maintenance respiration rate for protein turnover. Other important turnover processes for which costs are largely unknown are membranes – their proteins and lipids, and the macromolecules of DNA, chlorophylls and hormones.

Table 5.8 Assimilation (A) and respiration, either in the light ($R_{m (light)}$) or after 16 h darkness ($R_{m (dark)}$) and the ratio ($A/R_{m (light)}$) of a range of C_3 and C_4 species of cultivated plants. All units are $\mu mol\ CO_2\ m^{-2}\ s^{-1}$ (after Byrd *et al*. 1992).

Plant	A	$R_{m (light)}$	$R_{m (dark)}$	$A/R_{m (light)}$
C_3 species				
Triticum aestivum	27.9 ± 1.0	2.0 ± 0.1	0.5 ± 0.1	14.0
Panicum boliviense	18.3 ± 2.1	2.0 ± 0.4	0.8 ± 0.3	9.2
Mean	23.1 ± 1.6	2.0 ± 0.3	0.7 ± 0.2	11.6
C_4 species				
Sorghum bicolor	41.1 ± 2.8	1.8 ± 0.1	1.0 ± 0.2	22.8
Zea mays	35.3 ± 3.0	2.9 ± 0.5	0.9 ± 0.2	12.2
Panicum maximum	37.6 ± 4.0	2.1 ± 0.2	1.0 ± 0.1	17.9
Mean	38.0 ± 3.3	2.3 ± 0.3	1.0 ± 0.2	17.6

As protein turnover is metabolically expensive, it might be hypothesised that respiratory costs are lower in C_4 species, such as maize, compared with C_3, such as wheat and soybean. The reasoning behind such conjecture is that Rubisco is used more efficiently in C_4 photosynthesis. C_3 plants also allocate more of their soluble protein to Rubisco than do C_4 plants. The generally higher N and thus protein concentration in the leaves of C_3 as opposed to C_4 plants (Sinclair and Horie 1989) should mean that respiration costs are higher in C_3 than in C_4 species. Byrd *et al*. (1992) examined a range of CO_2 exchange parameters for several cultivated and wild C_3 and C_4 species (Table 5.8) and found that, while assimilation rates were higher as expected in C_4 species, respiratory losses of CO_2 in both the light and dark differed little between C_3 and C_4 plants. As a result, the ratio of assimilation to respiration was higher in the C_4 species. Further results by Byrd *et al*. (1992) showed that for both C_3 and C_4 plants A and R_m increased with specific leaf nitrogen (g N g^{-1} leaf) and this response dwarfed any difference between the two plant types. It seems to be the case that high N and thus protein levels increase both assimilation and respiration in plants, echoing the general points made earlier, and that the composition of leaves and the age of respiring tissue are more determining of protein turnover and thus the costs of maintenance. Thus, young expanding leaves with high levels of meristematic activity and mainly protein synthesis but low dry weight will lose CO_2 *via* growth respiration, whereas mature leaves with high dry weight, low meristematic activity but high protein turnover will expend energy in maintenance. It is also the case that the relative costs of other energy-demanding processes such as phloem translocation are also affected by leaf age and composition.

Uptake of ions, as active ion transport or to counteract membrane leakage, and the maintenance of intra-plant ionic gradients represent the largest respiratory cost after protein metabolism. Maintaining ionic gradients can require up to 30% of the total respiratory costs of young roots with the rest being allocated to protein turnover and growth-related ion uptake. Nitrogen uptake was mentioned above as being metabolically about twice as expensive for NO_3 than for NH_4. Uptake of other cations is thought to have the same cost as NH_4 of 1 ATP per ion. Using the stoichiometry mentioned above of 30 mol ATP per mol glucose and assuming an average cation relative molecular mass of 40, gives 0.06 g C (0.22 g CO_2) (g mineral uptake)$^{-1}$. In reality, cations vary in relative molecular mass and those taken up will depend on their availability in the soil and many other plant physiological factors, such as cell vacuole biochemistry. The concentration of non-N minerals is about 4% of plant dry weight and thus the demand for them can be calculated as the product of their concentration and plant dry weight, net of senesced plant parts. Maintenance of ionic gradients has been investigated by measuring effluxes and influxes of mineral ions under steady-state conditions. At this point, respiration cost can be calculated as the product of the specific ion uptake cost and the fraction of active ion influx that compensates for ion loss or efflux (Bouma and de Visser 1993), in turn calculated as the increase in the electrical conductivity of demineralised water when plant roots are placed in it. Results (Bouma and de Visser 1993) show that the balancing influx of ions and thus the maintenance of ionic gradients may consume 25–50% of the total respiratory costs associated with ion uptake, the rest being associated with growth as *de novo* ion uptake. Total ion fluxes may account for 25–50% of the total maintenance costs of roots, the rest being mainly protein turnover, although many of the estimates are uncertain.

In summary, respiration associated with growth, nitrate reduction, symbiotic nitrogen fixation, N-uptake, other ion uptake and phloem loading, can be calculated because reasonable estimates exist of the specific unit respiratory costs and the rates of these processes, the biochemistry well known, and the pathways are mainly unidirectional. Growth respiration leads to about 0.8 g C appearing in new biomass per g of C substrate utilised. Values of the specific unit respiratory costs of other processes are less certain, but estimates are:

- for nitrate reduction; 1.7 g C respired from a glucose substrate (g nitrate N)$^{-1}$;
- for symbiotic N_2 fixation; 4.6 g C respired from a glucose substrate (g N_2 reduced to ammonia)$^{-1}$;
- for NO_3-N uptake; 0.34 g C respired from a glucose substrate (g nitrate N)$^{-1}$ and about half this for NH_4-N uptake;
- for the uptake of other ions; 0.06 g C respired from a glucose substrate per g mineral taken up;
- for phloem loading; 0.06 g C respired from a glucose substrate (g C loaded)$^{-1}$.

At present, it is less easy to estimate accurately the specific unit respiratory costs and/or the rates of protein turnover, the maintenance of cell ion concentrations and gradients and all forms of 'wastage' respiration (Lambers 1997), the specific costs and significance of which are extremely speculative (Cannell and Thornley 2000).

5.7   Environmental effects on respiration

Long-term exposure of roots of crop plants to drought reduces respiration in line with a general decline in carbon assimilation under such conditions (Chapter 4).

Leaves also show a decline in respiration as their water potentials decline, but factors such as the speed of imposition of drought seem to affect the level of response. Although it is the case that the respiratory costs of coping with saline environments are small in salt-adapted species, it is unclear what the physiological and yield consequences would be of trying to adapt the major food crops to cope with salinity, although this is one of the major goals of crop breeding by conventional and biotechnological methods. Since ion uptake and the maintenance of ionic gradients are important consumers of respiratory C in roots, it is possible to predict that increasing nutrient concentrations would lead to increased respiration, leading to increased growth, and this seems to be the case. Conversely, a low supply of nutrients will tend to reduce root respiration but may increase the level and thus the costs of N_2 fixation in legumes, with the inference that the respiratory costs per unit of growth and maintenance must increase at low nutrient levels. Respiration rates of shoots are less influenced by levels of nutrient supply (van der Werf *et al.* 1994).

As noted above, temperature is an important stimulant of respiration *via* its effect on the rate of enzymatically catalysed reactions and thus the demand for ATP, protein turnover and biosynthesis. The effect of temperature on respiration is described by the temperature coefficient of respiration (Q_{10}, Equation 5.2), the value of which describes the degree of increase in respiration for a 10°C increase in temperature and which has been discussed above.

The elevation of the CO_2 level of the atmosphere presents crop physiologists with a challenge to understand and predict the effects of climatic change on the processes of yield formation for the major crops. Many factors influence the rate and amount of mitochondrial respiration and one of the more important ones is the level of CO_2 in the atmosphere, which has risen on average by about 100 μmol mol^{-1}, following the industrial revolution and the use of fossil fuels in Western countries. In C_3 crops such as wheat and soybean there is clear evidence that R_m is reduced per unit plant dry weight by 15–18% (Drake *et al.* 1997, 1999) when the current CO_2 concentration of about 360 μmol mol^{-1} is doubled. Such a response is not seen in C_4 crops such as maize, with the inference that the rate of photorespiration is reduced as one of the affected processes. Recent studies have provided explanations for the phenomena observed in the above cropping studies. Thus, there seem to be two respiratory responses to CO_2: a direct, reversible effect where respiration rates temporarily drop but thereafter rise to pre-exposure levels when the CO_2 level is restored to ambient; and an indirect, irreversible effect observed only in C_3 plants. The direct effect is brought about by the end-product inhibitory action of CO_2 on respiratory enzymes such as cytochrome oxidase and/or competition between CO_2 and other small anionic substrates for the binding sites of respiratory enzymes such as succinate dehydrogenase. Other mechanisms have also been postulated.

Exposure of young photosynthetic tissue to elevated CO_2 in the presence of adequate nutrients increases growth as do higher light levels, as seen above. However, as tissue ages, elevated CO_2 seems not to stimulate respiration but to reduce it, at least in C_3 species (Drake *et al.* 1999). The inference is therefore that higher concentrations of substrate formed under higher levels of CO_2 and light in actively growing material leads to higher rates of respiration in agreement with the experimental evidence shown above (Figure 5.4). Reduction in the rate of CO_2 loss at elevated CO_2 levels, known as acclimation, argues for the conclusion that more maintenance, for example protein turnover, rather than growth activities are being supported by respiration in acclimated tissue. The respiratory costs of lowered protein turnover and transport in expanded rather than actively growing tissues provide a testable hypothesis of respiratory acclimation and, in terms of the McCree formulation of respiration, are interpreted as the shift from growth to maintenance respiration, as seen in Figure 5.6. The current empirical formulation of the McCree model and derivatives are starting to account for the ontological shift from growth to maintenance respiration and need to be linked to plant and crop processes and plant chemical composition. What is required ultimately is a mechanistic model of respiration that incorporates the elegance of the Farquhar, von Caemmerer and Berry biochemical model of photosynthetic CO_2 assimilation in leaves of C_3 species, presented in Chapter 4.

5.8   Crop respiration in the future

The foundation established by crop scientists studying respiration in the 1970s and 1980s has basically endured but in more refined ways that now take account of the type of dry matter that is grown and maintained and the processes that underpin the accumulation and loss of environmental resources. The respiration paradigm has extended from the simple conceptualisation of respiration into growth and maintenance components to incorporate the notion of wasteful respiration and the stochiometry of the major groups of biochemical products. The sophistication with which crop models can calculate the respiratory losses associated with growth and maintenance metabolism of a variety of crop types contrasts favourably with many natural ecosystem models that treat respiration more simplistically. Furthermore, there is a unity of approach between studies of crop production in relation to radiation interception and RUE as seen in Chapter 4 and the notion of the conversion efficiencies of the direct products of photosynthesis into the more sophisticated biochemical structures needed for plant and crop survival and reproduction. Targets for the future will include updating models, as more detailed biochemical and genomic knowledge becomes available, and improving methods of respiration measurement at the large scale.

Finally, the agricultural fluxes of CO_2 and other greenhouse gases, such as methane (CH_4) and nitrous oxide (N_2O), from agricultural systems are being recognised as elements in regional greenhouse gas balances. An example of such fluxes for a range of cropping systems is shown in Figure 5.9 and the global warming implications of such emissions are presented in Table 5.9.

Taken overall, such fluxes show that CO_2 and N_2O represent the major fluxes from a range of agricultural systems and that the largest potentials for reducing greenhouse gas fluxes from cropping systems arise via the use of perennial legumes and biomass crops such as poplar. These arise because these cropping systems store C in the soil and fix small amounts of methane, the effects of which are large enough to overcome associated N_2O emissions.

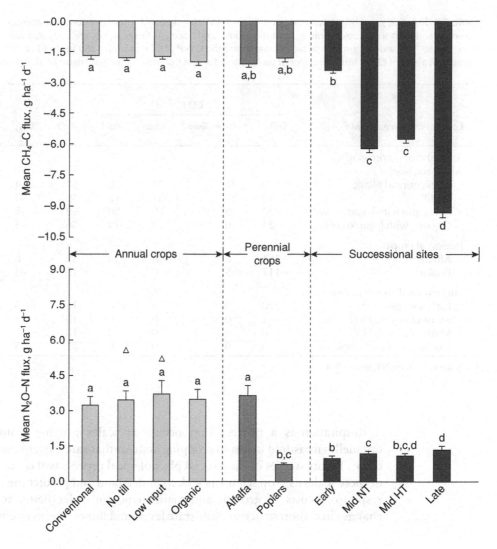

Figure 5.9 CH_4 (top) and N_2O production (bottom) in annual and perennial cropping systems and unmanaged systems. Annual crops were managed as conventional cropping systems, as no-till systems, as low-chemical input systems, or as organic systems (no fertiliser or manure). Successional systems were either never tilled (NT) or historically tilled (HT) before establishment. All systems were replicated three to four times on the same or similar soil series; fluxes were measured over the 1991–99 period. There are no significant differences ($P \geq 0.05$) among bars that share the same letter on the basis of analysis of variance. Triangles indicate average fluxes when including a single day of anomalously high fluxes in the no-till and low-input systems in 1999 and 1991, respectively (after Robertson *et al.* 2000).

Table 5.9 Relative greenhouse warming potentials (GWP) for different agroecosystem management systems based on soil carbon sequestration, agronomic inputs, and trace gas fluxes. Units are CO_2 equivalents (g m^{-2} year^{-1}) based on conversion factors weighting the GWP of CO_2 as unity, 280 for N_2O and 56 for CH_4. Negative values of GWP indicate a global warming reduction potential (after Robertson *et al.* 2000).

Ecosystem management	CO_2				N_2O	CH_4	Net GWP
	Soil C	N fertiliser	Lime	Fuel			
Annual crops (corn–soybean–wheat rotation)							
Conventional tillage	0	27	23	16	52	−4	114
No till	−110	27	34	12	56	−5	14
Low input with legume cover	−40	9	19	20	60	−5	63
Organic with legume cover	−29	0	0	19	56	−5	41
Perennial crops							
Alfalfa	−161	0	80	8	59	−6	−20
Poplar	−117	5	0	2	10	−5	−105
Successional communities							
Early successional	−220	0	0	0	15	−6	−211
Midsuccessional (HT)	−32	0	0	0	16	−15	−31
Midsuccessional (NT)	0	0	0	0	18	−17	1
Late successional forest	0	0	0	0	21	−25	−4

HT, historically tilled; NT, never tilled.

Respiration is a process that occurs at scales ranging from subcellular organelles measured in μm to cropping systems that range over tens or hundreds of m^2. Future studies of this crucial physiological process will occur at both ends of these scales as crop scientists seek to understand and predict the physiological respiratory costs of growth and maintenance in order better to predict the changes in carbon structure, substrate levels and fluxes at ecosystem levels.

Chapter 6

The partitioning of dry matter to harvested organs

> *. . . it is one of the important tasks before crop physiology to analyse the basis of the competing power of an organ for assimilates, because it is on empirical selection for this that past increases in yield have largely been based.*
>
> (Evans 1975)

Chapters 3 to 5 have dealt with the first three components of crop yield, following the analysis presented in Chapter 1: the supply of solar radiation over the lifetime of the crop (Q); the proportion of the incident radiation intercepted by the leaf canopy (I); and the efficiency of conversion of the intercepted solar energy to chemical potential energy in the form of fixed carbon (or plant dry matter) (ε). However, not all of the biomass produced ($Q \times I \times \varepsilon$) is located in harvested parts, and this chapter considers how dry matter is distributed among the organs of the crop, culminating in the fourth component of crop yield: the harvest index (the fraction of the dry matter of the mature crop that is partitioned to the harvested organs; Hay 1995). It begins with the physiology of assimilate partitioning within the individual plant, as an understanding of these processes is necessary in, for example, the interpretation of grain filling from storage and from current photosynthesis.

6.1 The processes and pathways of assimilate partitioning

Figure 4.3 describes the conversion of glyceraldehyde 3-phosphate (triose phosphate), the primary product of the photosynthetic Calvin cycle, into glucose, by a series of enzyme-regulated steps in a photosynthetically active leaf. In the *cytosol*, the enzymes of sucrose synthesis combine one molecule of glucose with one molecule of fructose (an intermediate in the synthesis of glucose) to give a molecule of sucrose, which is by far the most important compound in assimilate transport and partitioning within higher plants (Figure 6.1). Several crop plants, including the major cereal species, have some capacity to store sucrose in leaf cell vacuoles, whereas others, notably soybean, do not (Huber *et al.* 1992). Under

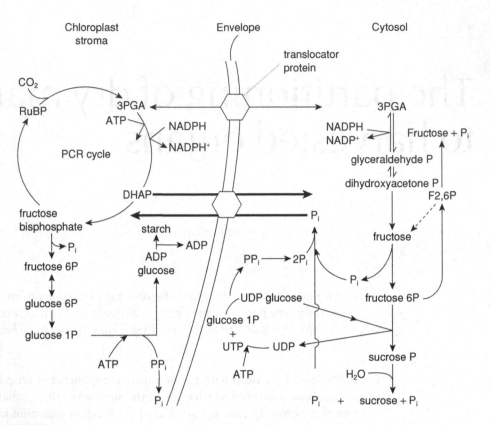

Figure 6.1 Pathways of carbon metabolism in the chloroplasts and cytosol of a photosynthetically active leaf, leading to the synthesis of starch and sucrose (from Lawlor 1993).

conditions where the concentration of sucrose in leaf cells rises (e.g. high irradiance and low temperature, inhibiting export and use), temperate grass and cereal species can also store soluble fructans (short chain polymers of fructose) in the vacuoles of mesophyll cells (Pollock and Cairns 1991). In contrast, within the *chloroplast*, fructose is converted to glucose which, in turn, is polymerised to give insoluble starch (Figures 6.1, 6.2).

These are the components of the system that buffers the level of sucrose in the cytosol of leaf cells, by exchange with sucrose (and/or fructans) in the vacuole, and by mobilising and depositing starch in the chloroplast. The key controlling process is the exchange of inorganic phosphate for phosphorylated intermediates (PGA and DHAP; Figures 4.3, 6.1) by the 'phosphate translocator' located in the chloroplast envelope (Lawlor 1993). Turnover can be rapid, measured in hours (Farrar 1999): during periods of rapid photosynthesis, sucrose and starch reserves can be replenished whereas, in the dark or under intense demand for assimilate (for local use or transport to distant storage), vacuolar sucrose can be mobilised rapidly. Subsequent hydrolysis of starch, and transport of the resulting phosphorylated intermediates across the chloroplast membrane (Figure 6.1) can meet demand over a prolonged dark period (Komor 2000).

These processes in the photosynthetic leaf are matched by a similar series at sites of storage (e.g. a developing cereal grain or potato tuber) (Figure 6.2). The

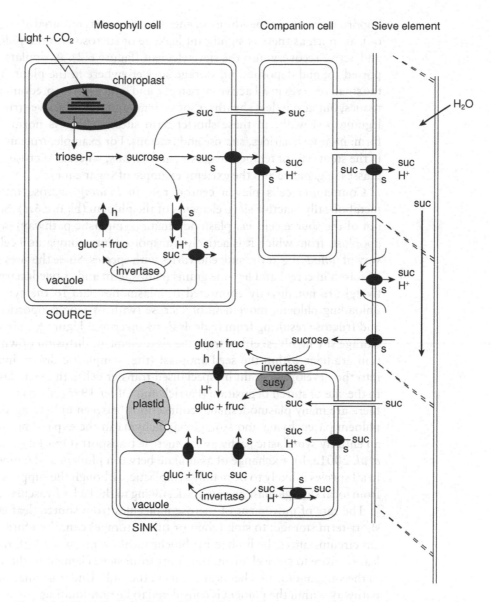

Figure 6.2 Sugar fluxes in higher plants: photosynthetic assimilates are transported *via* the phloem, mainly in the form of sucrose, to sink tissues, with loading and unloading occurring either symplastically through plasmodesmata, or apoplastically by sucrose transporters (*s*). Sucrose export or leakage and recovery occur along the length of the phloem. In sink tissues, invertase and sucrose synthase activities, and hexose transporters (*h*), establish a sucrose gradient between source and sink (from Hellman *et al.* 2000).

levels of sucrose in the cytosol are kept low by enzymatic hydrolysis to glucose and fructose, and by the incorporation of the resulting sugars into polysaccharides (normally starch) stored within plastids. The resulting differences in sucrose concentration generate a gradient in turgor pressure within the lumina of the phloem elements, providing the driving force for the mass flow of sucrose from the 'source' organ (net exporter of assimilate) to the 'sink' organ (net importer). Further expenditure of energy is, however, required along the pathway of transport, as there is significant leakage of sucrose into the phloem apoplast and active recovery into the sieve element (Figure 6.2). Assimilate is also transported to, and deposited in, storage sites elsewhere in the plant. These include the small reserves in all active meristems, and larger stores, predominantly in stem tissues, but also in leaf sheaths, root systems and reproductive structures such as legume pod walls; in these shorter-term sites, storage is normally in soluble form: monosaccharides, sucrose and fructans. For example, fructans predominate in the stem cells of temperate grasses and cereals, whereas sucrose is stored in C_4 grasses (e.g. maize and the extreme example of sugar cane).

Companion cells play a central role in *loading* sucrose into the largely metabolically inactive sieve elements of the phloem (Figure 6.2). Sucrose moves out of the source cells *via* plasmodesmata (symplastic pathway) or into the leaf apoplast, from which it is actively transported into companion cells. The pathway of *unloading* at the sink end varies with species. Since the sites of deposition of starch in cereal and legume grains (endosperm and cotyledonary cells respectively) are not directly connected by plasmodesmata to the symplasm of the unloading phloem, movement of sucrose (with varying proportions of glucose and fructose resulting from hydrolysis by invertase; Figure 6.2) from phloem to storage site involves: efflux from the sieve element; diffusion down a concentration gradient within the seed apoplast (the 'symplastic discontinuity'); uptake into the developing grain by specialised transfer cells; and symplastic transport to the site of starch deposition (Patrick and Offler 1995). By contrast, in potato, there are many plasmodesmatal connections between unloading sieve elements, phloem parenchyma and storage parenchyma in the expanding tuber, offering an entirely symplastic pathway for sucrose transport (Oparka *et al.* 1992; Viola *et al.* 2001a, b). Exchange of assimilate between phloem and temporary storage in all species is likely to be mainly symplastic, although the supplies to meristems from local storage must be apoplastic, owing to the lack of vascular development.

The rate of movement of sucrose molecules from source (leaf chloroplast or short-term storage) to sink (grain or tuber storage) can, therefore, under different circumstances, be limited by: biochemical events in the leaf; transport from leaf or store to sieve element; transport from sieve element to the site of storage in the sink; and/or biochemical events in the sink. Under normal conditions, the pathway within the phloem is considered to be non-limiting.

6.2   Ontogeny and assimilate partitioning: a survey of source/sink relationships

From unfolding or appearance (Sections 3.1, 3.2) to fully-expanded area, an individual leaf changes from being a sink for assimilate to a source, as its photosynthetic capacity increases during ontogeny (Figure 6.3a, b). However, the process is complex, with more mature tissues exporting to less mature tissues *within* the leaf, and there is the potential for simultaneous export from, and import to, a developing leaf (Figure 6.3a, b; Turgeon 1989). For example, in squash (*Cucurbita pepo*), the tip of a developing leaf stops importing and starts

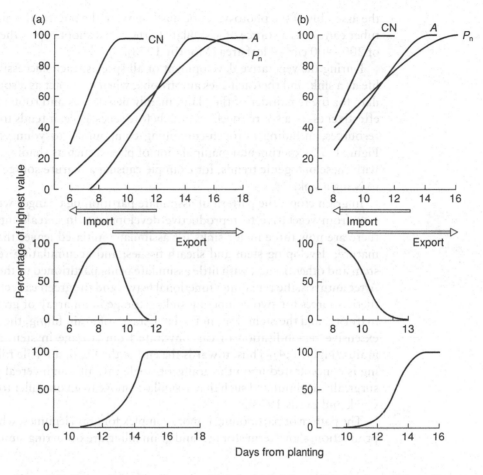

Figure 6.3 Time courses of cell number (*CN*), leaf area (*A*), net photosynthesis (P_n), and import and export of carbon from (a) primary leaves of bean (*Phaseolus vulgaris*) and (b) the second leaf of barley (from Dale 1985).

exporting when the lamina has achieved only 10% of its final area; such exports supplement the supply of assimilate into the leaf from mature leaves. By the time that the lamina has reached 35% of its final area, it has become a net exporter of assimilate, although concurrent import continues up to 45% (Turgeon and Webb 1973, 1975). In cereals and grasses, where cell division and expansion are concentrated in the extension zone near the leaf base (Section 3.1), net export of assimilate from mature regions to other organs of the plant takes place *through* immature tissues.

During plant ontogeny, the status of each plant organ tends to vary between sink and source, but not necessarily from sink to source. When the dormancy of endospermic seeds or potato tubers is broken, they act as sources of assimilate for construction of the early leaf tissues of the emerging plant which, in turn, pass through the transition from sink to source as their potential for photosynthesis increases (Figure 6.3). Where storage in the seed is in epigeal cotyledons (e.g. *Phaseolus vulgaris*), the pattern is more complex: once the cotyledons have expanded into aerial organs, they become temporary *consumers* of resources for the assembly of the photosynthetic machinery. In the potato, the planted mother tuber can act as a source of assimilate to the aerial shoots up to the achievement of 200–400 cm^2 of leaf area (Moorby 1978).

During the vegetative development of all species, each successive leaf begins life as a sink, and then achieves autotrophy, when it can act as a source of assimilate for the remainder of the plant, mainly new leaves and roots. As it ages, its effectiveness as a source declines and, before senescence, it tends to export other resources, including nitrogen-containing compounds, to younger tissues (e.g. Figure 6.5). Experimental manipulation of plants, such as shading, can interfere with these ontogenic trends, for example causing a mature source leaf to revert to being a sink.

In grain crops, the pattern of assimilate partitioning changes with the transition from vegetative to reproductive development. In cereals, up to anthesis, there are now three major sinks for assimilate: initiated leaves that are not yet mature; developing stem and sheath tissues; and accumulating reserves in the stem and other tissues, with little assimilate being partitioned to the root system. After anthesis, the remaining functional leaves and the green ear act as photosynthetic sources for two competing sinks (storage in an array of grains of similar maturity, and the stem) but, in the later stages of grain filling, there is normally extensive remobilisation of carbohydrate from storage in stem and leaves to grains (Figure 6.4). Thus, towards the end of the life of a fertile tiller, partitioning is concentrated upon the grains of single ear, although cereal plants can be surgically manipulated such that assimilates move from one tiller to another (e.g. Cook and Evans 1978).

The pattern of partitioning is more complex for grain legumes, where assimilate production, short-term storage, and grain filling are occurring simultaneously at

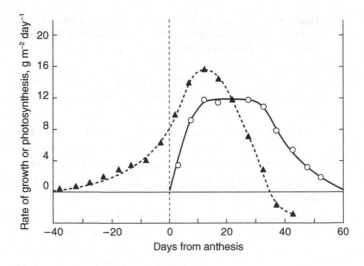

Figure 6.4 The rate of growth of the grains of a wheat crop (Skandia III) (o) and the estimated rate of photosynthetic production of dry matter eventually harvested in the grains (▲). The lack of correlation between the curves shows that photosynthate that is surplus to immediate requirement for grain filling can be stored and released later for grain filling when the rate of photosynthesis declines (from Stoy 1980).

several nodes on a single stem, but the nodes are at different stages of development (Flinn and Pate 1970). In indeterminate varieties, new vegetative phytomers can also be developing. In potato crops the initiation of tubers during the phase of rapid expansion of the leaf canopy (Figures 2.11, 6.15) ensures that there is competition among several sinks, with partitioning to new leaves, stem tissues and tubers being closely regulated (van Heemst 1986) (Section 6.8). In grass swards, where the harvested material is both source and sink, the partitioning of assimilates is crucial in a different way: here, successful regrowth after cutting depends upon the mobilisation of assimilate stored mainly in the lower stem nodes to generate the new canopy (Section 6.9).

The fine control of the partitioning of assimilate among competing sinks remains one of the unresolved areas in plant science (Farrar 1992). In the simplest analysis, the competitiveness of a given sink depends upon its relative strength (capacity and the rate of uptake of assimilate) and its distance from the source along the phloem pathway (e.g. in wheat, Cook and Evans 1978). Thus, under certain circumstances, a weak sink may draw assimilate from an adjacent source, to the detriment of a stronger but more distant sink, although the geometry of the phloem can play an important part in directing resources to a particular sink (Wardlaw 1990). At least three groups of growth regulators have been implicated: auxin, gibberellins and cytokinins. Detailed molecular analysis of assimilate partitioning in a range of model bacteria and higher plants (Hellman *et al.* 2000) has indicated that, at a given capacity, the strength of a sink depends upon the expression of a suite of genes involved in the transport and metabolism of sugars; this expression is, in turn, regulated *directly* by sucrose and other sugars acting as signals.

In summary, the partitioning of assimilate to the harvested organs of a crop is the net result of a complex series of physiological processes, and their interactions with the environment and crop management. The control and integration of these processes are still poorly understood. Competition for resources among the grains, tubers, tillers or other harvested parts of a given crop is considered in the following sections.

6.3   Time courses of dry matter partitioning: harvest index

The net effects of assimilate partitioning can be illustrated by the time course of dry matter distribution in a soybean crop (Figure 6.5a). For the first 50 days from crop emergence, most of the plant dry matter was partitioned to the leaf canopy but, after the initiation of the first flower, an increasing proportion was devoted to stem and petiole tissues, supporting an expanding population of flowers. Once grain (bean) filling started, virtually all of the increase in dry matter was partitioned successively to pods and grains; and, in the last days of the green crop, all new assimilate and stores located in the stem and pods were channelled to the grains. Meanwhile, there had been an accelerating loss of dry matter from the crop, owing to losses of senescent leaves and petioles, with the result that, by maturity, the grains constituted a high proportion of standing crop dry matter (here a harvest index of approximately 0.5). There was an even

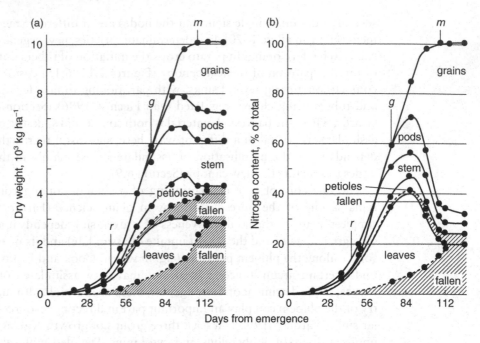

Figure 6.5 Time courses of the distribution of (a) dry matter and (b) nitrogen in soybean plants, where *r* indicates the initiation of reproductive development, *g* the start of grain (bean) filling, and *m* crop maturity (from Shibles *et al.* 1975).

greater concentration of plant nitrogen in the grains (Figure 6.5b). Time courses for the partitioning of dry matter in cereals are broadly similar, except that mature leaves remain attached to the parent plant after senescence.

Strictly speaking, harvest index is the economic yield of a crop expressed as a decimal fraction of total biological yield but, for practical reasons, it is normally measured in terms of above-ground dry matter production. Where the time course of root biomass has been charted (e.g. for soybean, Taylor *et al.* 1982), partitioning below ground has been small compared with that devoted to leaves and stems, and unimportant by early flowering.

Standard methods of measuring harvest index are now widely accepted, facilitating comparisons across seasons and varieties. For annual seed crops, care must be taken to ensure that the sample is sufficiently large and representative; the crop is cut at maturity by hand at ground level, dried to constant weight, threshed, and the resulting grain weighed (Hay 1995). This procedure minimises losses of senescent tissues and ensures that the same proportion of the plant is harvested (e.g. no variation in stubble height). Nevertheless, variation in the degree of shedding of senescent parts makes comparisons among species difficult; for example, if approximately half of the leaves of the soybean plants in Figure 6.5a had remained attached (as in temperate cereals) then the measured value would have been of the order of 0.35 rather than 0.5. There are even greater difficulties in making meaningful measurements of harvest index for vegetative species: in the potato, the practical problems involved in measuring

Table 6.1 Representative values of harvest index for selected temperate crop species, measured by applying standard methods to high yielding crops of modern varieties, mainly in the 1980s (sources in Hay 1995).

Species	Type	Area	Harvest index
Bread wheat	Winter	UK	0.43–0.54
	Spring	USA	0.31–0.51
		Canada	0.38–0.41
		Australia	0.37–0.47
Barley	Winter	UK	0.43–0.57
	Spring	UK	0.55–0.63
		Canada	0.33–0.49
Triticale		UK	0.45–0.47
Maize	Hybrid	USA	0.42–0.49
		Canada	0.47–0.57
		Nigeria	0.36–0.46
Soybean		USA	0.35–0.53
Oilseed rape	Winter	Germany	0.22–0.38
Potato	Maincrop	Canada	0.47–0.62

tuber dry weight as a proportion of crop dry weight (excluding roots) include the tendency for the haulm to disintegrate if permitted to senesce naturally; the risk of chemical degradation of tubers during drying; and the need to recover all stolon and tuber materials below ground.

Table 6.1 presents a set of representative harvest indices, for a range of species, determined according to standard procedures, which allows limited comparisons across species. In general, values around 0.5 can be expected for high-yielding cereals grown in the absence of stress (see Sections 6.6, 6.7). The values for grain legumes and oilseeds tend to be significantly lower (if due allowance is made for the complete defoliation of legumes), partly because of the greater metabolic costs of synthesising and storing proteins and oils, rather than starch (Sinha *et al.* 1982; Chapters 5 and 9). The concept of harvest index can be extended to quantify a particular component as a proportion of total yield: for example, sucrose yield as a proportion of the total dry biomass of a sugar cane crop.

6.4 Limitation of yield by source or sink

During at least two decades from around 1960, one of the major aims in plant physiology was to establish whether yield under a given set of circumstances was limited by the capacity of the leaves to generate assimilate (source limitation) or by the capacity of the harvested organs to store it (sink limitation). The results of the many experimental manipulations (leaf removal, selective shading, selective removal of organs etc.) have been authoritatively reviewed by Evans (1993) who concluded that this approach was, in general, inappropriate as source and sink are not independent: since sinks are constructed out of assimilate generated by sources, crop sink capacity may be determined by the availability of assimilate. Furthermore, many of the experimental manipulations caused disruptions to the intrinsic coordination of the plant.

More recent evidence for source or sink limitation of yield, from a crop physiological standpoint (i.e. considering the crop stand in the field as a whole), is most fully developed for cereal crops. One of the earliest observations of crop scientists applying the principles of yield components (Section 2.1.6) was that, for a particular variety of maize or temperate cereal, individual grain weight was the most stable component, unless grain filling was disrupted by stress or disease (e.g. in temperate cereals, Hay and Walker 1989). For example, in an extensive review of more than 50 winter wheat crops grown in England under favourable conditions, Gales (1983) showed that individual grain weight had the lowest coefficient of variation of the three components, and was not significantly correlated with grain yield. The coefficients of variation of ear population density and grain number per ear were generally higher, and the highest correlation with grain yield was achieved by combining them into a single component: grain population density ($r = 0.8–0.89$, $P < 0.001$). Most of the variation in the grain yield of maize crops grown under favourable conditions can also be explained in terms of grain population density (e.g. Table 6.2), although individual grain weight can fall sharply at very high plant population densities (e.g. from 290 mg per grain at 4×10^4 plants ha^{-1} to 219 at 13×10^4; Tollenaar *et al.* 1992).

The consensus that, in the absence of stress, yields of maize, temperate cereals and grain legumes are determined by grain population density (e.g. Shibles *et al.* 1975; Fischer 1985; Hay and Walker 1989; Egli 1998; Chapman and Edmeades 1999; Andrade *et al.* 2000a) appears, at first sight, to support the concept of sink limitation of yield. However, this ignores the events around anthesis. In wheat,

Table 6.2 Correlation coefficients ($P < 0.05$) between grain yield and grain population density, or individual grain weight, for four maize hybrids grown in Argentina at densities from 5 to 14.5 plants m^{-2} (experiment 1) or from 3 to 18 plants m^{-2} (experiments 2, 3) (from Echarte *et al.* 2000).

Experiment	Hybrid	Grain population density	Individual grain weight
1	DKF880	0.79	ns
	M400	0.99	ns
	DK4F37	0.73	ns
	DK752	0.99	ns
2	DKF880	0.96	ns
	M400	0.92	ns
	DK4F37	0.56	ns
	DK752	0.70	ns
3	DKF880	0.96	−0.89
	M400	0.81	ns
	DK4F37	0.65	ns
	DK752	0.84	ns

ns, no significant correlation.

the developing ear of a fertile tiller initiates a limited number of spikelets but each spikelet can, in turn, initiate up to 8 or 10 florets (Section 2.2.1; Figure 2.7). As few spikelets in the ears of commercial crops will carry more than four grains, there is a substantial overproduction of florets. Similarly, the axillary maize ear produces an excess of single-floret spikelets. The earliest formed florets in the middle of the ear for each species, and the first (basal) florets formed in each wheat spikelet, command precedence over later-formed florets. Thus the empty spikelets in a harvested ear are normally concentrated at the apex and base (Figures 6.8, 6.13). There is a similar overproduction of flowers and pods in soybean (Section 2.1.5).

In wheat, the magnitude of grain population density has been correlated with a series of indices of assimilate availability: the quantity of radiation intercepted by the crop (Fischer 1985); ear dry weight at anthesis (Thorne and Wood 1987; Slafer *et al.* 1990); and biomass produced by, or in the interval around, anthesis (Bindraban *et al.* 1998; Figure 6.6). In maize, similar correlations have been found with photosynthetic rate and crop growth rate (Edmeades and Daynard 1979b; Kiniry and Ritchie 1985; Tollenaar *et al.* 1992; Andrade *et al.* 1999) and with intercepted radiation (Otegui and Bonhomme 1998; Andrade *et al.* 2000b; Ritchie and Alagarswamy 2003; Figure 6.7), in each case during the critical period around anthesis. Further support for the concept that grain yield in wheat is dependent upon grain population density which, in turn, is determined by biomass at anthesis, is provided by the constancy of the rate of change of harvest index during grain filling (Moot *et al.* 1996; Bindi *et al.* 1999). There are similar findings for other grain crops (e.g. correlation between grain population density and crop growth rate at flowering in soybean; Egli and Zhen-wen 1991).

There is, therefore, a very substantial body of evidence showing that, in the absence of stress or other factors affecting grain set (see Section 6.5), cereal grain yield is effectively source limited: each grain has a defined capacity for assimilate, and the available biomass (from current photosynthesis or from reserves) is allocated to grains up to this limit, in order of maturity. For example, Fischer (1985) proposed that one wheat floret would survive for each 10 mg of ear dry weight at anthesis. Relatively straightforward rules of this kind, which are deployed in a range of cereal simulation models (Chapter 9), appear also to hold in the more complex partitioning environment of grain legumes. For example, in pea there is a common pool of assimilate within each stem, partitioned according to sink strength; resources are distributed preferentially to the earlier-formed grains because of their greater capacity, not because of their position (Jeuffroy and Devienne 1995; Jeuffroy and Ney 1997). Partitioning rules of this kind are used in the leading simulation model of soybean (Boote *et al.* 1998; Chapter 9).

In summary, this analysis shows that the high heritability of the harvest index of most grain crops (i.e the lack of influence of variation in the environment on the overall partitioning of biomass by a given genotype; Hay 1995; Vega *et al.* 2000) is fully consistent with overall source limitation of yield. Nevertheless, as explained in the following section, the yields of crops exposed to stress during reproductive development are commonly sink-limited and at a lower harvest index; the effects on yield can be large, permitting a clear distinction between source and sink limitation.

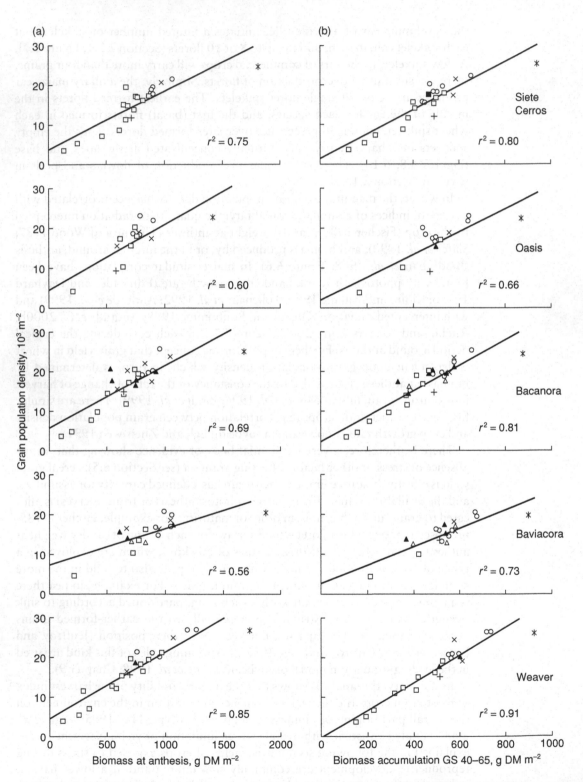

Figure 6.6 Relationships between the grain population densities of wheat crops (five varieties, with different crops indicated by symbols; two seasons 1993–5) in Mexico and (a) crop biomass at anthesis and (b) crop biomass accumulated between GS40 and GS6.5 (anthesis) (from Bindraban *et al*. 1998).

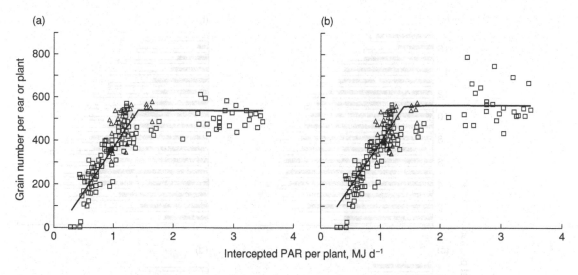

Figure 6.7 Relationships between grain number per ear (a), or per plant (b), of maize crops (hybrids DK 636, 639) grown in Argentina, and PAR intercepted in the period from 10 days before silking to 20 days after silking (from Andrade *et al.* 2000b).

6.5   Sink limitation of yield in cereals – physiology of ineffective grain setting

Winter wheat crops in the uplands of Nepal, where soils are deficient in boron, provide a clear example of sink limitation of yield. Here vegetative development and reproductive development up to floret initiation are relatively unaffected by boron supply, and merely slowed by low temperature, but pollen development and germination are strongly inhibited by both stresses (boron deficiency and temperatures lower than 8°C). In extreme cases, where there is complete failure to set grain, the crop consists of a source (the leaf canopy) without a corresponding sink to fill (i.e. harvest index = 0); in practice, such an imbalance can be corrected to a varying extent by the use of appropriately-tolerant varieties (Subedi *et al.* 1998; 2001). Detailed analysis of the responses of sensitive and tolerant varieties demonstrated that boron deficiency had a progressively greater inhibitory effect on pollination and grain set in subsidiary tillers (Subedi *et al.* 1998); and that low temperature increased floret sterility at all positions in the ear without affecting spikelet initiation (Figure 6.8). In response to each stress factor, there was a partial compensation for reduction in grain number per ear by increased individual gain weight. For example, under temperature stress (8°C day/2°C night around anthesis) the number of grains per ear of a temperature-sensitive variety (NL-683) fell from 60 to 13 (−78%) but grain weight rose from 33.3 to 44.9 mg (+35%). The corresponding responses for the tolerant variety Annupurna-3 were 58 to 40 (−31%) and 33.7 to 41.6 mg (+23%) (Subedi *et al.* 1998). Field crops showing such responses would be limited by sink capacity and have harvest indices lower than those protected from stress. Although the effects of boron deficiency and low temperature appear to be broadly additive, the pattern of response can vary considerably among varieties (Subedi *et al.* 2001).

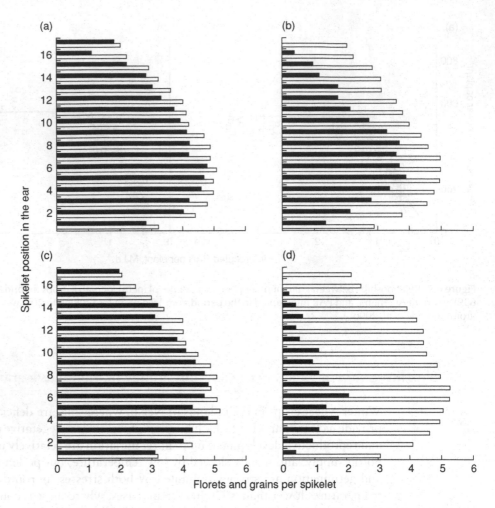

Figure 6.8 Numbers of florets (□) and grains set (■) in each spikelet of ears of wheat plants of varieties Annapurna-3 (a, b) and NL-683 (c, d) grown under controlled conditions with (b, d) or without (a, c) cold treatment. All plants received an adequate supply of boron (from Subedi *et al.* 1998).

This is a particularly dramatic example of the effect of the environment on the most sensitive phase of reproductive development in wheat, but the success of pollination can be affected by a range of other factors, particularly water stress and high temperature (affecting both male and female fertility; Saini and Aspinall 1982). There are concerns that the rise in temperature associated with global climatic change could have serious implications for grain set in temperate regions. For example, Ferris *et al.* (1998) found that, in a field-grown spring wheat crop in southern England where temperature treatments were imposed for 12 days around anthesis, there was a negative correlation between thermal time over 31°C during this period, and grain population density. Thus, at the time when the availability of assimilate was determining the size of the crop sink, a relatively brief period of high temperature stress could result in substantial losses of potential grains and grain yield (here up to 50%).

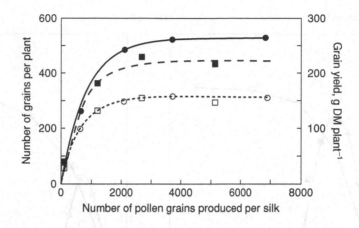

Figure 6.9 Relationships between grain yield (open symbols) or number of grains per plant (closed symbols) and pollen availability per silk, for maize crops of two varieties (P3978 circles; P3925 squares) grown in Minnesota (from Westgate *et al.* 2003).

Crops that rely on external pollination are potentially more vulnerable to environmental stress during reproduction. Thus, as outlined in Section 2.2.2, if maize plants are subjected to shading, drought or high temperature around anthesis, the development of the tassel is relatively unaffected unless the stress is severe, but the appearance and extension of silks can be delayed. The resulting increase in the anthesis–silking interval (ASI; Section 2.2.2) can lead to poor pollination and grain set.

Studies of maize pollen fluxes under favourable conditions have begun to establish thresholds below which grain set is impaired. For example, Westgate *et al.* (2003) showed that the number of fertilised florets (grains) per plant of isolines of Pioneer 3925 were unaffected until the pollen flux fell below 3000 pollen grains per silk (Figure 6.9), well below normal field estimates of 10^4 to 10^5 pollen grains per silk. However, the time course of pollen shedding is also crucial. For example, the shading stress associated with increase in the plant population density of plots of two Dekalb maize varieties from 3 to 9×10^4 plants ha^{-1} had three important consequences: substantial reduction in total pollen supply per tassel (40–60% fall in peak rate of supply); no change in the timing of pollen shedding; but a delay of 1–2 days in the achievement of 50% silking (increase in ASI from 0.5 to 2.0 days and 0.5 to 1.0 days for the two varieties; Figure 6.10). Comparing Figure 6.10 with the idealised relative time courses in Figure 2.5, shows that, even at low plant population densities, these Dekalb varieties ran out of viable pollen well before the completion of silk emergence, and this is reflected in published values of the proportion of potential grains set (50–60%; Uribelarrea *et al.* 2002). Consequently, the disruption of pollination caused by shading in this experiment did not have an *additional* effect on grain set, as pollen supply remained above the threshold during the main period of silk emergence. Breeding and selection of modern maize varieties for intensive grain production have concentrated on such resistance to the stresses associated with high plant population density (Russell 1991).

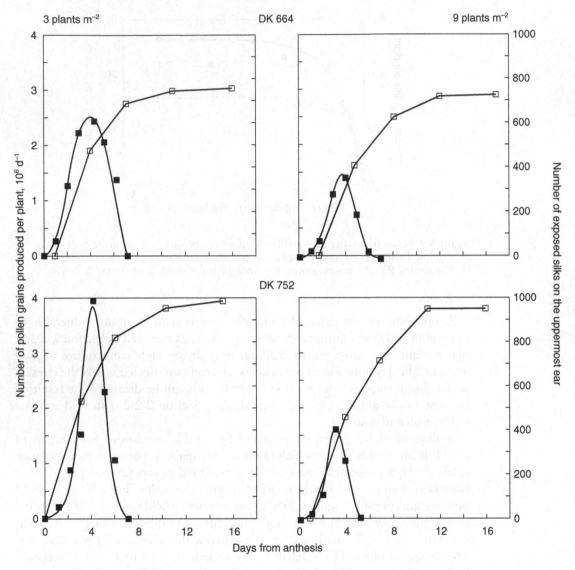

Figure 6.10 Time courses of daily pollen production (■) and cumulative number of exposed silks on the uppermost ear (□) of maize plants of two varieties (DK 664, 752) grown at two plant population densities (3 and 9 plants m⁻²) in Argentina (from Uribelarrea *et al.* 2002).

Interpreting the effects of drought on grain set is more complex as there are at least two other important mechanisms at work in addition to delay in silk emergence (Edmeades *et al.* 2000). First, water stress influences grain population density directly though its influence upon crop growth rate in the period round anthesis (Section 6.4; e.g. in terms of ear biomass at anthesis, Bolaños and Edmeades 1993; assimilate supply to individual florets, Schussler and Westgate 1991). Second, drought is also associated with reduction in the receptiveness of silks, as shown by their responses to the artificial application of viable pollen (e.g. Otegui *et al.* 1995). By far the most comprehensive account of the

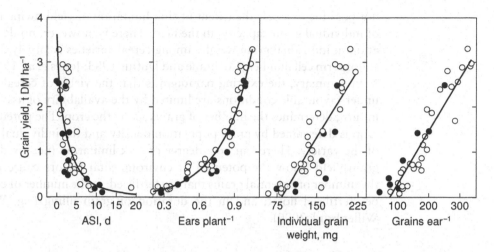

Figure 6.11 Relationships between grain yield of six tropical maize populations grown under a range of water regimes in Mexico, 1986–90, and anthesis–silking interval (ASI), number of ears per plant, individual grain weight, and number of grains per ear (from Bolaños and Edmeades 1996).

relationship between ASI, grain yield and its components, is for selections of the Mexican variety Tuxpeño Sequia (e.g. Figure 6.11; see also Figure 2.8). Here grain yield fell very sharply from 3.5 to 0.5 t ha^{-1} as the ASI increased from 0 to 5 days, in response to variation in water stress, and this was associated with a linear decrease in the number of grains per ear, as well as more modest increases in plant sterility (number of ears per plant falling from around 1 to 0.8). There were also substantial effects upon individual grain weight, at least partly as a consequence of shorter durations of grain filling under water stress (Westgate 2000, and see below). A clear curvilinear relationship between ASI and harvest index demonstrated the overall influence of water stress in reducing assimilate partitioning to the ears (Bolaños and Edmeades 1993).

The hypothesis that cereal crops are normally source-limited is founded on the observation that individual grain weight, for a given variety, is relatively invariable (Section 6.4), but there are numerous reports of lower grain weights in crops exposed to stress (e.g. for maize, Kiniry and Otegui 2000). Some of these effects can be attributed to the curtailment of the grain filling period (e.g. under water stress, Westgate 2000). However, there remains the possibility that exposure to drought, high temperatures or shading after anthesis, could also inhibit cell division and development in fertilised florets (completed over a period of 20 days from fertilisation in maize, Jones *et al.* 1996; or up to 30 days in small grain crops, Cochrane and Duffus 1981). Restriction in the capacity of the endosperm to accommodate assimilate could convert the crop from potential source-limitation to sink-limitation, resulting in individual grain weights that were lower than the potential for the variety. Experimental reductions in the number of cotyledon cells in grain legumes (e.g. by defoliation, depodding and shading of soybean, Egli *et al.* 1989) have resulted in such effects on individual grain weight. However, until there are more comprehensive datasets covering the effects of stress on all aspects of floret fertilisation, grain set, grain development and grain filling, it is not possible to assess the extent of sink limitation of yield owing to limitations of individual grain capacity, in the field. There is, however, no doubt that variation in individual grain weight among cereal varieties is largely determined by endosperm cell number (Cochrane and Duffus 1983; Jones *et al.* 1996).

In summary, the existing paradigm is that the yields of cereal crops grown under favourable conditions are limited by the availability of assimilate which, in turn, determines the number of grains set by the crop. The potential yield of a crop is determined by grain population density and the individual grain weight of the variety. There can be a degree of sink limitation (i.e. a reduction in the grain yield below the potential) if environmental factors cause reductions in the number of potential grains that are fertilised; in the number of cells produced per fertilised floret; and in the duration of grain filling (e.g. Wardlaw and Willenbrink 2000).

6.6   Assimilate partitioning and crop improvement: historic trends in harvest index of wheat and barley

Throughout most of the twentieth century, a primary breeding objective for small-grain cereals was to prevent lodging; unless the crops remained standing to harvest, the potential increases in grain yield resulting from application of fertiliser nitrogen could not be exploited. There were progressive increases in stem strength and stiffness and decreases in stem length, and these developments intensified with the incorporation of dwarfing genes, such as the *Rht* genes from Japanese Norin 10 wheat. Most commercial cereal varieties grown in the UK are now either semi-dwarf (0.7–0.9 m) or dwarf (0.5–0.7 m), a development that has been made possible by the control of weeds that would have a competitive advantage over such short crops. In association with these changes, grain yield of intensively managed crops has increased by at least 50% (winter wheat, Austin *et al.* 1980b; spring barley, Riggs *et al.* 1981).

However, the increased grain yield potential of newer varieties could not be explained simply in terms of the increased biomass of crops receiving more nitrogen fertiliser. When series of historic varieties of UK wheat and barley were grown at a common high level of husbandry, and physically prevented from lodging, there was no evidence of consistent change in the potential for biomass production across the century (Austin *et al.* 1980b; Riggs *et al.* 1981). For example, reduction in barley straw length from 1.40 m (Plumage, introduced in 1900) to 0.75 m (Triumph, 1980) had little effect on crop dry weight at harvest; indeed some varieties, such as the high-quality malting barley Golden Promise (1966), actually showed a significantly lower potential. There is supporting evidence for this pattern of response from small grain cereal crops in other countries (e.g. Argentina, Slafer *et al.* 1990), although some reports are less clear cut, also showing significant increases in biomass production by more recent varieties (Hay 1995). Thus, breeding and selection for straw characteristics have resulted in strong (but unconscious) selection for increased partitioning of plant dry matter to the ear: under the same conditions, varieties of different vintages produce similar amounts of biomass, but more of the biomass is harvested in grain

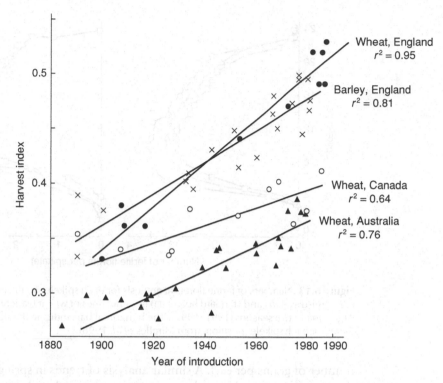

Figure 6.12 Relationships between harvest index and date of introduction of wheat varieties in England (●), Canada (○) and Australia (▲) and of barley varieties in England (×), from field experiments in which all varieties were grown under the same conditions (from Hay 1995).

in more modern varieties. The resulting progressive increases in harvest index of cereal varieties in England, Canada and Australia are documented in Figure 6.12.

Analysis of the components of yield of historic UK varieties of winter wheat (e.g. Table 6.3) established that, although there was a tendency for individual grain weight to be higher for varieties introduced between 1953 and 1972, most of the increase in grain yield potential could be explained in terms of grain population density (some increase in ear population density but a greater effect on

Table 6.3 Components of yield of varieties of winter wheat of different vintages grown under uniform conditions at Cambridge, UK, 1984–1986 (adapted from Austin *et al.* 1989).

Year of introduction	Biomass yield (t DM ha^{-1})	Grain yield (t DM ha^{-1})	Harvest index	Ears m^{-2}	Grains per ear	Grains 10^3 m^{-2}	Weight per 10^3 grains (g)
1830–1907	15.00	5.05	0.34	393	32.2	12.7	50.9
1908–1916	15.41	5.57	0.36	406	34.5	14.0	49.8
1953–1972	14.84	6.69	0.45	402	33.4	13.4	59.7
1981–1986	15.88	8.05	0.51	447	42.0	18.8	51.5

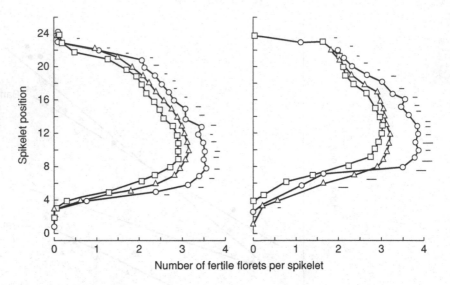

Figure 6.13 Numbers of fertile florets at anthesis for each spikelet on the main shoot of dwarf (○), semi-dwarf (△) and standard height (□) isogenic lines of wheat (variety Maringa) grown in Argentina in two seasons (1991, 1992). The horizontal bars indicate the standard error of the mean for each spikelet position (from Miralles *et al.* 1998).

number of grains per ear). A similar analysis of trends in spring barley indicated that the effects on grain population density were caused predominantly by increased ear population density (Riggs *et al.* 1981), presumably because there is less scope for increases in ear size in a species that produces only one floret per spikelet.

The effects of alteration in plant stature on the physiology of assimilate partitioning have been investigated in detail using contrasting varieties, and near-isogenic lines, of wheat, carrying different combinations of *Rht1*, *Rht2* and *Rht3* alleles (e.g. Fischer and Stockman 1986; Siddique *et al.* 1989; Youssefian *et al.* 1992; Flintham *et al.* 1997). The emerging consensus is that it is reduced demand for assimilate for the growth and expansion of the stem in semi-dwarf and dwarf cereals, rather than any increased competitiveness of the ear, that results in greater partitioning to the ear, and significantly greater ear weights at anthesis. This, in turn, leads to a higher proportion of surviving florets and grains set, predominantly at more distal sites in the central spikelets of the wheat ear (Figure 6.13); in general, these dwarfing genes do not have a significant effect on the *production* of spikelets or florets (i.e. dwarfing affects the culm but not the ear; Youssefian *et al.* 1992; Flintham *et al.* 1997). In some investigations, there have also been effects on mean individual grain weight, owing to incomplete filling of the distal grains (e.g. at position 4; Miralles and Slafer 1995); such effects indicate a degree of sink limitation, possibly caused by the curtailment of grain filling.

Most of these investigations have concentrated upon the ratio of ear to stem weight without considering the effects of dwarfing upon the proportion of stem weight represented by non-structural carbohydrate, and the implications of variation in storage for grain filling. In wheat and barley crops growing under favourable conditions, ear weight increases linearly with time, after an initial lag period, up to the cessation of grain filling. This behaviour, and the related linear increase in harvest index with time, are achieved by the coordination of assimilate supply to the ear from current photosynthesis in the ear and surviving green tissues, and from storage of pre-anthesis assimilate, particularly in the stem (e.g. Figure 6.4). Since supply from storage plays a progressively greater part as grain filling proceeds, it tends to make a greater contribution to grain yield when stress causes the acceleration of canopy senescence. Thus, assimilates from up to 5 days after anthesis contributed 54–61% of grain dry weight in barley crops in the classic drought year of 1976 in England, but only 14–15% in the following, less stressful, year; these difference were associated with a 55% lower yield in 1976 (Austin *et al.* 1980a). Across a wide range of wheat varieties, seasons and locations, the contribution of storage to grain yield has varied from 6 to 100% (Blum 1998).

A restricted range of measurements indicates that taller varieties do have a greater capacity for storage of non-structural carbohydrate, predominantly in the form of fructans, but that this difference does not normally play an important part in determining yield (e.g. Borrell *et al.* 1993): the role of *Rht* alleles in influencing ear weight at anthesis is more important, and dwarfed varieties also have a shorter lag phase before the linear phase of grain filling is achieved (Moot *et al.* 1996). Nevertheless, there are records of lower yields in dwarfed varieties under severe drought, which can be interpreted in terms of restricted grain filling from storage (e.g. Austin *et al.* 1980a).

6.7 Assimilate partitioning and crop improvement: historic trends in harvest index of maize

In contrast to other cereal crops, the harvest indices of commercial varieties of maize in North America were already high in the early 1900s. Figure 6.14 shows that the open-pollinated varieties planted in the USA up to the 1920s had values of around 0.45, compared with around 0.35 for wheat and barley in the UK (Figure 6.13), and this had risen slightly to around 0.5 in the latest hybrid varieties introduced in the 1980s. The overarching breeding objective for maize during the century was not dissimilar from that of wheat and barley: increased yield by breeding and selecting for tolerance to more intensive cultivation. This was primarily increased tolerance to nitrogen application and resistance to diseases and pests in both cereal types, but also increased tolerance to higher population density in maize, which cannot respond by variation in branching. As discussed in Section 6.5, this last attribute involved the maintenance (at least) of plant prolificacy (number of ears per plant) and a short anthesis–silking interval under intensive cropping. Thus, there was an (unconscious) selection for maintenance of high harvest index under intensive cultivation.

The result has been a greatly increased production of maize crop biomass for two reasons: first, increased inputs but, second, increased potential for biomass production per plant. For example, in the comparison of a range of varieties of differing vintages under the same conditions by Russell (1991) (Figure 6.14), biomass production had increased by 40% at a relatively constant harvest index,

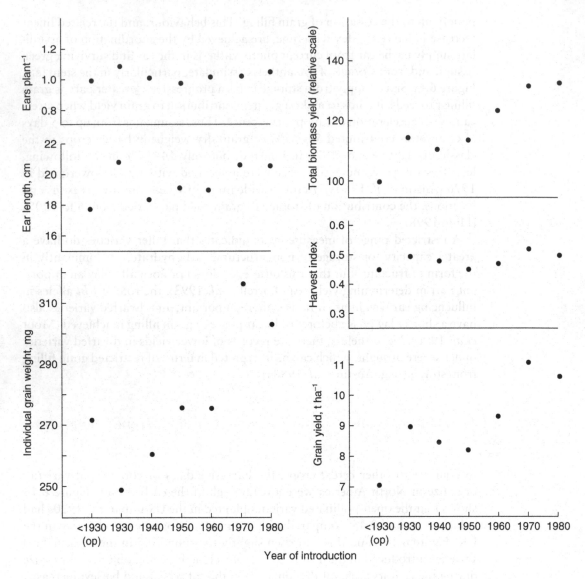

Figure 6.14 Yield and its components for maize varieties, introduced into the USA between 1930 and 1980, grown under uniform conditions (op refers to open-pollinated varieties) (from Russell 1991).

leading to an improvement in grain yield of similar magnitude. Since plant prolificacy had remained very stable at around one ear per plant, and the number of grains per ear (ear length) had varied less than 10% in varieties from 1930, the major effect was on individual grain weight.

Intensive breeding and selection of maize in North America have, therefore, been highly conservative in terms of overall assimilate partitioning. This is not the case for maize varieties elsewhere, particularly in the tropics, where harvest indices can be much lower, and the partitioning of assimilate dominated by events at anthesis (Hay and Gilbert 2001).

6.8   Assimilate partitioning to potato tubers

In temperate zones, where the interaction of daylength and temperature is favourable for the rapid initiation of tubers (Section 2.2.4), the first are formed on stolons growing from the basal nodes of mainstems during the rapid phase of expansion of the leaf canopy. The maximum number of tubers is normally determined within a few days (Figure 6.15); for example, non-destructive measurement of potato plants growing in pots in the Netherlands, showed that stolon initiation had begun by 15 days after planting, and all of the tubers that contributed to yield were formed between 25 and 40 days after planting (Struik *et al.* 1988). Later in the season, numbers can fall slightly owing to resorption or loss of small tubers, but there is normally no further recruitment.

The number of tubers initiated per stem is primarily under genetic control, with the extremes of UK maincrop varieties represented by Pentland Crown (typically 2–4 per stem), which can produce few, outsize, tubers under the highest-yielding conditions (Figure 6.17), whereas King Edward (5–8) can give a high proportion of unsaleable small tubers (chats) under less favourable conditions (Allen 1979). The expression of this genetic potential is, in turn, affected by management and environment (see below).

Under normal circumstances, each stem is independent, and the assimilates it generates are partitioned only among its tubers. Experimental manipulation has shown, however, that there are pathways for the exchange of assimilate among the tubers of the different stems of a plant *via* the mother tuber, as long as

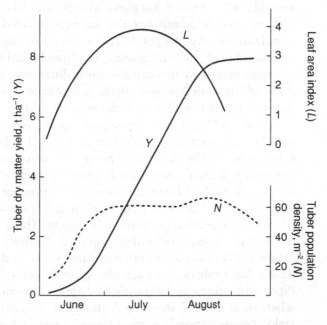

Figure 6.15 Time courses and interrelationships of leaf area index, tuber numbers and tuber yield for a model early-maturing crop of maincrop potatoes in England. In a cooler area, the canopy would persist into September (from Scott and Wilcockson 1978).

it retains functional integrity (Oparka and Davies 1985). Partitioning among the tubers of a stem is determined by the order of initiation (priority to the earliest formed; Gray 1973; Engels and Marschner 1986), and the pattern of vascular connections between source leaf and corresponding sink tuber (Oparka and Davies 1985), although there is scope for reallocation among tubers (e.g. Engels and Marschner 1987). Fine control at the tuber level appears to be exerted by the regulation of sucrose and starch metabolism (e.g. Zrenner *et al.* 1995; Appeldoorn *et al.* 1999).

The net results, at the plant and crop levels, are normal distributions of tuber size and weight whose means and standard deviations both increase as the season progresses: although the individual weights of most of the tubers increase progressively, a tail of slowly growing small tubers remains (Figure 6.16). This outcome contrasts with that for cereals and grain legumes where, although priority is also given to the first formed organs, the grains are filled to a pre-determined weight and low individual grain weights tend to occur only when the duration of grain filling is curtailed by stress (Sections 6.4, 6.5). In the potato, the sink capacity of a given tuber (number of starch-storing parenchyma cells) can increase throughout the life of the crop by cell division, whereas, in the cereal grain, the number of storage endosperm cells is fixed by 20 to 30 days from floret pollination. The size distributions of the tuber yields of potato crops have important implications for crop management. For example, in the UK, the ware fraction, for normal domestic consumption, is determined by a pair of riddles with 45 mm and 80 mm square holes. Figure 6.16 shows that there can be significant differences among varieties in the proportion of small unsaleable tubers (<45 mm), which represent the population of slow-growing tubers (e.g. around 20% in Record compared with around 10% in Désirée).

Tuber size distribution is also strongly affected by crop management. The population density of a potato crop can vary through differences in the spacing, size (number of potential sprouts) or physiological age of seed tubers; although variation in these factors can also cause differences in the spatial arrangement of the resulting stems (clumping in the case of larger seed tubers), the most practical index of population density is the number of (main) stems per unit area (Allen and Wurr 1992). As stem population density rises, the total yield of tubers rises rapidly to a sustained plateau (e.g. at 1.5 to 2×10^5 stems ha^{-1}, Figures 6.17, 6.18), whereas the number of tubers per unit area continues to rise, even at densities well above commercial optimum ranges (Figure 6.18). This response, which reflects the relative constancy of the number of tubers initiated per stem (e.g. in tubers of different physiological age; Haverkort *et al.* 1990), results in a progressive reduction in mean tuber weight. Thus, increase in plant population density, particularly beyond the optimal density for total yield (Figure 6.17), causes a shift in size distribution towards smaller tubers, and the proliferation of poorly filled tubers. For example, approximately half of the yield of a Maris Piper crop grown at 10^5 stems ha^{-1} took the form of larger tubers (>5.1 cm) whereas, at 4×10^5 stems ha^{-1}, this grade constituted less than 10% of total yield. The corresponding proportions of unsaleable chats (<3.8 cm) were 11% and 40% respectively. Figure 6.17 also shows that there are distinct differences in response to population density among varieties (Pentland Crown being less

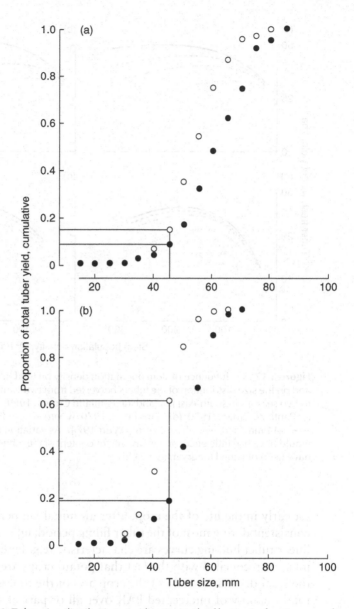

Figure 6.16 Tuber size distributions (within 1 month of harvest) of maincrops of two potato varieties, (a) Désirée and (b) Record, grown in 1984 in the east of Scotland under irrigation (●) or drought (○) (adapted from MacKerron and Jefferies 1988).

affected than the more prolific Maris Piper) and seasons (the effects more pronounced in years of higher yield potential).

The marked influences of nitrogen application, and particularly its effect on tuber fresh weight yield through changes in tuber water content, are considered in detail in Section 8.4. Figure 6.16 indicates how the incidence of drought can alter the time course, but not the shape, of tuber size distribution, owing principally to reduction in the availability of photosynthate.

Nevertheless, although the demography of the tubers of a potato crop can be complex and dynamic, the overall partitioning of crop assimilate to the tubers is

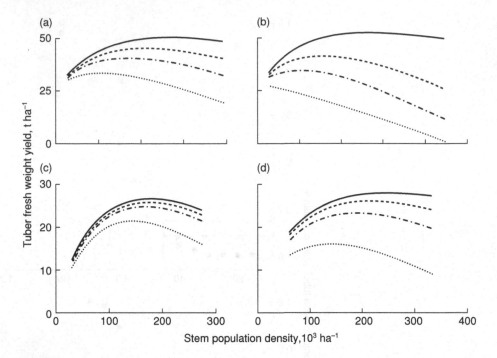

Figure 6.17 The influence of stem population density on total tuber fresh weight yield (——)
and on the size distribution of the tubers harvested from two varieties of potatoes grown
in two seasons in south-east England: (a) Pentland Crown, 1969; (b) Maris Piper, 1969;
(c) Pentland Crown, 1970; (d) Maris Piper, 1970, where ---- indicates yield >38 mm,
–·– >44 mm, and ······ >51 mm (from Wurr 1974). As variation in plant population density
would have had little effect on the dry matter content of the tubers, similar responses would
have been obtained for dry matter yield.

set early in the life of the crop, after an initial lag period, and tends to remain
consistent during most of the tuber filling period, unless affected by stress. Stable
linear tuber bulking curves are characteristic (e.g. Figures 2.11, 6.15), and these
must be reconciled with the fact that potato crops are generally source limited:
the total dry matter yield of the crop and of the tubers is linearly related to the
total quantity of intercepted PAR over all or part of a season (Allen and Scott
1980). Here the rate of transport of assimilate to the tubers remains constant
even though the daily input of radiation can vary considerably.

The explanation, at least in part, is that bulking rates are normally calculated
from destructive harvests at intervals of at least one week. There is, therefore, a
degree of statistical smoothing of the data since, during the midsummer months
(May to July in the UK), the total incident radiation per week will vary much
less than from day to day. If the canopy is intercepting more than 90% of the
incident radiation during this period, an approximately linear relationship
would be anticipated. However, the observation that linearity can continue into
the early autumn (September) suggests that, as in cereals (Section 6.4), there is
an overall control of partitioning (van Heemst 1986) governing the release of
assimilate from storage in stems, which maintains the rate of bulking, as the rate

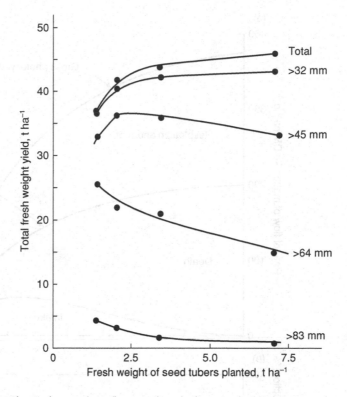

Figure 6.18 Classic data on the influence of seed tuber population density on the total tuber fresh weight yield, and on the size distribution of the tubers harvested from main crops of potatoes in southern England (from Holliday 1960).

of production of current photosynthate declines with the ageing of the canopy and the seasonal reduction in incident PAR.

6.9   Assimilate partitioning in grassland: implications for management of grass yield

For grain and tuber crops, it is possible to discriminate sensibly between source and sink for assimilate, as long as allowance is made, for example, for the photosynthetic activity of the green ear. Such a clear distinction is impossible when dealing with grass swards: green leaves, sheaths and stems are the sources of photosynthate but, as the harvested product by cutting or grazing, they are also the sinks. Even when the sward is permitted to flower (for example when managed for hay), it is the above-ground biomass rather than the grain that is harvested; and, since the sward is perennial, assimilate must be available for the re-establishment of the photosynthetic canopy from the residual stubble.

The dynamics of harvesting grass swards are crucial (Parsons and Chapman 2000). To maximise the productivity of an individual grass leaf, it should be permitted to pass from sink to source status (Section 6.2) but, if harvesting is delayed too long, the leaf will be lost to senescence. Furthermore, the nutritional

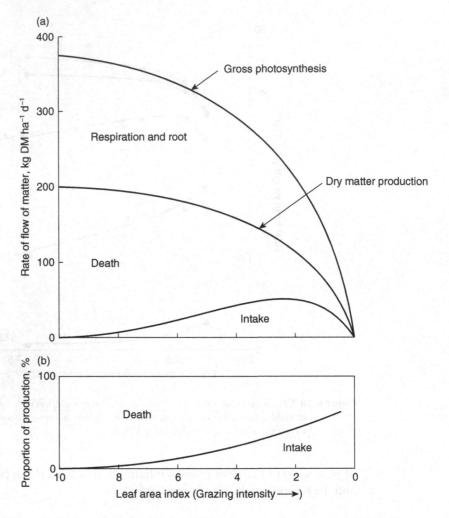

Figure 6.19 Model of the physiological limitations to sward production under continuous sheep grazing: (a) the relationship between production and loss of dry matter in swards maintained at different leaf area indices; and (b) the same data presented to describe the proportions of the dry matter production that are harvested by the grazer or lost to senescence (from Parsons *et al.* 1983; Parsons and Chapman 2000).

quality of older leaves declines with the remobilisation of nitrogen. At the same time, the frequency of harvesting also influences yield by determining the rate of stem branching (e.g. population densities in perennial ryegrass swards varying from 3×10^3 tillers m^{-2} when cut at 10-week intervals to 20×10^3 and 60×10^3 under lenient or hard continuous grazing, respectively, in southern England; Parsons *et al.* 1983; Hay and Walker 1989).

The management of grass swards for optimal yield must, therefore, take into account a range of opposing processes. The basic physiological principles can be understood from Figure 6.19, which presents a model of sward productivity under *grazing* (here by sheep), formulated from intensive programmes of research on perennial ryegrass swards over many seasons, principally in the UK and New Zealand. The model assumes that there are no other limiting factors at work, such as nitrogen or water supply. There are three possible fates for assimilate produced by the canopy: intake of green tissue by the grazing animal, loss as senescent tissue, or loss in sward respiration (combined here with dry matter partitioned to the root system). Changes in the intensity of grazing cause alterations not only in the *proportions* represented by these three components, but also in the *total quantity* of dry matter to be partitioned. In a heavily grazed sward at a low L around 1, gross photosynthesis is low, owing to incomplete interception of incoming radiation, but the grazing animals remove at least one-third of the crop dry matter production, with similar proportions being lost to respiration and death. At $L = 3$, the gross amount of assimilate produced has increased threefold, owing to increased interception, but, at the stocking density required to maintain the canopy at this L, the grazers leave a higher proportion of the crop to senesce. Thus, although the dry matter intake reaches its peak at this intensity of grazing, it has declined as a percentage of production; more leaf area that has functioned as source for assimilate has not been harvested by the grazing animal. Finally, under lax grazing ($L > 6$), gross photosynthesis is higher but most of the sward dry matter is lost to senescence.

In a complete analysis of the productivity of such swards it would also be necessary to quantify the intake per grazing animal because the yield of the system of exploitation would be expressed in terms of number of grazing days to reach a marketable carcass weight. For example, the regime in Figure 6.19 where an L of 3 was maintained by 24 sheep was intrinsically more efficient at producing finished animals than that maintained at $L = 1$ by 47 sheep (Parsons *et al.* 1983).

The physiologies of grass swards cut for silage or hay differ in several important ways. Depending upon the management system (intensity of grazing, frequency of cutting), cut swards tend to have lower stem population densities but larger culms, with more leaves developing to full photosynthetic potential but more leaves lost to senescence. Meaningful comparisons between the two types of harvesting are difficult to make, not least because of the need to monitor herbivores in grazing treatments, and the variable losses in dry matter during the processing of silage and hay. Grazing livestock also tread and foul pastures. The productivity per leaf, per tiller, or per unit of L, of cut swards can be similar to that of grazed swards, or higher, between cuts (e.g. Figure 6.20; Parsons *et al.* 1988), but the cut sward experiences periods of very low productivity after defoliation (Figure 6.21; Richards 1993). In general, continuously grazed swards show less pronounced seasonal variation, because they maintain a photosynthetic canopy, intercepting radiation throughout the growing season, and because of reduction in the shading of newly appearing leaves and tillers by the removal of tall reproductive culms (Parsons and Chapman 2000; Section 2.2.5).

Rapid recovery from complete defoliation by cutting depends upon the existence of branch (tiller) meristems carrying initiated and partly expanded leaves, which can re-establish photosynthetic area rapidly by leaf expansion (Sections 3.1, 3.2), and the mobilisation of the seasonally varying reserves of non-structural carbohydrates, predominantly in the lower stem rather than the root system (Pollock and Jones 1979; Richards 1993). For example, when perennial ryegrass plants, grown under controlled conditions, were cut to a

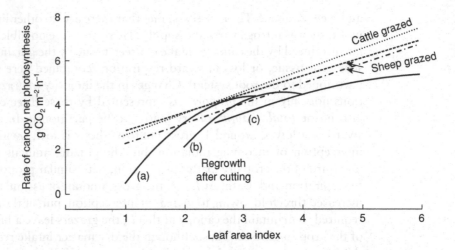

Figure 6.20 Relationships between canopy net photosynthesis (at 320 W m⁻²) and leaf area index for swards, dominated by perennial ryegrass, which were defoliated in different ways: continuous grazing by sheep or cattle; (a) cut weekly to 2 cm; (b) cut every 3 weeks to 4 cm; (c) cut weekly to 4 cm. Data from southern Scotland (from Hodgson *et al.* 1981).

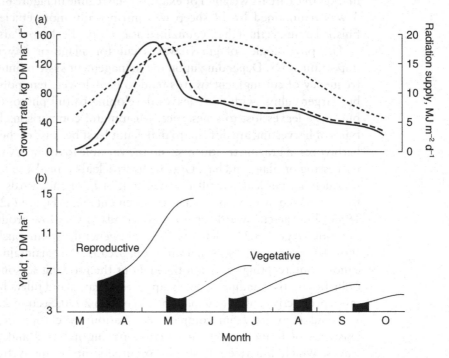

Figure 6.21 (a) Seasonal patterns of radiation receipt (upper curve), and of dry matter production, based on frequent cutting, of swards of S24 (────) and S23 (- - - -) perennial ryegrass in the South of England. (b) Dry matter production of infrequently cut swards of S24 in the south of England during regrowth; shaded areas indicate periods of incomplete interception of radiation (from Robson 1981).

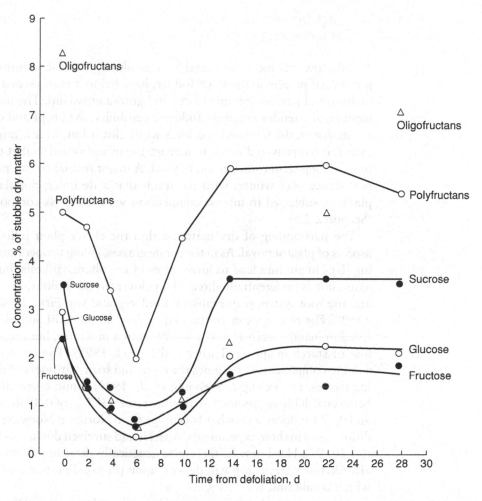

Figure 6.22 Time courses of contents of different storage carbohydrates in stubbles of perennial ryegrass after defoliation (from data of Gonzalez *et al.* 1989).

height of 4 cm, the sugar (glucose, fructose, sucrose) and low-molecular weight fructan reserves in the remaining stubble were immediately drawn upon, and largely exhausted over 6 days, but the higher molecular weight fructans were not mobilised until day 2 (Figure 6.22). Of the 17.6% of stubble dry weight lost by day 6, fructans had contributed three-quarters. By day 10, stores of all components were being recharged as the leaf canopy resumed its role as source, and pre-cutting levels were re-established by day 28. Detailed study of defoliated ryegrass plants in mini-swards has confirmed that there are strict rules for the partitioning of stored assimilate, with the re-establishment of leaf area taking priority, followed by root growth and, finally, production of new tillers (Donaghy and Fulkerson 1998). Consequently, reduction in the amount of non-structural carbohydrate stored in the sward, for example by frequent severe defoliation, will influence productivity first through reductions in tiller population density (i.e. by reduced radiation interception and, ultimately, plant death).

6.10   Assimilate partitioning in grassland: implications for the overwintering and early growth of white clover

Trends towards more 'sustainable' agriculture, and the substitution of plant for animal protein in livestock fodder, have led to a reawakened interest in the inclusion of pasture legumes in cut and grazed grassland. The incorporation of legumes also tends to improve fodder digestibility. In Europe and other cool temperate zones, the favoured species is white clover but, over a range of environments, it has proven difficult to manage the mixed sward so that the legume can make a long-term contribution to yield. A major feature of this problem is poor persistence over winter, with the result that a declining population of clover plants is subjected to intense competition with the grass component in spring (Section 2.2.5).

The partitioning of dry matter within the clover plant influences different aspects of plant survival. As in temperate grasses, falling temperatures and shortening days in autumn lead to lower rates of net photosynthesis, but the resulting assimilate is preferentially directed, in clover, to the stolons, at the soil surface, and the root system (e.g. Frankow-Lindberg and von Firks 1998; Corbel *et al.* 1999). The reserves can contain equivalent amounts of soluble carbohydrate (predominantly sucrose) and insoluble starch in winter, but there is a rapid net loss of starch in spring (Turner and Pollock 1998). The concentration of the soluble components determines the cold- and frost-hardiness of the overwintering tissues. For example, Svenning *et al.* (1997) found correlation coefficients between soluble sugars and lethal temperatures (LT_{50}) of 0.87 for cv. AberHerald and 0.92 for Bodø, a very hardy ecotype from Northern Norway; under the conditions used in these experiments, AberHerald survived down to −13°C and Bodø to −20.3°C. Hardiness is, therefore, genetically determined but the protection afforded to a given crop will depend upon pre-conditioning and the severity of winter conditions.

During winter, starch reserves can be exploited to maintain the level of soluble sugars (Turner and Pollock 1998) and to meet the respiratory demand of the tissues. For example, Collins and Rhodes (1995) found that, for Swiss varieties growing in Wales, the non-structural carbohydrate reserves fell from 46% of total dry weight in November to 12% in April; for cv. Grasslands Huia from New Zealand, they fell from 37% to 10% in the same period. Other analyses have shown that the sharpest decline in reserves occurs just before the resumption of active growth (e.g. in February in maritime Wales; Turner and Pollock 1998). It would be reasonable to predict, therefore, that success in overwintering and regrowth in the following spring would depend upon the size of reserves at the start and finish of winter. Support for this hypothesis comes from a series of experiments in which early growth and annual yield of clover in a sward were related to the number or quantity of stolons surviving winter (e.g. Harris *et al.* 1983; Collins *et al.* 1991), but other work has failed to confirm such a relationship.

The factors determining overwintering and spring growth of white clover have now been evaluated thoroughly in an experiment at 12 sites in Europe (including temperate maritime, subarctic maritime, lowland continental, and alpine zones) over 7 years, using similar planting materials (mainly cvs. AberHerald and

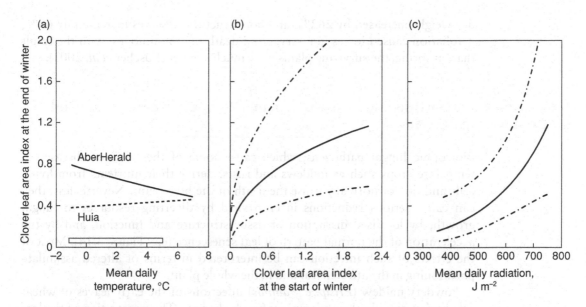

Figure 6.23 One of the sub-models arising out of the European multi-site investigation of the overwintering of white clover, showing the relationships between leaf area index at the end of winter and (a) winter temperature (note the variety/temperature interaction); (b) leaf area index at the start of winter; and (c) mean daily radiation during winter. Data for both varieties are combined in (b) and (c). The horizontal bar in (a) indicates the range over which the differences in the two curves are significant. Dashed lines indicate 95% confidence limits (from Wachendorf *et al.* 2001b).

Grasslands Huia) and similar agronomic practices, in perennial ryegrass swards (Wachendorf *et al.* 2001a, b). Using a novel modelling approach, it has been possible to relate survival and subsequent productivity to a limited set of parameters: clover cultivar, clover leaf area index at the start and end of winter, temperature and mean daily radiation receipt (Figure 6.23) (as well as those other factors determining the degree of competition offered by the grass component: grass tiller density and mean daily precipitation). Overall, analysis of the factors affecting clover production throughout the year emphasised the importance of clover leaf area index, but there were no consistent effects of stolon parameters.

These findings can be reconciled with the concept that stolon and root reserves are crucial by considering the marked decline in carbohydrate reserves observed in clover plants with little leaf area during winter (e.g. Collins *et al.* 1991; Collins and Rhodes 1995; Frankow-Lindberg *et al.* 1997). Where leaf area is maintained throughout winter, the reserves can be supplemented, at a low rate, by net photosynthesis (e.g. the observations of Turner and Pollock (1998) which showed that photosynthate was directed to the formation of new stolon phytomers). Thus, in the European multi-site experiment, the early growth of the clover component was dependent upon the photosynthetic activity of the overwintering leaves, and the rapid reconstruction of the canopy using reserves that had been maintained by continued partitioning to the stolons over winter. These conclusions were confirmed by the results of a single-site experiment in the Swiss Alps where clover plants produced 4–7 new leaves over winter, stolon dry weight increased by 262%, and non-structural reserves increased by 68%; defoliation caused losses in reserves and death of stolon nodes, with the result that, in spring, the surviving plants were much smaller (Lüscher *et al.* 2001).

6.11   Assimilate partitioning in diseased plants: temperate cereals affected by biotrophic fungal pathogens

Biotrophic fungal pathogens, which cause some of the common diseases of temperate crops such as mildews and rusts, derive their nutrients from living cells and do not normally cause the death of the host plant. Nevertheless, they can cause serious reductions in crop yield by diverting resources to fungal growth, by localised disruption of tissue structure and function, and by the acceleration of the normal pattern of leaf senescence (e.g. Figure 3.14). Most of the effects of such infection can be interpreted in terms of altered assimilate partitioning in the affected leaf and in the whole plant.

Powdery mildew (*Erisiphe graminis*) infections of the early leaves of wheat and barley provide clear examples of such changes. The penetration of the leaf by fungal hyphae and the development of haustoria within the epidermal cells (Szabo and Bushnell 2001) are associated with a rapid increase in invertase activity in the mesophyll (e.g. three times the level in control leaves by 7 days after infection of barley leaves; Scholes *et al.* 1994). This leads to the hydrolysis of sucrose (Figure 6.2) and the accumulation of glucose and fructose (e.g. a 10-fold increase in the combined concentration of monosaccharides by day 7 after infection of barley leaves; Scholes *et al.* 1994), which are then available for uptake by the haustorial complex. The pathogen is, thus, intercepting the flux of sucrose from the mature leaf to the rest of the plant, but the metabolic disturbance is more complex than a simple diversion of product: in spite of being actively hydrolysed, sucrose also accumulates in the infected leaf and there is an increase in local storage of starch. The accumulation of soluble carbohydrate results in a feedback inhibition of the Calvin cycle in the leaf mesophyll cells (signalled by a progressive decrease in Rubisco activity) and an associated decline in chlorophyll content. A very similar pattern of response has been observed in early wheat leaves infected with powdery mildew (Wright *et al.* 1995a, b).

The net result of such infection is the reduction of sucrose export from the leaf to the developing plant and the immobilisation and loss of starch in prematurely senescing leaves. Such a failure of infected leaves to act as a source of assimilate to the remainder of the plant is a characteristic feature of biotrophic infections (Farrar and Lewis 1987), although unaffected leaves of infected plants can compensate, to a degree, by enhanced rates of net photosynthesis, and resistance to the pathogen (e.g. Walters and Ayres 1983; Rooney and Hoad 1989).

Lower grain yields from crops exposed to disease during the early growth stages are, therefore, caused primarily by reduction in crop biomass, and in the number of grains set per unit area, but there can also be effects on harvest index. In particular, infection of flag leaves can lead to the premature senescence of the photosynthetic area (Figure 3.14), without compensatory increases in the exploitation of stem reserves. As grain population densities are determined shortly after anthesis, individual grain weights can be reduced considerably by fungal infection. For example, inoculation, with *Septoria nodorum*, of the flag leaves of previously uninfected wheat plants at anthesis, caused a very rapid senescence of the leaf canopy; measurement of mainstems at harvest showed that infection had caused decreases of 21% in culm dry weight; 13% in grain number per ear; and 27% in grain dry weight per ear (Rooney 1989). Taking the opposite approach, Gooding *et al.* (2000) found that application of fungicides to protect the flag leaves of 21 crops of winter wheat (10 varieties, 8 seasons) from a range of biotrophs resulted in an increase in grain yield of 11.8%, which was almost entirely caused by increased individual grain weight (+9.9%).

Thus biotropic infections can cause yield losses in a series of ways: by reduced interception of radiation through early canopy senescence; by reduced rates of net photosynthesis through feedback inhibition of the Calvin cycle and loss of chlorophyll; and by reduced partitioning of assimilates from infected leaves, and to the ear at grain filling.

This chapter has explored the factors determining the harvest index of a range of crops, showing that the same principles of assimilate partitioning can be applied under different circumstances. This completes the analysis of the physiology of crops whose yield is ultimately determined by the amount of solar radiation captured during the growing season. Although this approach is appropriate for high-yielding crops in temperate environments, other factors can become limiting under other circumstances. This is the subject of the following chapter on limiting factors and the achievement of high yield.

Chapter 7

Limiting factors and the achievement of high yield

. . . there is a heightened need to understand farming systems in terms of their capture of energy, carbon, nitrogen and water. Interest in 'sustainability' has accented the requirement not only to maximize efficiencies but to ensure that they balance.

(Scott *et al.* 1994)

The preceding chapters of this book have been founded on the guiding principle, generally held by crop physiologists dealing with temperate and intensive cropping, that the primary factor limiting crop yield is solar radiation (Chapter 1). Other factors can be considered as secondary in the sense that they influence the interception of solar radiation, the conversion of solar energy to chemical potential energy, or the partitioning of the resulting dry matter to harvested parts.

Thus, the nitrogen status of a cereal crop can influence grain yield by affecting canopy size and duration (area and longevity of leaves; branching; Section 3.2) and radiation use efficiency (of individual leaves; Section 4.3.2). The water status of the crop can also influence leaf size and longevity (Section 3.2), the rate of net assimilation (directly *via* stomatal aperture and mesophyll resistance, and indirectly *via* changes in leaf temperature etc.; Section 4.5), and the duration of grain filling (Sections 6.5, 6.6). Similarly, biotrophic infections can cause yield losses by reduced interception of radiation through premature canopy senescence; by reduced rates of net photosynthesis through feedback inhibition of the Calvin cycle and loss of chlorophyll; and by reduced partitioning of assimilates from infected leaves, and to the ear at grain filling (Sections 3.3.2, 6.11).

This guiding principle has provided a valuable analytical approach to the complexities of crop physiology, permitting the effects of a range of environmental and management factors to be brought together in coherent models (Chapter 9). It is consistent with the observation that, under intensive cropping where moderate to severe stress is avoided or prevented, yield is proportional to the amount of radiation intercepted. Nevertheless, the approach has several drawbacks. First, it encourages a non-mechanistic approach to the effects of some environmental factors; for example, the focus is on above-ground tissue nitrogen and water status, and there is a tendency for the below-ground processes that determine them to be neglected. The emphasis is on leaves rather than roots. Second, by concentrating upon dry matter production, it ignores the fact that the levels of nitrogen and other nutrients in the harvested organs can be crucially important in determining the quality of the product (see Chapter 8). Third, and perhaps most importantly, in areas where stress is common, the limiting factors are more likely to be supplies of water, nitrogen or phosphorus; high temperature; or pathogen damage. Under less intensive cropping, and in the tropics in particular, interception of solar radiation may be less important than the avoidance of high temperatures and drought, or the optimisation of nitrogen recovery. Finally, there is an increasing need worldwide to conserve resources and prevent pollution. For all of these reasons, other analytical approaches can be taken, as discussed below.

This chapter takes a crop physiological approach to the determination of crop yield, building on the more plant physiological analysis of assimilation in Chapter 4. More detailed background on plant water and nutrient relations can be found in Fitter and Hay (2002).

7.1   Limitation by water supply

For decades, there has been a consensus that, where water is limiting, the accumulation of crop biomass is linearly related to cumulative transpiration (reviewed by de Wit 1958; Ludlow and Muchow 1990; Figure 7.6). Consequently, by analogy with the approach outlined in Chapter 1, the yields of crops limited by water supply can be described by the equation:

$$Y = Q_w \times A \times \varepsilon_w \times H \tag{7.1}$$

where

Q_w is the total quantity of water potentially available to the crop over the growing season,
A is the fraction of Q_w that is absorbed by the crop,
$Q_w A$ is the total quantity of water transpired by the crop,
ε_w is the overall photosynthetic efficiency of the crop in terms of the total plant dry matter produced per unit of water transpired (i.e. the water use efficiency),
H is the harvest index of the crop, and identical to the value used in analyses based on the interception of radiation (analysed in detail in Sections 6.2–6.7).

Thus, for a given quantity of water transpired by a crop canopy, the harvested yield can depend upon the efficiency of use of the water supply in maintaining the photosynthetic machinery, and upon the partitioning of the resulting assimilate to harvested organs. The magnitude of each of these components can, in principle, be affected by drought; the effects of anaerobic stress, caused by *excess* of water, are not considered here.

7.1.1   Acquisition of water

For crops relying solely on stored water, Q_w is the soil water content at sowing, less the content when the entire soil profile has reached permanent wilting point (PWP, corresponding to a standard soil water potential of −1.5 MPa, when the water in all soil pores of diameter greater that 2 μm has been removed). To this total must be added any precipitation received during the life cycle of the crop. Q_w is, therefore, determined by environmental factors; although there are differences among crop species in the extent to which they can extract water from drying soils (i.e. the PWP is at least partly a plant characteristic; see the discussion on osmotic adjustment below), these usually have a modest effect on Q_w because there are only limited quantities of water in soil pores narrower than 2 μm.

In contrast, *A* varies considerably among species, varieties and crops. Liquid water cannot move from soil to root, at sufficient rates to support rapid rates of transpiration, over distances of more than a few mm, especially in drying soils. Consequently, to tap the water resources of a soil volume effectively, the crop must place roots within that volume, and these roots must be capable of absorbing water and transmitting it to the above-ground parts. Bänziger *et al.* (2000) estimate that maize crops require densities of 0.5–1.5 cm of active root cm^{-3} of soil to extract available water.

Differences among species or varieties in the extent to which they explore the soil profile, although well-documented, are not normally considered to be of great importance in temperate cropping, where soil water is regularly recharged from the soil surface. However, where crops are growing principally on stored water, the rate and extent of exploration of each successive soil horizon, and the maximum rooting depth, can be critical. For example, the ability of sunflower crops to extract more water than other dryland crops from a soil profile arises, largely, out of the deeper penetration of a rapidly-extending tap root system: in Figure 7.1, the superiority, in terms of water extraction, of sunflower over

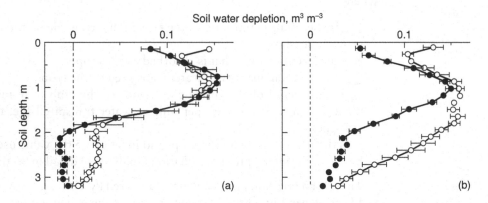

Figure 7.1 Soil water depletion (28 June to 6 September 1987) by crops of sunflower (Triumph 585) (open circles) and grain sorghum (Triumph TWO 54-YG) (closed) grown in a deep loess soil in a semi-arid zone of Kansas, at two levels of irrigation: (a) 100 mm in November and 75 mm in June; (b) 400 mm in November and 75 mm in June. The crops also received 450 mm of precipitation, distributed throughout the growing season (from Stone *et al.* 2002).

the fibrous-rooted sorghum, was associated with maximum rooting depths of 2.8–3.0 m, compared with 2.3 m for the cereal crop. The sunflower root system was exploiting a greater volume of soil.

Variation in *A* can also be caused by differences in the ability of roots to extract the water per unit volume of soil, but such differences can be explained mainly in terms of the physiology of the crop *leaves*. The transpirational loss of water from a crop canopy causes a lowering of the water potential of the leaf cells, thus setting up a gradient in water potential within the plant, down which water moves from root to shoot. Roots extending into successive soil horizons can take up water and transport it to the leaf canopy only if the water potential of the root xylem can be lowered by transpiration below that of the surrounding soil.

The extraction of soil water by a crop canopy can be illustrated by Figures 4.17, 7.2 and 7.3. The potential rate of transpiration by the canopy (e.g. the maximum rate in Figure 7.2) is determined primarily by the evaporative surface area (the leaf area index), whose magnitude, in turn, depends upon the plant and environmental factors reviewed in Section 3.3. For a canopy exploiting an unrestricted water supply (e.g. day 1 in Figure 7.2), there is a diurnal variation in the rate of transpiration, determined by the diurnal rhythm of stomatal aperture (opening progressively in the morning and closing in the afternoon; note the relationship

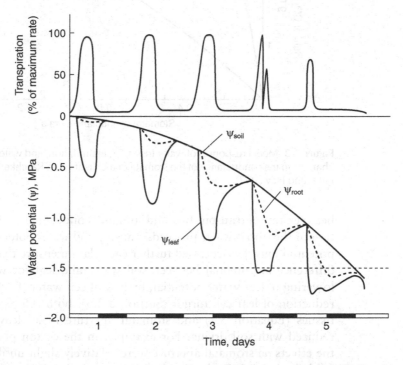

Figure 7.2 Schematic representation of the changes in leaf, root surface, and bulk soil water potentials, and in the rate of transpiration, associated with the exhaustion of the available soil water over a five-day period. A difference in water potential of at least 0.5 MPa between leaf and root is necessary to overcome the hydraulic resistance of the soil–plant system, and maintain the maximum rate of flow of water from soil to leaf (from Slatyer 1967, adapted by Fitter and Hay 2002).

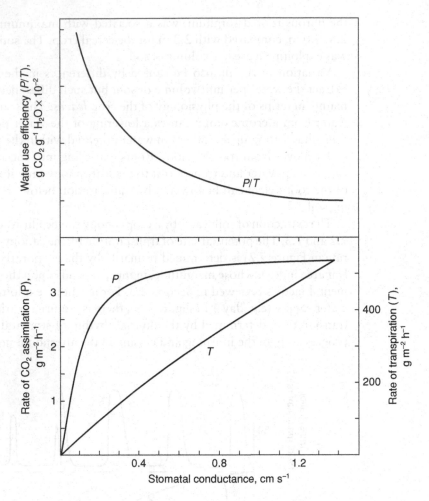

Figure 7.3 Model responses of water loss, CO_2 assimilation, and water use efficiency to changes in the conductance of the stomata of a C_3 leaf (from Raschke 1976, adapted by Fitter and Hay 2002).

between rate of transpiration and stomatal conductance in Figure 7.3). As depletion of the stored water proceeds, causing soil water potential to fall, leaf water potential must be depressed further each day to ensure that there is a sufficient difference in potential between soil water and leaf for water uptake. As the lowering of leaf water potential, by loss of cell water (Figure 4.17), results in a reduction of leaf cell turgor (Section 3.2.3), both cell expansion in immature tissues (Equation 3.1) and stomatal aperture in all leaves are progressively reduced with soil drying. For example, in the cotton plants in Figure 4.17a, the effects on stomatal aperture were relatively slight until leaf water potential fell below −1.0 MPa, but further water loss caused a rapid increase in stomatal resistance. In the model canopy in Figure 7.2, leaf water potential had fallen to −1.5 MPa on the morning of day 4, causing the loss of leaf turgor, complete stomatal closure, and the cessation of transpiration. Some turgor was re-established by the equilibration of water relations within the plants when the stomata were closed, but transpiration ceased on day 6 with the exhaustion of all available soil water.

In summary, stomata close completely and transpiration ceases finally in most temperate crop species when leaf water potential reaches −1.5 to −2.0 MPa (corresponding to permanent wilting point; soil water potential −1.5 MPa); any further loss of leaf water would lead to irreversible damage, although the leaves of species from drier environments can normally tolerate lower values, down to around −4 MPa. However, in addition to the relatively simple responses illustrated in Figure 7.2, partial or complete stomatal closure can occur over a range of leaf water potentials, in response to internal signalling of the onset of water shortage, modulated by changes in the balance of chemical regulators (principally, but not exclusively ABA (abscisic acid); Davies *et al.* 2002). Stomatal aperture can also depend upon the degree of osmotic adjustment achieved by leaf cells: crop canopies that can maintain leaf cell turgor for longer periods, and at lower leaf water potentials, by secreting solutes into the vacuole, have the capacity to continue to transpire, and absorb more of the available water in the soil explored by its root system.

Wheat and sorghum, originating in dry zones, have a proven capacity for, and genetic variation in, osmotic adjustment (Ludlow and Muchow 1990). For example, comparison of a range of wheat varieties exposed to controlled drought in the field in Australia revealed that leaf water potential at zero turgor varied between −2.3 MPa and −3.3 MPa. Varieties showing greater degrees of adjustment would experience less inhibition of photosynthesis by daytime closure of stomata; and there would be an effective lowering of PWP, and a modest increase in the amount of available water in exploited soil horizons. (Such results show the limitations of separating Q_w and A in Equation 7.1.) As a result, there were linear relationships between the degree of osmotic adjustment and the amount of water taken up, grain yield and harvest index (Morgan and Condon 1986). Because osmotic adjustment occurs in roots as well as leaves, there can also be benefits in terms of continued root function and extension (Ludlow and Muchow 1990).

The potential for osmotic adjustment in maize may have been underestimated in the past. For example, laboratory studies of the leaves of Argentinian maize varieties showed that a water potential of −2.0 MPa (the lower limit for survival of the sensitive genotypes) corresponded to leaf relative water contents (RWC) varying from 58.2% to 87.5%. Parallel field studies confirmed that varieties with a higher capacity for osmotic adjustment extracted more water from the soil profile (higher A); for example, in Figure 7.4, variety LP125R, with a high capacity (80.3% RWC at −2.0 MPa in the laboratory), absorbed 40% more water from the same soil volume under conditions of drought than did LM017 (lower potential, 66.8% RWC), and 10% more under irrigation. Such large effects under drought are, presumably, mainly the result of prolongation of uptake of soil water, and access to greater soil volumes.

It should be noted that, through its role in shading, leaf area index plays a further part in determining A and its time course by influencing the proportion of Q that is lost by direct evaporation from the soil surface and, therefore, not available to the plant.

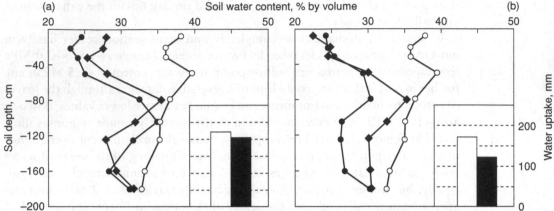

Figure 7.4 Soil water extraction at 30 days after anthesis by maize crops (closed circle LP125R; closed diamond LM017) grown under (a) irrigation and (b) drought in Argentina. The curves to the right (open circle) show the soil water content at the start of crop growth. The inset histograms show that total water uptake by LP125R (white) was greater than that of LM017 (black) under both conditions, the difference being significant ($P > 0.1$) under drought (from Chimenti *et al.* 1997).

7.1.2 Water use efficiency

The quantification of Q_w and A for a crop variety tends to be fairly arduous owing to the need to make extensive measurements, preferably over several seasons and at more than one site, down to maximum rooting depth, which may be several metres in dryland crops and pastures. As the necessary above-ground measurements are more readily collected, more is known about water use efficiency (ε_w, Equation 7.1), although many studies refer to evapotranspiration (including water loss by the soil surface) rather than strictly to transpiration. To ensure clarity, the term transpiration efficiency can be used.

Pioneering measurements of ε_w in terms of the amount of water transpired per unit of (above-ground) dry matter produced revealed major differences between species and groups of species (Figure 7.5). These observations were the first indications of the marked differences between plants employing the C_3 and C_4 pathways of photosynthesis (Table 4.2), notably in the physiology of their stomata. As explained in Chapter 4, water molecules leaving a transpiring leaf traverse the substomatal cavity, the stomatal pore and the boundary layer outside the leaf before reaching the bulk air. CO_2 molecules diffusing in the opposite direction traverse the same pathway but an additional resistance is encountered between the substomatal cavity and the site of fixation: the poorly characterised mesophyll or residual resistance (Figure 4.7). The diffusive resistance offered by the leaf to CO_2 uptake is, therefore, greater than that offered to water loss, and any increase in the overall resistance to gaseous diffusion will have a greater proportional effect on water loss than on CO_2 uptake (i.e. causing an increase in water use efficiency at the molecular level: Fitter and Hay 2002). Thus, in Figure 7.3, increasing stomatal resistance (reduced conductance) causes a greater reduction in water loss than in assimilation of CO_2, with the result that

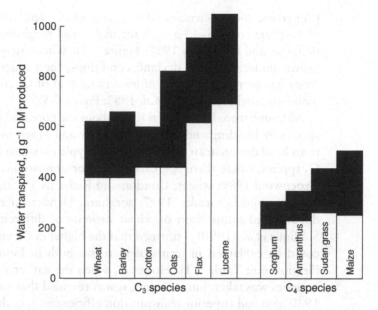

Figure 7.5 Relationships between water use and dry matter production for ten crop species grown in the open in pots in the dry rangelands of Colorado, in two contrasting seasons (unshaded columns: dry season, evapotranspiration = 848 mm; unshaded plus shaded columns: wetter season, 1198 mm). Note that the values are the inverse of water use efficiency (from data of Shantz and Piemesal 1927).

water use efficiency (g CO_2 fixed per g H_2O lost) rises progressively. Better use of water is achieved, but at the expense of the rate of net photosynthesis.

Although the behaviour shown in Figure 7.3 is common to all crop species, the effects are much greater in those employing the C_4 pathway of assimilation (Figure 7.5). Because of the capacity of C_4 plants to concentrate CO_2 in the leaf bundle sheaths (Section 4.4), the gradients in CO_2 concentration between the intercellular spaces and the bulk air are generally much steeper than in C_3 plants. For example, the mean value of p_i/p_a in *Sorghum bicolor* crops grown under controlled conditions and in the field in Australia was 0.33 compared with typical values of 0.7 for C_3 species (Henderson *et al.* 1998). Such differences permit optimal fluxes of CO_2 into C_4 leaves at higher stomatal resistances (smaller stomatal apertures/lower conductances) than in C_3 leaves, resulting in significantly fewer water molecules lost per molecule of CO_2 fixed (i.e. C_4 leaves have a higher intrinsic efficiency than C_3).

Many of the more reliable measurements of ε_w (transpiration efficiency) have come from experiments in large pots where surface evaporation has been reduced by sealing the soil surface; others have arisen out of direct measurement of transpiration in the field. In general the efficiency of C_3 plants receiving adequate water falls within the range 1–5 g DM kg^{-1} water transpired; under drought stress, these values can increase by up to 10%, presumably owing to partial stomatal closure (as illustrated in Figure 7.3) (e.g. in a range of cereals, grain legumes and oilseeds reviewed in Ehleringer *et al.* 1993; Knight *et al.* 1994). Significantly higher efficiencies have, however, been measured in some C_3 crops (e.g. 8–11 g kg^{-1} in potato; Vos and Groenwold 1989). In line with their stomatal properties, the efficiencies of C_4 crop plants tend to be higher than those of C_3 plants (e.g. 5–13 g kg^{-1} for maize and sorghum; Donatelli *et al.* 1992; Bänziger and Edmeades 1997; Figure 7.5). Where adapted varieties have been grown under the same dryland conditions, the transpiration efficiency of C_4 crops has been shown to be at least twice that of C_3 crops (e.g. comparison of maize and soybean, Angus *et al.* 1983; Figure 7.5).

Although there have been doubts about the potential for improving ε_w within species by breeding and selection (Ludlow and Muchow 1990), some investigations have demonstrated significant genotypic variation in efficiency, not only in C_3 species, where there is greater scope for improvement (e.g. potato, Vos and Groenwold 1989; wheat, Condon and Richards 1993), but also in C_4 (maize, Bänziger and Edmeades 1997; sorghum, Henderson *et al.* 1998). In a well-documented comparison of wheat varieties of different vintages in Australia, Siddique *et al.* (1990) confirmed that the higher grain yields of modern varieties could be explained in terms of increases both in biomass and harvest index (Section 6.6; Table 7.1). However, when the earlier maturity of more recent varieties was taken into account, it was revealed that varieties introduced since 1930 also had superior transpiration efficiencies (i.e. they generally transpired less water than old varieties owing to a shorter growing season but produced more biomass; Table 7.1). This physiological effect appears to have been a step change, with little variation among the more modern varieties. Other comparisons of genotypes have shown more limited variation, particularly in C_4 plants (e.g. among 46 lines of *Sorghum*, including five species; Figure 7.6a).

Until the 1980s, attempts to improve water use efficiency of crop varieties by breeding and selection were hampered by the lack of appropriate tools for the rapid screening of large populations. This changed with the introduction of techniques for the routine measurement of stable isotopes (Ehleringer *et al.* 1993). Of the two naturally occurring stable isotopes of carbon, ^{12}C predominates over ^{13}C, representing 98.9% of the carbon atoms in atmospheric CO_2. The proportion of the lighter isotope is even higher in plant tissues because of discrimination against ^{13}C during the diffusion of CO_2 into the leaf and in the fixation of CO_2

Table 7.1 Yield and water use of spring wheat varieties growing in a mediterranean climate in W. Australia (from Siddique *et al.* 1990).

Variety	Year of introduction	Days to maturity	Yield (kg ha^{-1}) Biomass	Grain	HI	Transpiration to maturity (mm)	Transpiration efficiency (kg biomass mm^{-1})
Purple Straw	1860s	167	4824	1162	0.25	122	39.5
Nabawa	1915	159	4794	1342	0.28	129	37.2
Bencubbin	1929	159	5326	1619	0.30	119	44.7
Gamenya	1960	154	4781	1674	0.35	107	44.9
Tincurrin	1978	148	5264	1870	0.36	121	43.7
Miling	1979	159	5286	1826	0.34	122	43.5
Gutha	1982	148	5266	1888	0.36	114	46.1
Kulin	1986	148	5122	1892	0.37	119	43.2

HI, harvest index.

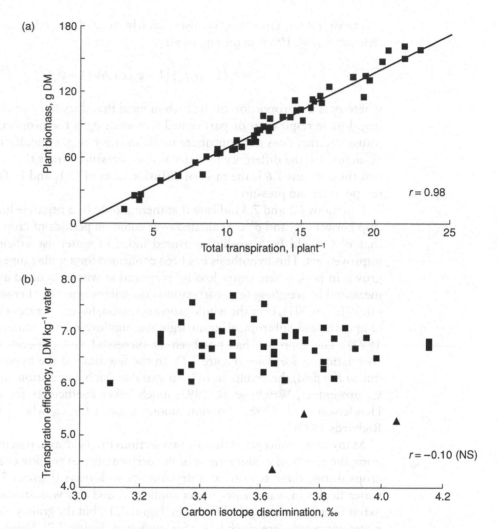

Figure 7.6 Comparison of the water relations of 49 lines of sorghum, including five species, grown on stored water in pots up to the flag leaf stage: (a) relationship between biomass production and total cumulative transpiration; (b) relationship between transpiration efficiency and carbon isotope discrimination. Each symbol represents a sorghum line (from Hammer *et al.* 1997).

by Rubisco. Thus, for plants relying on the C_3 photosynthetic pathway, the discrimination against ^{13}C can be expressed by

$$\delta^{13}C = a + (b - a)p_i/p_a \qquad (7.2)$$

where a is discrimination against ^{13}C owing to diffusion, b is the net discrimination owing to carboxylation, and p_i, p_a are the partial pressures of CO_2 in the intercellular space and the bulk air, respectively (O'Leary 1988). Because discrimination is weaker during carboxylation by PEP carboxylase, the value of b used for C_4 plants must take into account the proportions of CO_2 fixed by the two different enzymes.

Transpiration efficiency is also linearly related to p_i/p_a, as formalised (Farquhar *et al.* 1989) in the equation:

$$\varepsilon_w = p_a(1 - p_i/p_a)\,(1 - \phi_c)\,/\,1.6v\,(1 + \phi_w) \qquad (7.3)$$

where ϕ_c is the proportion of the carbon fixed that does not contribute to yield (e.g. lost in respiration or partitioned to roots); ϕ_w is the proportion of water transpired that does not contribute to assimilation (e.g. cuticular transpiration at night); v is the difference in water vapour pressure across the stomatal pore; and the constant 1.6 is the ratio of diffusion rates of CO_2 and H_2O at standard temperature and pressure.

Equations 7.2 and 7.3 indicate that there should be a negative linear relationship between ε_w and $\delta^{13}C$ evaluated over common periods of crop growth, and that $\delta^{13}C$ could be a readily measured index of water-use efficiency in crop improvement. This hypothesis has been confirmed for a wide range of C_3 species grown in pots, where water loss by evaporation was minimised and water use measured by weighing (e.g. correlation coefficients generally between −0.6 and −0.98 ($P < 0.001$), over the whole growing season, for temperate cereals, potato, temperate and subtropical grain legumes, sunflower and tomato; Hall *et al.* 1994). The approach has not been as successful in C_4 species (e.g. lack of correlation in *Sorghum*, Figure 7.6). In the few tests of the hypothesis carried out *in the field*, the results have been variable (high correlation coefficients for C_3 groundnut, Wright *et al.* 1993; much lower coefficients for C_4 sorghum, Henderson *et al.* 1998; variation among seasons for C_3 wheat, Condon and Richards 1993).

Many of the concepts in the last two sections can be drawn together in considering the results of a comparison of the performance of sunflower and sorghum crops during three seasons at a dryland site in Kansas (Figures 7.1, 7.7). Soil water depletion was greater under sunflower, and this was attributed to more extensive rooting at depths below 1m (Figure 7.1), but the grain yields per unit of water extracted were much lower for sunflower (Figure 7.7). Using the available data, it is not possible to partition this difference into effects caused by differences in ε_w (C_4 sorghum > C_3 sunflower) and H (sunflower < sorghum owing to the extra resources required for the synthesis of seed lipids); but what is striking is that the overall efficiency of use of water in the production of grain dry matter was effectively constant across seasons and treatments (water-stressed and irrigated) for each species. Reports of this kind reveal the lack of basic physiological data for rigorous analysis of water-use efficiency in field crops, using the approach summarised by Equation 7.1.

7.1.3   Crop yield where water supply is limiting

Separating out the relative importance of water uptake, water-use efficiency and assimilate partitioning in determining the yields of crops grown with limited water supply is complicated by the fact that these components are not strictly independent. In particular, A and ε_w are both highly dependent upon stomatal

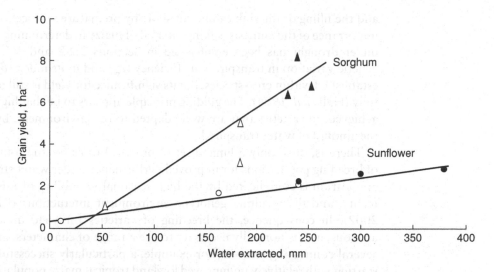

Figure 7.7 Relationships between grain yield and soil water extracted for crops of sunflower (Triumph 585) and sorghum (Triumph TWO 54-YG) grown with full (closed symbols) or partial (open symbols) irrigation over three seasons in Kansas (see Figure 7.1) (from data of Stone *et al.* 2002).

physiology, but the effects can be opposed: thus, the maintenance of cell turgor and open stomata by osmotic adjustment, leading to the extraction of a higher fraction of available soil water, may be associated with lower transpiration efficiency (Figure 7.3).

The phenology of the crop is also very important, perhaps even more important than under limitation by radiation interception. There are two principal seasonal patterns of water availability in productive dryland farming (Ludlow and Muchow 1990): crops grown on stored water only (or on precipitation received during a short period at the start of growth) experience terminal water stress; whereas crops grown in rainfed environments, where drought is less predictable, can experience intermittent stress at any phase of development. Where terminal stress occurs in grain crops, the stored water must be used economically in the construction of the canopy and in the initiation of grains so that sufficient water is left for grain filling (e.g. Section 6.5). Thus, earlier initiation of reproductive development in cereals is commonly associated with higher yield under water stress (Ludlow and Muchow 1990; e.g. wheat in a Mediterranean climate, Perry and D'Antuono 1989), and genotypes with inappropriate phenologies can be screened out at an early stage of breeding programmes. An alternative approach, applied to wheat with limited success in terms of grain yield, has been to limit the amount of water used in the vegetative phase by selecting for higher hydraulic resistance in the seminal roots (Richards 1987).

For crops of appropriate phenology that are limited by water supply, biomass accumulation is linearly related to water uptake/cumulative transpiration (reviewed by Ludlow and Muchow 1990); and the factors determining the harvest index, as laid out in Sections 6.3 to 6.7, are common to crops limited by radiation or water. For example, the capacity of the sink (related to the biomass at anthesis for cereal crops) can be reduced by water stress during canopy generation;

and the filling of the sink can be cut short by premature senescence. The central importance of the anthesis–silking interval of maize in determining harvest index under drought has been emphasised in Sections 2.2.2 and 6.5. A degree of genetic variation in transpiration efficiency (ε_w) and in its index ($\delta^{13}C$) has been established within crop species, but its significance for yield is still to be explored fully (Hall *et al.* 1994). The guiding principle appears to be that high yield can be achieved, in varieties that are well adapted to the environment, by maximising the amount of water transpired.

There is, thus, only a limited set of practical tools for the routine screening of breeding populations for improved performance under water stress, and their application is complicated by the high seasonal variability of rainfed environments, and the resulting genotype–environment interactions (Richards *et al.* 2002). In consequence, the breeding of varieties for yield under drought is arduous, as it is normally necessary to use a range of characters, and select over several contrasting seasons. For example, a particularly successful programme, starting with relatively unimproved lowland tropical maize populations, involved a weighted combination of several relevant characters (grain yield under water stress and irrigation; short anthesis–silking interval; low canopy temperature and reduced leaf senescence under drought; and high rates of tissue expansion; Edmeades *et al.* 1999). This regime of recurrent selection generated increases in grain yield of 4–13% per cycle of selection over up to eight cycles, with accompanying increases of up to 1.5% per cycle when the materials were grown under irrigation. However, such a demanding programme of improvement is unlikely to be attractive to commercial breeders.

This discussion has concentrated on drought *avoidance*, the most appropriate approach to terminal stress. The breeding and selection of crop varieties to yield well specifically under intermittent stress is even more challenging because characters conferring the necessary drought *tolerance* or *resistance* during vegetative or reproductive development are commonly associated with low yield. Thus species native to semi-arid and arid environments tend to divert a significant proportion of their assimilates to protective functions, and their lifestyles tend to be conservative, geared to survival rather than to the production of biomass or economic yield (Fitter and Hay 2002). Nevertheless, dryland crop varieties can show tolerance to drought. For example, in comparisons of four grain legumes growing on stored water in pots, soybean plants died within 2 days of exhausting the supply of available water but black gram (14 days), pigeonpea (18 days) and cowpea (>24 days) could survive much longer owing to lower stomatal conductances and a higher degree of tolerance at the cellular level (Sinclair and Ludlow 1986). Judged on the basis of water extraction alone, the potential yield of soybean was higher but it did not have the capacity to convert this to harvestable yield either under terminal or intermittent stress.

Although this section has concentrated upon productive dryland crops, a high proportion of the world's food is grown in semi-arid environments at a subsistence level. Here, the emphasis is on yield stability rather than high productivity; it is important to obtain some yield every year rather than to endure violent fluctuation in response to the seasonal pattern of precipitation. The main guiding principles are the same as in intensive irrigated agriculture except that the subsistence farmer may choose varieties conservatively. For example, where the phenology of cereal varieties can be expressed in terms of the days of available water required to complete the life cycle, the choice will tend to lie with varieties well below the mean duration of available water at the planting site; stability can be gained at the expense of potential yield in an above average year (but see Hay 1981, where nutritional aspects predominated). The analysis has also concentrated upon the existing climate, but the progressively increasing concentration of CO_2 in the Earth's atmosphere will tend to act to improve the ε_w of most crops by increasing the gradient of CO_2 between bulk air and leaf mesophyll, permitting the same inward flux at a lower leaf conductance (Figure 7.3; Jarvis *et al.* 1999). As shown by Magliulo *et al.* (2003), where other resources are not limiting, this offers the prospect of higher yields of dryland crops. These issues are explored further in Chapter 10.

7.2 Limitation by nitrogen supply

Following the approach adopted for limitation by the supply of radiation or water, the yields of crops limited by nitrogen supply can be described by the equation:

$$Y = Q_N \times A \times \varepsilon_N \times H \tag{7.4}$$

where

Q_N is the total quantity of nitrogen potentially available to the crop over the growing season,
A is the fraction of Q_N that is taken up by the crop,
$Q_N A$ is the total quantity of nitrogen taken up by the crop,
ε_N is the overall photosynthetic efficiency of the crop in terms of the total plant dry matter produced per unit of nitrogen taken up (i.e. the nitrogen use efficiency),
H is the harvest index, and identical to the value used in analyses based on the interception of radiation (Sections 6.2–6.7) or the uptake of water (Section 7.1).

Thus, for a given quantity of nitrogen taken up by the root systems of a crop, the harvested yield depends upon the efficiency of use of the acquired nitrogen in constructing and maintaining the photosynthetic machinery, and the partitioning of the resulting assimilate to harvested organs. One important difference from limitation by water supply is that the nitrogen taken up by the crop remains within the plant tissues and much of it can be remobilised in the later stages of the crop life cycle.

7.2.1 Acquisition of nitrogen

For crops relying on the soil solution for their supplies of nitrogen, Q_N is the sum of the nitrate and ammonium ion contents of the soil profile at sowing, and the quantities of nitrate and ammonium released to the soil solution by mineralisation and nitrification, and applied as fertiliser, during the lifetime of the crop. In certain areas, it is also necessary to include nitrogen input from the atmosphere (Goulding 1990). Q_N is, therefore, determined solely by environmental and management factors; these include not only soil temperature, moisture content and biotic activity before and during the growing season, but also the history of management of the soil, which determines the quantity and quality of the soil organic matter. Q_N can be measured by frequent (and laborious) sampling of the soil or soil solution throughout the soil profile (e.g. Armstrong *et al.* 1986; Devienne-Barret *et al.* 2000), or evaluated by modelling the mineralisation of organic matter and the fate of applied fertilisers (e.g. the DAISY model, Hansen 2002). There have been relatively few measurements of Q_N in the full soil profile (e.g. Ridge *et al.* 1996). As a quantitative reference point, high-yielding temperate wheat crops will normally contain up to 200 kg N ha^{-1}, with highly variable proportions contributed by fertiliser and soil nitrogen (Vaidyanathan 1984).

The term Q_N is less appropriate for grain and pasture legumes, as they also have the potential to exploit the nitrogen in the soil atmosphere *via* symbiotic relationships with nitrogen-fixing bacteria. The quantity of nitrogen supplied to the crop by root nodules depends upon the presence of appropriate species and strains of rhizobia in the soil, the establishment of a functional association between host plant and bacterium, and favourable soil physical conditions. For example, the profitable cultivation of soybeans cannot advance into cool temperate zones without the introduction of strains of *Bradyrhizobium japonicum* that are tolerant of low temperatures (Zhang *et al.* 2002). Similarly, the success of soybean production on very infertile soils in Brazil has depended upon the introduction of appropriate bacterial strains (Alves *et al.* 2003). Effective rhizobial associations are capable of supplying 100 kg N ha^{-1} to annual grain legume crops, although there are measurements of up to 450 kg N ha^{-1} for soybean (Unkovich and Pate 2000).

The fraction of Q_N acquired by a crop (A) depends upon the distribution of roots within the soil profile, and the capacity of the root system to absorb nitrogen-containing ions from the soil solution. The first products of the mineralisation of soil organic matter are ammonium ions, but these are rapidly transformed to nitrate by nitrifying bacteria in most agricultural soils. Since the principal form of available nitrogen is nitrate, which does not bind to soil particles, the soil solution of the profile is a single source or pool; nevertheless, nitrate ions do not move more than a few millimetres from the bulk soil towards root axes, and it has been estimated for maize that a root density of 1 cm cm^{-3} of soil is required for the complete exploitation of soil nitrate (Bänziger *et al.* 2000). The fact that all of the soil nitrate is in solution also means that there can be significant losses to leaching when the soil water content exceeds its field capacity. Typical concentrations in the soil solution are up to 25 mM for nitrate and 5 mM for ammonium, and unless the soil is anaerobic or acidic, the proportion of available nitrogen in the form of ammonium ions, which can bind to soil particles, will be low.

Extensive studies of isolated root axes and root systems have established that the rate of uptake of nitrate depends upon the concentration of the ion in the soil solution; the kinetics of uptake, as analysed by the Michaelis–Menton equation, show that most species possess both low affinity (operating over the mM range) and high affinity (μM to mM) systems for nitrate (Forde and Clarkson 1998). However, there is still considerable uncertainty about the factors controlling uptake of nitrogen into crops *in the field* (Grindlay 1997), not least the interaction with soil water content (Robinson 1994). In several field experiments (e.g. Lemaire and Salette 1984, pasture grasses; Greenwood *et al.* 1990, C_3 and C_4 species; Justes *et al.* 1994, wheat), the rate of uptake of nitrogen has been found to match the growth rate of the crop (determined by interception of radiation), allowing for the normal ontogenic decline in the nitrogen content of plant tissues. This relationship can be described by the equation:

$$\%N \text{ in shoot} = aW^{-b} \tag{7.5}$$

where W is shoot dry matter content and a and b are constants that vary between groups of species. For example, Greenwood *et al.* (1990) found that values of $a = 5.7$ and $b = 0.5$ were common to a range of C_3 species; the corresponding constants for C_4 species were 4.1 and 0.5, suggesting a higher nitrogen use efficiency for C_4 plants (see below). This analysis implies that demand from the shoot determines the rate of uptake, possibly by genetic control of inducible uptake systems, irrespective of the availability of nitrate in the soil solution; the associated nitrogen dilution curves (e.g. Figure 7.8) have, conversely, been used to identify when crops are experiencing nitrogen deficiency (when the shoot content falls below the critical minimum, e.g. Colnenne *et al.* 1998).

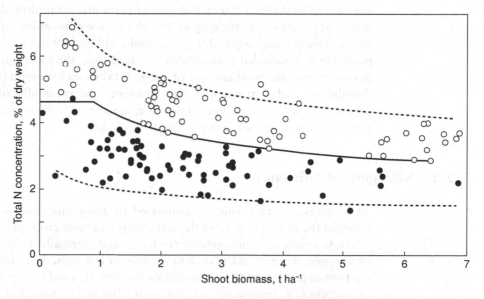

Figure 7.8 Dilution curve (continuous line) for winter oilseed rape (cv. Goeland) constructed from a dataset of 142 measurements of crops grown in northern France (two sites, two growing seasons, and six nitrogen fertiliser regimes). The open symbols indicate crops that were classed as not limited by nitrogen supply, and the closed symbols indicate crops that were limited (from Colnenne *et al.* 1998).

An alternative position, favoured by crop modellers, considers nitrogen uptake, *before canopy closure*, in terms of a positive feedback loop: increased uptake of nitrogen leads to enhanced interception of solar radiation (larger leaves, more branches; Section 3.2), higher rates of growth, and greater demand for nitrogen.

A common problem with these approaches is that most crop species have the potential to accumulate levels of nitrogen, as organic nitrogen or nitrate, that are significantly in excess of the critical minimum required for maximal growth rate under a given combination of environmental factors (Millard 1988). Similarly, Figure 7.8 shows that, under certain circumstances, shoot growth can continue even though the nitrogen content of the leaves falls well below the dilution curve. Devienne-Barret *et al.* (2000) have suggested a possible reconciliation of the existing findings (root uptake described by Michaelis–Menton kinetics; shoot uptake governed by crop growth; existence of luxury uptake and deficiency) using a new parameter, nitrate uptake rate index (ratio of the actual and the critical minimum uptake rates). Values of the index below 1 lead to nitrogen deficiency, whereas values greater than 1 lead to accumulation of nitrogen (points above the nitrogen dilution curve; Figure 7.8). Plots of the index against soil solution concentration for a series of wheat, oilseed rape and maize crops can then be interpreted in terms of Michaelis–Menton kinetics (Figure 7.9).

This analysis suggests that the nitrogen supply to the shoot is in fact controlled by the rate of uptake at the root surface. Although an internal balance is normally struck between nitrogen uptake and growth (controlled by radiation interception), this can be disturbed to give either a deficiency or an excess of nitrogen (i.e. points below or above the dilution curve; Figure 7.8). Much remains to be done to clarify this area of plant and crop physiology, including wider application of techniques for detailed examination of relationships between supply and demand (e.g. Ingestad and Lund 1986); nevertheless, it can probably be concluded that, in well-fertilised soils, the most important factors in determining the total amount of nitrogen taken up by a crop ($Q_N A$) are root distribution, and crop management to maximise the availability of nitrogen, and minimise losses. The concepts in this section have yet to be evaluated in agricultural systems where nitrogen is normally limiting.

7.2.2   Nitrogen use efficiency

After uptake, nitrate ions are converted to ammonium by nitrate reductase either in the root system or in the leaf canopy; in most crop species, the conversion takes place predominantly in the leaves and, normally, there is little storage of nitrogen as nitrate in the vacuoles of root or leaf cells. As ammonium ions are phytotoxic at relatively low concentrations, they are used rapidly in the synthesis of, successively, glutamine and glutamate. The newly assimilated amine group can then contribute to the synthesis of other amino acids by transamination reactions; subsequent biochemical pathways lead to the formation of other nitrogen-containing compounds including nucleic acids, pigments and alkaloids. In the case of legumes, nitrogen fixed in root nodules is supplied to the shoot

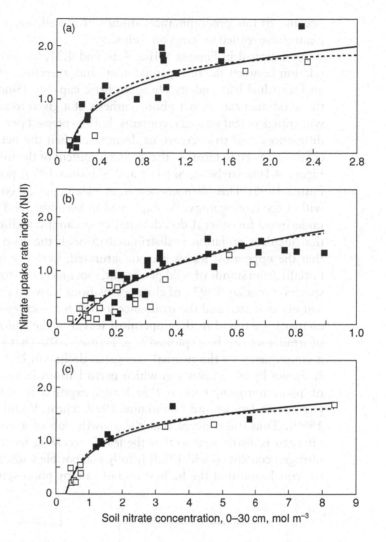

Figure 7.9 Relationships between crop nitrate uptake rate index and soil nitrate concentration (0–30 cm) for (a) three wheat crops, (b) six oilseed rape crops and (c) one maize crop grown with varying nitrogen fertiliser inputs in France. The closed symbols indicate crops that were classed as not limited by nitrogen supply, and the open symbols indicate crops that were limited (note that this is the reverse of Figure 7.8) (from Devienne-Barret *et al.* 2000).

in the form of ureides, whose reduced nitrogen is incorporated into amino acids *via* urea.

The enzyme Rubisco is by far the predominant nitrogen-containing compound in crop plants, representing 50% of the soluble leaf protein of C_3 species and up to 25% in C_4 (Chapter 4; Sinclair and Horie 1989); and leaf nitrogen content is normally linearly related to Rubisco content (Figure 4.21; Evans 1989). These observations have led to the suggestion that the enzyme may play a storage role in addition to its central part in photosynthesis (Huffaker and Peterson 1974). It is perhaps only in tobacco, amongst major crops, that tissue concentrations of nicotine, a nitrogen-rich alkaloid, reach comparable levels.

Because of this predominance, studies of the efficiency of use of nitrogen by plants have tended to focus on Rubisco.

As indicated in Figures 4.10a, 4.20 and 4.21, for most crops there is a correlation between the nitrogen content (and, therefore, the Rubisco content) of an individual leaf and its photosynthetic capacity (Sinclair and Horie 1989): the maximum rate of net photosynthesis at a given irradiance increases over a wide range of leaf nitrogen contents, both expressed per unit area. Such marked differences tend to decrease or disappear when the net photosynthesis of the crop is assessed in terms of the nitrogen content of the full canopy (e.g. ryegrass, Figure 4.10b; soybean, Sinclair and Shiraiwa 1993; potato, Vos and van der Putten 1998). This phenomenon is, at least in part, a consequence of gradients within the crop canopy. As explained in Sections 3.4.3 and 3.4.4, it has been understood for several decades that crop canopy architecture tends to ensure that incident radiation is distributed through the crop stand, with the result that the uppermost leaves are not saturated. Evidence has been accumulating, initially from stands of wild plants but, more recently from a wide range of crop species (Grindlay 1997), of close correlations between the nitrogen content of a leaf per unit area and the irradiance to which it is exposed; higher leaf nitrogen contents are found in the uppermost leaves of the canopy where higher levels of irradiance can be exploited (e.g. Figure 7.10). That this effect is not simply a consequence of the normal ontogenic decline in leaf nitrogen with leaf age is shown by experiments in which perturbations have caused the reallocation of plant nitrogen, even *within* leaves exposed to differences in irradiance (e.g. maize, Drouet and Bonhomme 1999; wheat, Vouillot and Devienne-Barret 1999). Thus the concept that the growth rate of a crop is maximised when nitrogen is distributed so that the leaves receiving most PAR have the highest nitrogen contents (Field 1983) is fully compatible with, and complementary to, the conclusion that the highest rate of canopy photosynthesis will be achieved

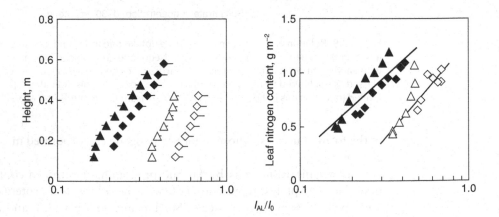

Figure 7.10 Gradients of radiation interception (I_{AL}/I_0: intercepted PAR per unit leaf area as a proportion of incident PAR above the canopy; 78 days after germination) by wheat stands, under controlled conditions, in terms of height within the canopy and leaf nitrogen content. Closed symbols indicate stands classed as not limited by nitrogen supply, and open symbols indicate stands that were limited (from Dreccer *et al.* 2000).

when all of the incident radiation is intercepted by leaf blades that are disposed in space in such a way that no leaf is more than saturated (Section 3.4.3).

In summary, therefore, nitrogen use efficiency, expressed in terms of dry biomass produced per unit of nitrogen taken up, depends upon:

- the proportion of the nitrogen resources of the crop allocated to Rubisco (no systematic information available for comparison among species or varieties, although there is widespread evidence of overproduction of Rubisco, in relation to its role in photosynthesis);
- the distribution of Rubisco within the canopy (increasing evidence that this is determined by the distribution of irradiance for all major crop species);
- the properties of Rubisco (no evidence of major differences in enzyme properties across species or varieties apart from the discontinuity between C_3 and C_4 species, with the latter achieving (higher) maximal rates of net photosynthesis than the former at lower leaf nitrogen contents; Section 4.3.3; Sinclair and Horie 1989); and
- the longevity of Rubisco/functional leaves.

Of these, only the last appears to have made a significant contribution to crop improvement through the introduction of 'stay green' characters in North American maize varieties (Section 3.2.5).

Published data on variation in nitrogen use efficiency are scarce, and interpretation is complicated by a spectrum of definitions. Practical agronomists use the term to express harvested yield per unit of fertiliser applied (Below 1995) or grain yield per unit of available nitrogen (van Beem and Smith 1997). The resulting data are of limited use here as they do not discriminate among uptake, use and allocation. Others have accepted definitions as vague as 'the ability of a genotype to produce superior grain yields under low soil N conditions in comparison with other genotypes' (Presterl *et al.* 2003). Garnier *et al.* (1995) propose the other extreme of exactitude: their 'photosynthetic nitrogen use efficiency' is the ratio of the rate of (net) photosynthesis and leaf nitrogen concentration.

In an unusually full investigation of a range of ten wheat varieties released by CIMMYT between 1950 and 1985, Ortiz-Monasterio *et al.* (1997) found that nitrogen use efficiency (here termed 'biomass production efficiency') declined with fertiliser application, particularly at 150 kg N ha^{-1} and above, in a 2-year experiment in Mexico. Analysis of the responses of varieties indicated that the significant effects (at $P = 0.05$) were limited to two introductions. In the unfertilised treatment, the oldest variety (Yaqui 50) had a much higher ε_N than the other varieties (139 kg kg N^{-1} compared with a mean of 119 kg kg N^{-1}, associated with the lowest harvest index of 0.30) but a similar value at 300 kg N ha^{-1} (93 compared with 86); in contrast, the 1970 introduction, Yecora 70, had the lowest efficiency at each level of fertiliser application (103 and 74 kg kg N^{-1} respectively), associated with the highest harvest index (0.43–0.45). Thus, within this collection of varieties, there was some scope for improvement in nitrogen use efficiency but this appeared to be linked to lower yield potential. As there are few comparable datasets for other species and locations, and no simple indices for use as tools in screening, it is not clear to what extent ε_N can play a part in future crop improvement programmes.

7.2.3　Crop yield where N supply is limiting

Although some aspects remain to be characterised, most of the evidence from temperate and intensive agriculture indicates that nitrogen influences crop yield primarily through its effects upon canopy expansion, and the survival and longevity of organs (Sections 3.2, 3.3) rather than through effects on canopy photosynthesis. Biomass yield is dependent upon how much nitrogen the crop can absorb, and how much leaf area can be constructed per unit of nitrogen taken up (a characteristic that can be constant across varieties and seasons; Figure 7.11; Lemaire and Gastal 1997). In practice, biomass yield tends to increase linearly with nitrogen application up to a maximum above which yield is maintained or depressed (e.g. in cereals, potatoes and grassland; Hay and Walker 1989). The effect of nitrogen status upon (carbon) harvest index varies considerably among geographic zones and crops, depending upon the extent to which nitrogen supply is critical for the duration of a functioning canopy and for the survival of harvested organs (e.g. little effect upon the harvest index of temperate grain crops; Hay and Walker, 1989; Hay 1999). The nitrogen harvest index (ratio of the amount of nitrogen in the harvested parts to that of the whole plant) is considered in detail in Chapter 8.

In areas of more subsistence cropping, predominantly on the tropics and sub-tropics, which tend to be chronically short of nitrogen, strategies for increasing production have tended to lay stress upon greater inputs of organic and fertiliser nitrogen. Indeed, in some environments, improved varieties, developed for intensive production, have proved to give higher yields under low fertility than apparently well-adapted landraces (e.g. maize in Mexico; Lafitte *et al.* 1997). However, in recent decades, major projects have begun to concentrate more on

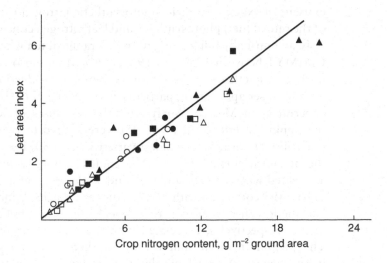

Figure 7.11 Relationship between the nitrogen content and green leaf area index of vegetative canopies of winter wheat grown in eastern England with 90 (circles), 200 (squares) or 350 (triangles) kg N ha⁻¹. Closed symbols indicate early sowing (September) and open symbols late (November) (from Scott *et al.* 1994).

optimising yield at lower levels of fertility. In the absence of appropriate tools for screening for relevant traits, breeding programmes have relied upon yield under low fertility, in some cases also assessing to what extent yield at higher fertility can be maintained (thereby developing varieties with a valuable degree of plasticity). Information on the characters that favour yield under nitrogen stress can therefore be sought by examining the physiology of varieties generated by such programmes.

For example, Lafitte and Edmeades (1994) subjected a population of Mexican maize (Across 8328), which had shown tolerance to low nitrogen, to three cycles of selection, using a weighted combination of increased grain yield with and without fertiliser application, and other characters under low nitrogen that might be involved in stress physiology (high chlorophyll content, delayed senescence, increased plant height). Table 7.2 shows that significant increases in grain yield were achieved by selection at low nitrogen fertility, without compromising crop potential under higher fertility. As harvest index was not affected by selection, increases at low and high N input were the consequence of greater biomass production, which was translated into higher grain population densities (owing to higher proportions of florets developing into grains), at unchanged numbers of ears per plant (Section 6.4). How this increase in biomass was achieved is not clear: higher yield was not associated with greater uptake of nitrogen, nor with systematic increases in translocation of nitrogen from stem and leaves to grains. As a consequence, the percentage of nitrogen in the grain tended to fall (see Section 8.1). Thus, in this experiment, differences in yield were not the result of

Table 7.2 The responses of a population of Mexican maize (Across 8328) to three cycles of selection (C_{1-3}), with (200 kg N ha^{-1}) and without the application of nitrogen fertiliser (see text for details) (from Lafitte and Edmeades 1994).

		C_0	C_1	C_2	C_3
Grain yield t ha^{-1}	Low N	2.62	2.65	2.80	2.81 *
	High N	5.76	5.86	6.12	6.13 ***
Biomass yield t ha^{-1}	Low N	7.33	7.23	7.80	7.56 *
	High N	12.78	13.09	13.70	13.52 **
Harvest index	Low N	0.36	0.37	0.36	0.37
	High N	0.45	0.45	0.45	0.45
Grains per ear	Low N	207	207	209	211 ns
	High N	349	358	368	372 **
Anthesis–silking interval, days	Low N	2.6	3.3	2.9	0.5
	High N	0.6	0.9	0.9	0.6
Total N at maturity g m^{-2}	Low N	5.3	5.1	5.4	5.3 ns
	High N	14.8	14.4	15.2	15.0 ns
Grain N %	Low N	1.14	1.13	1.10	1.10 ns
	High N	1.62	1.53	1.55	1.55 **
Grain N from mobilisation %	Low N	43	48	44	49 ns
	High N	42	48	45	49 ns

*, **, *** and ns indicate effects of selection that are significant at $P < 0.1$, 0.05 or 0.01, or non-significant.

increased Q_NA or H but in ε_N (more biomass produced per unit of nitrogen acquired).

One of the most important outcomes of the sustained programme of improvement of Mexican maize in the 1980s and 1990s was the finding that selection for yield under one stress could have a beneficial effect on yield when the resulting improved varieties were exposed to another form of stress. In particular, populations selected for yield under mid-season drought (predominantly with a shorter anthesis–silking interval, leading to higher harvest index; Sections 2.2.2, 6.5) also showed improvements in nitrogen uptake (Q_NA) in soils of low N status (Bänziger *et al.* 1999). Similarly, varieties selected on the basis of reduced plant height were more tolerant of high population density but also produced higher yields under less-intensive management (Johnson *et al.* 1986).

Responses of this kind are of great importance in areas of low yield potential because drought, nitrogen and other stresses commonly occur together; their relative severity and timing can vary among seasons and sites; and there are important interactions between the effects of each stress on crop physiology and yield (e.g. Bennett *et al.* 1989). This introduces a major preoccupation of all plant breeders: genotype–environment interactions. Account must be taken of the fact that the order of ranking of a set of crop populations or cultivars can vary considerably among seasons and sites, and that changes in management practices can elicit major differences in growth, development and yield (Evans and Fischer 1999; Slafer 2003).

7.3 Achieving high yield: resource capture and assimilate partitioning

The evidence evaluated in this chapter indicates that, with some important exceptions, crop biomass production depends upon resource capture rather than resource utilisation. Under intensive cultivation, biomass yield depends upon the amount of PAR intercepted by the crop canopy whereas, under limitation by water supply, it depends upon the quantity of water transpired by the crop. The situation for crops limited by nitrogen is not quite so clear, but many investigations have shown that total nitrogen uptake is more important than nitrogen use.

Crop phenology plays an important part in the translation of biomass yield into harvestable products: the resources acquired must be used effectively in the construction, survival and filling of harvestable organs. Thus, it is important for the ontogeny of the crop plants to match the availability of resources (e.g. rapid reproductive development in grain crops, ensuring that water is available for grain filling) or to be coordinated with the acquisition and storage of resources (e.g. the mobilisation and transport of stored and non-structural nitrogen from other organs to developing grains). The harvest index of the crop, therefore, depends primarily upon the interrelationship between crop phenology and the time course of availability of the limiting resource.

The implications of this analysis for breeding and managing high yielding crops can be explored through the extensive documentation of temperate cereals over the last century. For example, Figure 6.12 indicates that, for wheat crops in the UK, Canada and Australia, higher grain yield was achieved by *managing* well-adapted varieties to express the existing potential for biomass production (i.e. by increasing radiation interception through optimum nitrogen application, plant population density, sowing dates, and crop protection) and by *breeding and selecting* varieties in such a way that harvest index increased progressively throughout most of the twentieth century. Resource capture (interception of solar radiation) increased through improved management but, in many areas, there is little scope for further improvement of this kind; for example, winter cereals in Northern Europe already have significant leaf area index in spring, as PAR supplies begin to increase, and any extension of the crop canopy into the autumn would have serious consequences for harvest conditions and the planting of the following crop.

Figure 6.12 indicates that there is still potential for higher harvest indices in Canadian and Australian wheat varieties (0.35–0.4 compared with 0.55 for UK varieties). In the search for such improvement, the phenologies of the crops may play a part, as lower harvest indices are commonly caused by premature cessation of grain filling under stress (e.g. Perry and D'Antuono 1989). In the case of European wheat varieties, however, it has been estimated that the maximum harvest index is around 0.65, as a proportion of plant dry matter must be devoted to canopy expansion and physical support of the ear (Austin *et al.* 1980b). The most recent varieties are approaching this level (>0.6; e.g. Whaley *et al.* 2000; Brancourt-Hulmel *et al.* 2003), indicating that this source of yield improvement will shortly be exhausted. It seems inevitable, therefore, that any increases in the potential yield of wheat varieties for the UK must arise out of higher biomass through increased radiation use efficiency. There are indications that conventional selection for grain yield in Europe may already be delivering such improvement, but this may well lead to a shift to limitation of yield by other factors, principally water supply (Foulkes *et al.* 2001). Breeding for high yield potential in new varieties is, however, not the only strategy for increasing crop performance across a given area of arable land; since surveys of actual farm yields (e.g. Hay *et al.* 1986) invariably show a wide range of yield on apparently similar soils, action to reduce the 'tail' of poor yield by improved management, possibly aided by the techniques of 'precision agriculture' may be as important as breeding for higher yield.

The pattern of yield improvement in maize in North America has been slightly different. Figure 6.14 shows that changes in harvest index over the twentieth century have been small, and that yield increases have been a consequence of enhanced biomass production, mainly the result of longer canopy duration ('stray green' traits). However, this tends to obscure the fact that the breeding effort concentrated on producing varieties that could produce more biomass because they were tolerant of higher plant population densities and the associated increases in input of nitrogen, while maintaining the levels of plant fertility and harvest index. In parallel with temperate wheat, there have been few indications of major increases in radiation use efficiency, and it is important to ensure that comparisons between varieties of different vintage are valid. For example, the apparently higher efficiency of some newer hybrids may be explained in terms of the enhanced duration of their canopies rather than an all-season effect (Tollenaar and Aguilera 1992).

As there is a limit to the increases in biomass yield that can be achieved by following the approach of the twentieth century, there is increasing interest in improving radiation use efficiency, for example *via* genetic modification. Progress in this area is likely to be slow owing to the complexity of the photosynthetic system (Reynolds *et al.* 2000; Richards 2000) and the close inverse relationship between leaf area and the rate of net photosynthesis per unit area (Evans and Dunstone 1970; Pellny *et al.* 2004).

Lack of detailed understanding of resource capture and use efficiency hampers the formulation of strategies for the improvement of crops for environments where yield is limited by other resources, singly or in combination. From the limited evidence considered in this chapter, it has been concluded that resource capture is normally the primary limiting process where crops are grown under water or nutrient stress. It is possible that current fundamental studies of the quantitative and molecular genetics of ion uptake and drought tolerance will ultimately provide powerful tools in the breeding and selection of varieties with enhanced potential for resource capture and use (e.g. Hirel *et al.* 2001; Bray 2002). Furthermore, there appears to be slightly more inter-varietal variation in water- and nitrogen-use efficiency than in radiation-use efficiency (wheat, Siddique *et al.* 1990; maize, Lafitte and Edmeades 1994). Nevertheless, until greater understanding is achieved, progress will tend to be limited to the results that can be achieved by empirical approaches, selecting for yield and paying particular attention to genotype–environment interactions (Araus *et al.* 2002).

Chapter 8

Physiology of crop quality

. . . in contrast to the about 800 million children, women and men who suffer from protein-energy undernutrition, nearly 2 billion suffer from deficiencies of micronutrients like iron and iodine

(Swaminathan 1998)

Consumers' expectations are high. They want food to offer all the qualities needed to fit their lifestyle and values. This means good taste, convenient preparation and health enhancing properties, produced in a manner consistent with environment care

(Bruhn 2003)

Most of this book is about the quantitative aspects of crop yield: how the yield of dry matter is determined by environment and management, and the extent to which it is partitioned to harvested organs. The quality of crop products has always been important, but the consequences of the major increases in yield over the last 50 years, and concerns about the inter-relationships between diet and health, have enhanced its importance, to such an extent that quality has now become a primary breeding objective in many parts of the world, as explained in the following sections.

The quality of a crop product can never be absolute; it will always depend upon the requirements of the consumer. Quality can relate to the quantity of a major component (e.g. the protein content of bread wheat grain, the oil content of rapeseed, or the starch and water content of potato tubers), or to its composition (e.g. the lysine content of the protein fraction of feed maize; the degree of saturation of the triglyceride lipids in oilseeds). Elsewhere, it can relate to the quantity of a minor component (e.g. vitamin A and its biochemical precursors in rice grain; micronutrients, such as Fe, Cu, or I, particularly if the entire diet is harvested from an area of mineral deficiency). Some products are classified according to cooking or processing quality (e.g. malting grain for alcohol production), and the term can also refer to the dimensions of harvested organs (e.g. in ware potatoes where grades can attract different prices) and to the incidence of diseases and blemishes (particularly on vegetables). In addition to all of these, the consumer is very interested in the taste and texture of the resulting food.

There are two major physiological aspects of crop quality. First, across a wide range of crop species there is a tendency for quantity and quality to be inversely related. The most thoroughly explored example of this effect is the nitrogen content of cereal grains (Section 8.1): where dry matter yield varies owing to site, season or management other than nitrogen fertilisation, a clear inverse relationship between yield and protein content is found (Simmonds 1995; Feil 1997). However, high yield can be associated with high protein content if the timing and quantity of nitrogen fertilisation are carefully managed. (As shown in Section 8.2, soybean is a major exception to this rule, with oil content relatively unaffected by dry matter yield; Figure 8.3.) Another important example is associated with the 'green revolution' in countries such as India: here the use of cereal varieties that are more responsive to nitrogen fertilisation on soils that are deficient in trace elements can result in a 'dilution' effect (higher yields of lower-quality grain, leading to health problems, such as the association between low Fe content in grain and anaemia; Hurrell 2001). Second, the quality of crop products can be modified significantly by the incidence of stress during the crop life cycle.

Out of a very extensive literature on the subject, this chapter presents a limited selection of case histories of major arable crop species, exploring some important physiological principles. Although the physiology of the quality of horticultural crops is a particularly rich field for investigation (e.g. the factors determining the balance between acidity and sweetness in tomatoes), it is beyond the scope of this book.

8.1 Wheat: protein content

The highest prices are paid for wheat (*Triticum aestivum*) grain that meets the standards required for bread making. The physical properties of the grain must be suitable for milling into flour: the preferred 'hard' wheats give a clean separation of endosperm from embryo and bran whereas the products of milling 'soft' wheat tend to clog the machinery. This distinction is principally genetic and less affected by management or environment, although hard wheat varieties are generally better adapted to drier continental zones, where yields are modest and protein contents high (see below).

The quality of the resulting flour for bread making depends upon the protein content: the major fraction of the storage proteins in the endosperm, the water-insoluble glutenin, provides the flexible and cohesive framework of the dough, trapping air, and determining the degree of 'rise' and crumb structure of the resulting loaf. In the UK, a threshold grain protein content of 13% (equivalent to 2.1% nitrogen) is required for classification for breadmaking. Grain can, however, be rejected if there is an excessive content of α-amylase, leading to extensive hydrolysis of carbohydrate, and doughs of unacceptable viscosity. This problem is characteristic of cool moist growing areas where the content of the enzyme depends upon the rate and temperature of grain maturation, and the degree of sprouting in the ear (e.g. in England; Lunn *et al.* 2001). The α-amylase content is normally expressed in terms of the industry standard (Hagberg Falling Number, with a minimum of 250 seconds required for breadmaking).

Flours of inferior quality can be used in the manufacture of biscuits (<10% protein), and wheat grain is also used widely in livestock feeds (where the content and composition of protein, notably the lysine content, are critical). In general, therefore, the quality of wheat grain depends upon its protein content; only in the production of grain spirits (where the grain is primarily a source of carbohydrate), is a low protein content preferred.

The nitrogen content of harvested wheat grain, predominantly in the form of protein, depends upon the amount of nitrogen acquired by the crop (Section 7.2.1), and the nitrogen harvest index (H_N: in practice, the fraction of the nitrogen content of the above-ground dry matter at harvest located in the grain). Grain filling, therefore, involves two parallel sets of processes leading to the mobilisation of nitrogen and dry matter in the leaves, sheath and stems, and their translocation to the grain. There are also contributions from current photosynthesis and, although the roots are increasingly starved of photosynthate after anthesis, the nitrogen status of the grain can be enhanced by late soil applications of nitrogen fertiliser or foliar applications of urea (e.g. Wuest and Cassman 1992; Bly and Woodward 2003). As the photosynthetic activity of the canopy depends upon the integrity of protein-containing structures, it would be predicted that the flows of dry matter and nitrogen into the grain would be tightly linked, and related to the onset of senescence. The coordination of protein and starch deposition in the grain (Kavakli *et al.* 2000) would support this prediction.

Evidence from a range of environments, however, indicates that this hypothesis is not correct, and that carbon and nitrogen metabolism during grain filling are independent (Triboï and Triboï-Blondel 2002). Even though, in the absence of stress, the H of wheat is highly heritable (Section 6.6), H_N is both higher than H (typically >0.8 compared with <0.6 in cool temperate crops) and less variable (Löffler and Busch 1982). Values of H_N as high as 0.92 have been recorded where additional nitrogen has been applied at anthesis (Wuest and Cassman 1992). Thus, since most of the non-structural nitrogen in the crop around anthesis is mobilised and transported to the ear, whereas the proportion of crop dry matter partitioned is lower and more variable, there is scope for significant variation in grain protein content.

This can be illustrated by field experiments in which the management of the wheat crop has been varied to produce significant changes in biomass production but less effect on nitrogen uptake. For example, in the winter wheat crops documented in Figure 8.1, a wide range of sowing densities led to variation in crop biomass from 11 to 13.5 t ha^{-1} (23%), in grain yield from 6.9 to 8.6 t ha^{-1} (25%), but in protein yield in the grain from 960 to 1060 kg ha^{-1} (10%) (all values at 200 kg N ha^{-1} of fertiliser at sowing). There was a classic inverse relationship between grain yield and quality: protein content fell from 14% to 12.8%, with increasing plant population density. Application of very high levels of fertiliser nitrogen (350 kg ha^{-1}) did, however, prevent the dilution of protein content with increasing yield; the protein content remained above 14% at all plant population densities, but at the expense of H_N (see also Wuest and Cassman 1992, where H_N fell from 0.8 to 0.6 as fertiliser application increased to 250 kg N ha^{-1}).

Even more marked reductions in grain nitrogen content can be induced if biomass is increased by applying irrigation at constant fertiliser application. For

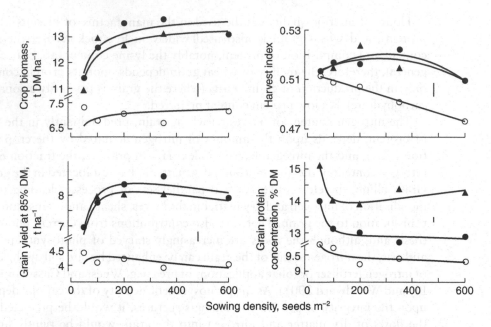

Figure 8.1 The effects of variation in plant population density on the yield and protein content of a crop of Hereward winter wheat grown in England receiving 0 (○), 200 (●) or 350 (▲) kg N ha^{-1}. Vertical bars indicate SEM (adapted from Gooding *et al.* 2002).

example, in a Mediterranean climate in Victoria, Australia, the grain yield increases resulting from irrigating a wheat crop receiving 150 kg N ha^{-1} varied between 72% and 80% (increases in grain number and individual weight), but the nitrogen content of the resulting grain fell by a quarter to 2.0% (Whitfield *et al.* 1989). Thus, the total nitrogen content of the grain is source limited, but reduction in the number of grains leads to increased protein content (Martre *et al.* 2003), whereas increase in the number and size of grains is associated with lower values.

Inverse relationships between grain protein content and yield are also a feature of wheat breeding programmes, with modern, high-yielding varieties tending to have higher yield but lower grain quality at a given level of nitrogen fertilisation (e.g. varieties bred for the UK, Austin *et al.* 1980b, 1993; maize in Mexico, Section 7.2.3). This effect is clearly demonstrated by measurements of wheat varieties, introduced to Argentina between 1920 and 1990, grown under uniform conditions (Calderini *et al.* 1995). There were increases in grain yield and in the total amount of nitrogen harvested in the grain, but the former effect predominated, with the result that grain nitrogen content declined from 2.8% to 1.8%. These results can be summarised in Figure 8.2, which shows a continuous decline in H_N in relation to H with later date of introduction.

The reverse of dilution of protein tends to occur when crops experience environmental stress, particularly during grain filling. Drought and high temperatures inhibit the deposition of starch in the grains more than that of protein, leading to lower individual grain weights but higher protein contents (Bindi *et al.* 1999). For example, Triboï and Triboï-Blondel (2002) found that an increase in

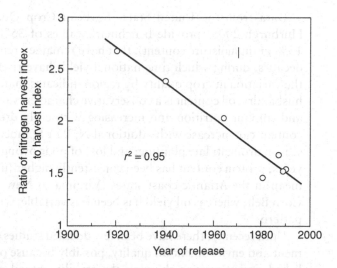

Figure 8.2 Changes in the ratio of nitrogen harvest index to harvest index for Argentinian wheat cultivars introduced at different dates, but grown under uniform conditions. Vertical bar indicates SEM (adapted from Calderini *et al.* 1995).

temperature from <20°C to <30°C during anthesis resulted in a 27% reduction in individual grain weight, associated with a 25% increase in grain protein content, and there were similar effects under a series of imposed drought regimes. The implications for grain quality are further complicated by the observation that stress after anthesis can also cause alterations in the proportions of the protein components (Jamieson *et al.* 2001).

In summary, there is a very substantial body of physiological knowledge underpinning the choice of variety and management system for the production of wheat grain to a standard suitable for breadmaking.

8.2   Soybean: oil and protein contents

Soybean crops are dual purpose, producing oil for direct human consumption and a protein-rich residue for feeding livestock. Soybean products are also major components of the protein content of Western vegetarian diets. The quality of soybean grain is, therefore, complex, involving:

- oil content;
- oil composition (proportions of saturated, monounsaturated and polyunsaturated fatty acids);
- protein content;
- protein composition (particularly the proportion of sulphur-containing amino acids);
- carbohydrate content;
- carbohydrate composition (sucrose, starch but also the indigestible polysaccharides raffinose and stachyose);
- anti-nutritional factors.

Data from the United States Soybean Crop Quality Survey (Brumm and Hurburgh 2003) provide benchmark values of 35% protein and 19% oil (at 13% grain moisture content), that have remained remarkably stable over recent decades, during which time national yields have risen to a plateau. Analysis of the variation in crop quality by region indicates that, with current varieties and husbandry, oil content is a conservative character, varying only with phosphorus and sulphur nutrition and increasing with early drought. In contrast, protein content can increase with additional N, P or S; inoculation with Rhizobia; late season drought; late planting; and loss of pods owing to insect damage. Over 20 years, protein content has been consistently higher (up to 2%) than the national mean in the Atlantic coast states (Virginia to New Jersey) and in the Eastern Corn Belt; whereas oil yield has been less variable, without a consistent regional pattern.

Until recently, there have been few detailed studies of the influence of management and environment on quality, possibly because of the many, largely unpublished, investigations showing the stability of oil and protein contents with variation in yield (e.g. in Canada; Figure 8.3); and it is not known to what extent regional variations in quality relate to the maturity group of the predominant varieties in the area (Figure 2.9; Yaklich *et al.* 2002). In a very extensive study (1863 variety/location/season combinations involving 20 elite varieties from 10 maturity groups and 60 sites), based on the USA Soybean Uniform Tests, Piper and Boote (1999) showed that the relationship between grain oil content and mean air temperature could be described by a quadratic equation, with oil content increasing almost linearly from 14°C to a plateau at 28°C (from 14% to 22%). In contrast, although protein content was generally less variable, there was a pronounced minimum around 22°C (41%), with similar levels at 16°C and 28°C (43%). However, when the varieties were considered separately, it was seen that this was a 'composite response', with limited biological meaning: for example, the variety Evans, Maturity Group 0, grown at northern latitudes, showed a decrease in protein content ($P < 0.01$) with increasing temperature (16–25°C), compared with an *increase* at mid-latitudes for Forrest (V) ($P < 0.01$) (19–28°C), and very little effect for Hutton (VII) in the southern states (20–27°C) (see Figure 2.9). A significant ($P < 0.01$) promotive effect of temperature on oil content was found for all varieties and sites, but the more northerly crops (Groups 00 to II) were the most responsive.

The results of extensive breeding programmes have established that there is significant genotypic variation in oil and protein contents, but that there is a pronounced inverse relationship between these traits (e.g. Brim and Burton 1979; Burton and Brim 1981; Piper and Boote 1999; Figure 8.4). In the light of the relative stability of the *combined* oil and protein contents (Piper and Boote 1999; Morrison *et al.* 2000), the objectives of a breeding programme, and the choice of variety for a given area, must be to maximise the yield of the economically more important component, at any given time. Recent breeding objectives have included high protein, high yield cultivars (Wilcox and Cavins 1995; Cober and Voldeng 2000); and increased proportions of monounsaturated fatty acid (oleic) at the expense of polyunsaturated (linoleic, linolenic) and saturated (palmitic, stearic) acids (Rahman *et al.* 2001; Streit *et al.* 2001).

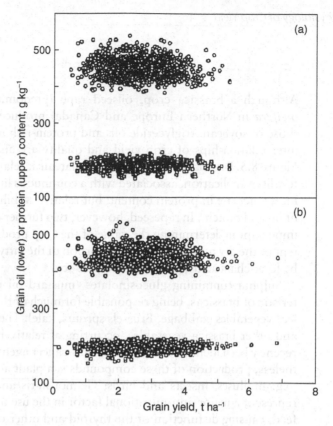

Figure 8.3 Relationships between grain yield and quality (g kg⁻¹ oil or protein) for soybean lines, originating from single (a) or back (b) crossing of a high-protein variety with a high-yield variety, grown under uniform conditions at Ottawa, Canada (from Cober and Voldeng 2000).

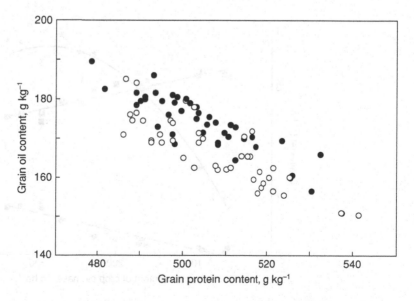

Figure 8.4 The relationship between grain oil and protein content for soybean lines, originating from single (o) or back (●) crossing of a high-protein variety with a high yield variety, grown under uniform conditions at Ottawa, Canada (from Cober and Voldeng 2000).

8.3   Oilseed rape: glucosinolates and erucic acid

Although a brassica crop, oilseed rape (predominantly *Brassica napa* ssp. *oleifera* in Northern Europe and Canada) produces very similar products to those of soybean: triglyceride oils and protein-rich meal for livestock feed. The inter-relationships of crop yield and quality are also similar, as illustrated by Figure 8.5. Dry matter yield reached a plateau at relatively high levels of nitrogen fertiliser application, associated with a continuous linear decline in oil content, a linear increase in protein content, but relative stability of the combined content of oil and protein. In rapeseed, however, two further quality aspects are critically important in determining the value of the crop products: the glucosinolate content of the dry matter, and the proportion of the fatty acids in the oil represented by long-chain acids, particularly erucic acid.

Sulphur-containing glucosinolates (mustard oil thioglucosides) are characteristic of brassicas, being responsible for much of the 'hot' taste of mustard and leaf vegetables (cabbage, Brussels sprouts, kale). The glucosinolates in calabrese and other brassica vegetables, occurring at relatively low concentrations, have recently been identified as potential anti-cancer agents in the human diet. Nevertheless, production of these compounds is a plant adaptation against herbivory (e.g. molluscs, insects and birds; Giamoustaris and Mithen 1995), and they represent a major anti-nutritional factor in the use of rapeseed meal in livestock feed, causing disfunction of the thyroid and other organs of a range of species from poultry to cattle. Such effects can be avoided if grain with glucosinolate levels below 25 µmol g^{-1} is used (Scarisbrick and Ferguson 1995).

Many of the physiological aspects of glucosinolate levels in oilseed rape can be explored through the results of a comprehensive study of crops growing in the

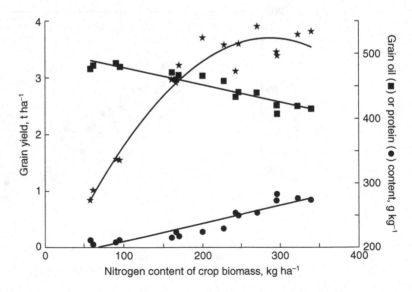

Figure 8.5 The effect of variation in nitrogen status on the interrelationships between crop yield and quality for an oilseed rape crop grown in France (from Triboï and Triboï-Blondel 2002).

north-east of Scotland (Walker and Booth 2003). In varieties with intrinsically high (Rafal) and low (Cobra) levels, there were similar ontogenic trends in the glucosinolate content of plant tissues. The content of vegetative tissues fell progressively during crop growth from around 10 μmol g^{-1} dry matter to around zero at harvest, and this trend was little affected by genotype or crop nutrition. Both varieties had generally higher levels in reproductive tissues, between 10 and 30 μmol g^{-1} in flowers, although pod levels were lower. The marked differences between the varieties were not expressed until the later stages of reproductive development, when Rafal grains contained 48–63 μmol g^{-1} compared with 12–17 μmol g^{-1} for Cobra. Increases in glucosinolate content caused by the application of 32 kg S ha^{-1} were inconsistent, and mainly confined to the early weeks of reproductive development.

In addition to these genetic and ontogenic effects, the glucosinolate content of the grain was also affected by interactions among nitrogen and sulphur application, and site (Figure 8.6). Rafal had higher levels than Cobra at each site and level of nitrogen application, but responded to sulphur application only at the site with low sulphur status. At the low sulphur site there was also a distinct dilution effect associated with high nitrogen application, giving a difference of 15–20 μmol g^{-1} between nitrogen treatments across the full range of sulphur application. By contrast, the glucosinolate content of Cobra showed little response to applied sulphur or nitrogen at either site and, even though the levels generally rose by around 10 μmol g^{-1} between the low and high sulphur status sites, they remained consistently below the threshold for use as animal feed (25 μmol g^{-1}). Associated data showed that the highest grain yield of Cobra at the low sulphur site required a combination of 32 kg S and 150 kg N ha^{-1}.

The results of this and other series of experiments indicate, therefore, that, as long as sulphur deficiency is avoided in this sulphur-demanding crop (20–30 kg S ha^{-1} year^{-1} compared with 5–15 kg S ha^{-1} year^{-1} for cereals), high yield and quality can normally be combined by using appropriate varieties rather than by specific agronomic practices.

However, the crops in Figure 8.6 were grown in a temperate environment without stress. Where oilseed rape crops are exposed to stress, there can be significant effects on grain quality. For example Triboï and Triboï-Blondel (2002) have documented responses of seed components to high temperature that parallel those already established for wheat (Section 8.1): decreased individual grain weight (−17%) was associated with decreased oil content (−12%) and yield per grain (−27%), and increased protein content (+38%). Uniquely, the yield of protein per grain also increased (+15%). Similar responses have been recorded in crops exposed to drought. There can also be significant effects of thermal and water stress on glucosinolate content. For example, in a fully-documented lysimeter experiment in Denmark, Jensen *et al.* (1996) found that, in a cool humid season, the glucosinolate content of the mature grain was generally around 10 μmol g^{-1}, rising to 17 μmol g^{-1} in only one out of four drought treatments. In the following hot sunny season, stressed crops produced grain with glucosinolate levels of up to 24 μmol g^{-1}, increasing above controls whenever leaf water potential fell below −1.5 MPa, and when plants had been exposed to more than 6 days of water stress. Although some of the increase in glucosinolate

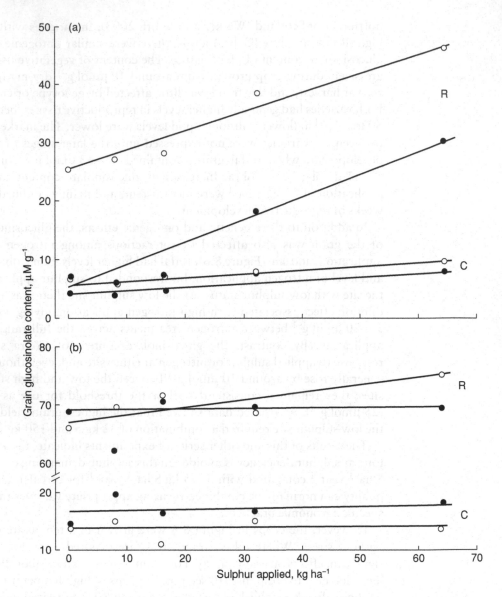

Figure 8.6 The effects of nitrogen (150 ○ or 250 ● kg N ha^{-1}) and sulphur nutrition on the glucosinolate content of two oilseed rape varieties (Rafel R and Cobra C) growing at sites in north-east Scotland of (a) low and (b) high soil sulphur status (from data of Walker and Booth 2003).

level could be attributed to poorer grain filling, there was a contribution from increased synthesis. Thus, even in cool temperate Europe, the quality of oilseed rape grain can be affected significantly by environmental stress.

The position in relation to the erucic acid content of rapeseed is altogether simpler. As dietary exposure of test animals to this 22C monounsaturated fatty acid has indicated that it can cause serious problems with heart function, rapeseed oils for human consumption must contain very low levels. Pioneering work on a series of *Brassica* species revealed that inheritance of the level of erucic acid in the oil was relatively uncomplicated, and the molecular genetics of erucic acid biosynthesis have now been worked out (Lühs *et al.* 1999). As the fatty acid composition of oils is relatively unaffected by environmental factors, the quality of oilseed rape crops for food and feed purposes has, since the 1980s, been achieved by breeding 'double low' varieties (i.e. low glucosinolate and erucic acid). Outside the UK, the predominant 'double low' varieties (erucic acid <2% of fatty acids) for human consumption are generally referred to as 'canola'. Recently, using detailed knowledge of the genetics of fatty acid metabolism in *Brassica* crops, oilseed rape varieties with very high erucic acid contents (>40% of oil) have been bred, particularly for Europe, to provide industrial feedstock for chemical industries.

8.4   Potato: tuber size and processing quality

In contrast to the activities of cereal breeders, whose principal breeding objective during the twentieth century was increased potential grain yield, potato breeders tended to concentrate on the introduction of disease resistance; the aim was to exploit the existing yield potential in the face of a wide range of pathogens, although, more recently, there has been a trend towards breeding for product quality (Caligari 1992). As a consequence, newer genotypes do not show consistent improvement in tuber yield over older varieties (Allen and Scott 1992); vintage varieties such as King Edward, Majestic and Bintje, introduced a century ago, still feature on recommended lists in Europe because of the desirable tastes and textures of their tubers. In general, growers select potato varieties primarily for their tuber quality rather than yield.

Potato crops can provide a wide range of tuber and processed products, depending upon the needs of the consumer. To produce high-grade planting material ('seed tubers'), the starting material, from tissue culture, and subsequent generations of crops in the field, must be managed to generate tubers of the correct size (25–60 mm in the UK) and of high health status (freedom from viral diseases; low levels of bacterial pathogens; and the absence of pests, such as potato cyst nematodes, on the tuber surface). Achievement of such high quality depends upon national certification schemes (e.g. Rennie 2001).

Where the tubers are for direct human consumption, quality criteria can range from shape (determining ease of peeling and degree of waste) and size (the extremes being for canning and baking), through cooking quality (e.g. the texture of the cooked tuber; degree of blackening) and taste, and the absence of diseases and blemishes, to protein content (of particular importance where the potato constitutes a large part of the diet). The increasing proportion of the world crop that is processed by frying (crisps, chips, French fries etc.), must have high dry matter content (minimising the absorption of cooking oil and the energy requirement to remove water) and low levels of reducing sugars (to prevent browning). Tubers for all uses need to be resistant to physical damage during handling and have appropriate storage qualities. All of these are predominantly varietal characteristics, but with important genotype–environment interactions (Storey and Davies 1992). This section concentrates upon two criteria (tuber size distribution and tuber dry matter content), for which the crop physiology is well understood.

As explained in Section 6.8, the size distribution of the tubers harvested from a potato crop is determined by the interactions among genotype (the number of tubers initiated per stem), the physiological age of the seed tubers, stem population density and the date of harvest. By reducing the supply of photosynthate, the incidence of stress can also affect mean tuber size, without affecting the overall pattern of assimilate partitioning (e.g. Figure 6.16). Thus large tubers for baking can be produced by choosing a variety that is less prolific in initiating tubers; using seed tubers that are physiologically young (tendency for apical dominance to be effective) and low planting densities (e.g. a target stem population density of around 10^5 ha^{-1} for Pentland Crown in Figure 6.17a, c); maintaining the crop canopy to maturity; and providing irrigation and mineral nutrition at the appropriate levels to meet the requirements of the crop. In contrast, where a crop is grown to produce tubers for canning (2–5 cm), a more profligate variety, in terms of tuber initiation, must be used (e.g. Maris Peer and Arran Comet in the UK) at a high stem population density (Figure 6.17), and the tubers harvested when the desired size distribution has been achieved.

Tuber dry matter content, normally expressed in terms of specific gravity, is primarily a varietal character. Pioneering work by Burton (1966) established that the mature tubers of traditional varieties could be divided into three classes: high dry matter content (>23%; e.g. Golden Wonder, Record); medium (21–23%; King Edward, Dr McIntosh, Majestic); or low (<20%; Arran Banner and other early varieties), and the ranking of varieties did not appear to vary with environment (Killick and Simmonds 1974). As these values are the culmination of an ontogenic increase in dry matter content (in some cases reaching a peak or plateau before maturity; Wurr and Allen 1974), accompanied by reductions in reducing sugars and sucrose, then tubers harvested before crop maturity will tend to have lower values. This is reflected in the waxy texture of early potatoes after boiling compared with the more floury texture of many maincrop varieties, grown to maturity.

Managing crops to maximise tuber fresh weight yield can lead to a deterioration in dry matter content. In field experiments investigating crop responses to nitrogen fertiliser in Europe, the first increment of nitrogen (50–100 kg N ha^{-1}) is usually sufficient to generate a leaf canopy that intercepts 95% of incoming solar radiation by the time of tuber initiation (Section 3.4.1; Figure 6.15), setting the tuber filling (bulking) rate at its highest attainable value; Section 6.8). Application of further increments, resulting in greater investment of dry matter in higher leaf area indices, can increase the interception of solar radiation slightly by extending the duration of the canopy, but most experiments show that dry matter yield reaches a plateau below 100 kg N ha^{-1} (Figure 8.7; Hay and Walker 1989; Harris 1992). However, it is not uncommon for the *fresh* weight yield of tubers to increase up to 200 kg N ha^{-1} (e.g. up to 150 kg N ha^{-1} in Figure 8.7; and higher rates of application in crops of Cara and Wilja documented by Harris 1992), and the effect can be enhanced by irrigation (e.g. Clutterbuck and Simpson 1978). As this increase in yield above 50–100 kg N ha^{-1} is effectively an increase in tuber water harvested per unit area, albeit associated with higher

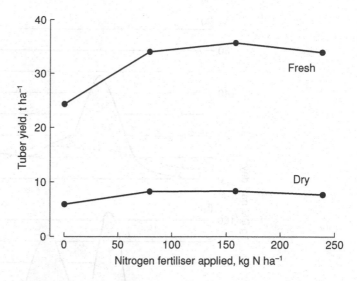

Figure 8.7 Classic potato yield data: the influence of nitrogen fertilisation on the mean fresh and dry weight yields of four varieties grown in the Netherlands in 1961 (adapted from Schippers 1968).

proportions of larger tubers, the grower must evaluate the relative economic importance of higher yield and lower quality. In the light of these well-established findings, and the tendency for tubers of low dry matter content to be susceptible to mechanical damage during handling, it is difficult to understand why farmers in Northern Europe are still being advised routinely to apply 150–220 kg N ha^{-1} to ware crops.

8.5   The quality of conserved forages: ontogeny and yield

Although the original ruminants adopted by man for agricultural production were adapted to grazing mixed pastures dominated by grasses, the diet in most intensive systems now involves at least a proportion of conserved forage, either as hay or silage. In the case of feedlot production in North America, the separation of livestock rearing from forage production is complete, but even in temperate systems based on grazing, the lack of synchrony between the seasonal pattern of dry matter supply by the sward (e.g. Figure 2.13), and demand by the livestock ensures that conservation must play a part in the annual cycle of nutrition. For example, in each of the (northern hemisphere) case histories shown in Figure 8.8, conservation of surplus dry matter produced during May to September would be necessary to supply demand in November to March when sward growth was inhibited by low temperatures and short days.

Evaluation of the quality of forage, to meet the many dietary requirements of ruminant livestock, is complex, and falls within the discipline of animal nutrition. Furthermore, the ultimate quality of the product depends upon the processes of drying or ensilage, and storage. Nevertheless, two important components of quality ('digestibility' – the proportion of the dry matter that can be

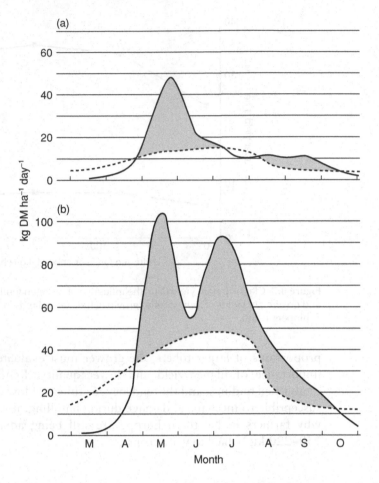

Figure 8.8 Relationships between grazing supply and demand: the dry matter production of irrigated swards of S23 perennial ryegrass (continuous lines) and the dry matter requirement of grazing ewes with twin lambs (broken lines). Model case histories from the UK: (a) sward receiving 173 kg N ha^{-1}, grazed by 3.34 ewes ha^{-1}; (b) sward receiving 690 kg N ha^{-1}, 10.9 ewes. The shaded areas indicate production that is surplus to current demand (from Spedding 1971).

metabolised by the animal; and nitrogen/protein content), are determined largely by sward physiology, notably the now-familiar inverse relationship between quantity and quality, and the relationships with crop ontogeny.

These features can be illustrated by data from perennial ryegrass, the principal species of sown swards in the UK and other parts of temperate Europe. The quality of the harvested herbage, the raw material for conservation, depends primarily on the proportions of total dry matter from cell contents (all readily metabolised) and cell walls (a proportion only, principally cellulose, digested and metabolised with the aid of rumen bacteria). Figure 8.9a shows that there is a progressive ontogenic deterioration in the digestibility of the herbage, as the proportion of cell wall material increases, reflecting the development of leaf sheaths and true stem tissues. This caused a 25% deterioration in the digestibilty

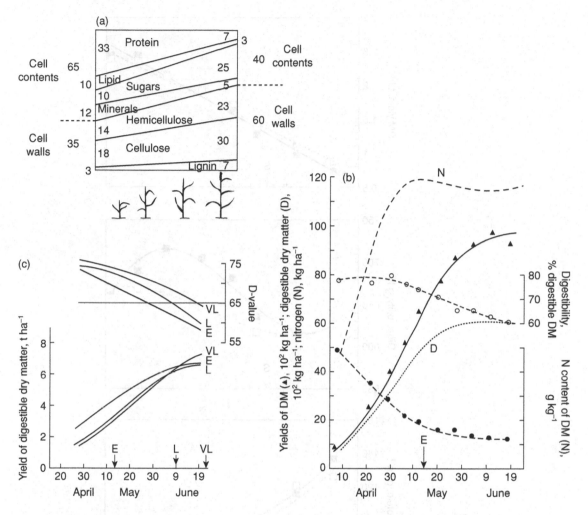

Figure 8.9 Seasonal tends in the quality of herbage produced by swards of perennial ryegrass in the UK: (a) schematic presentation of the ontogenic trends in herbage composition; (b) seasonal pattern of dry matter production and herbage quality for an early heading variety (at E); and (c) representative relationships between herbage digestibility and dry matter yield for early (E), late (L) and very late (VL) heading varieties (adapted from Spedding 1971; Beever *et al.* 2000).

(% digestible dry matter) of the herbage shown in Figure 8.9b but, because of the rapid increase in total dry matter production, the maximum yield of digestible dry matter from this sward occurred shortly after ear emergence (heading). Figure 8.9c presents a series of inverse relationships between quality (digestibility here expressed as the D-value; Beever *et al.* 2000) for varieties of contrasting heading date.

This analysis provides useful tools in the optimisation of sward management for conservation, but Figure 8.9 shows that concentration on dry matter production may compromise other quality attributes. In particular, where swards receive a single application of nitrogen fertiliser at the start of rapid growth in spring (e.g. Figure 8.9b), most of the uptake of nitrogen takes place well before

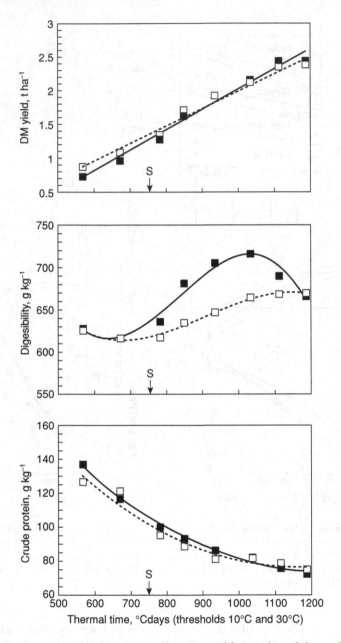

Figure 8.10 Ontogenic relationships among dry matter yield, crop digestibility and protein content in plots of forage maize (means of four hybrids, and two seasons) grown under intensive management (200 kg N ha^{-1}) in Wisconsin. The filled symbols indicate fresh material, the open symbols, after ensiling; and S indicates the date of 50% silk emergence (adapted from Darby and Lauer 2002).

heading, with the result that the total yield of nitrogen thereafter remains constant or decreases (see also the protein yield in Figure 8.9a). This results in a steady decline in the quality of the herbage, in terms of nitrogen content (for example throughout April, May and June in Figure 8.9b). There is, therefore, a conflict between management for maximum quantity of digestible dry matter or for protein content; concentration on the former may require the livestock farmer to enhance the conserved forage with protein concentrate.

Production of forage maize rather than the traditional pasture grasses for silage production has expanded rapidly in North America and Europe over the last two decades, partly because of the very high yields that are attainable (10–25 t DM ha^{-1}) but also because high yield is not linked to lower quality. For example, in the high yielding crop documented in Figure 8.10, the digestibility of the dry matter produced actually *increased* after silk emergence, and improvement continued almost to crop maturity. This pattern, which contrasts with that of pasture grasses, arises because the highly digestible ear constitutes an increasing proportion of the crop biomass as grain filling progresses; this is confirmed in the case of the crop in Figure 8.10 by the observation that the digestibility of the stover (stem plus remaining leaves) declined over this phase from 630 to 510 g kg^{-1} (Darby and Lauer 2002). Nevertheless, as for other forage crops, increase in biomass was associated with a decline in quality as measured by protein content (Figure 8.10). Lauer *et al.* (2001) have shown that, in common with a range of grain crops, the substantial genetic advances in forage maize yield over the twentieth century have been associated with a dilution of the protein content of the product, when grown under identical conditions.

In summary, this chapter has demonstrated that the plant and crop physiologist can provide a wide range of characters to assist in the breeding and selection of varieties that are superior in quality as well as in dry matter yield.

Chapter 9
The simulation modelling of crops

The invention of deliberately oversimplified theories is one of the major techniques of science, particularly of the 'exact' sciences, which make extensive use of mathematical analysis.

(Williams 1954)

9.1 Introduction

Humans have an intuitive ability and a need to make abstract or conceptual models of their observations and experiences. Such models do not substitute for reality but permit sense and predictions to be made of the world and beyond. Such a mental facility allows humans to understand, forecast and decide different actions given a range of current circumstances. This ability and the use of extensive verbal communication is the most important characteristic that identifies humans as conscious beings. The ability to reason and theorise is with humans from the earliest months of life. As they grow, models of the world are ever more important as a way of making sense of experience – hence a child's fascination with toys and games. Scaled representations of real things, possessing structure and form, can be manipulated and made to interact. Scientists and engineers use various sorts of model as tools to help them understand the phenomena they are studying. Physical models, such as a scaled-down representation of an aircraft in a wind tunnel, can mimic aspects of a real aircraft in flight and thus help in improving design. The construction of a scaled model, using chemical retort stands for support and cut out pieces of cardboard in the shape of the nucleotide bases, was central to revealing the three-dimensional structure of DNA and the birth of modern biotechnology.

Certain phenomena are made abstract in non-physical or mathematical models. For the smallest quantum and the largest parsec levels of space and time, we are unable to make physical models of events or structures that inhabit such domains. Here we resort to purely conceptual models, most often formulated in mathematics. The simplified representations of reality found within models are intrinsic to science, since conceptual models are synonymous with hypotheses that are at the intellectual cornerstone of science. Simulation models are logically and quantitatively constructed series of justifiable beliefs about how a system works (Porter 1985a). Not all conceptual models are quantitative; however, models gain an added significance when they provide quantitative, and thus mathematically formulated, descriptions of processes. Simulation models are also reductionistic; that is, they seek to represent the explanation for an event or process at a more detailed level than the event or process itself. However, they are also integrationist in that they reconstruct a picture of the behaviour of the whole system or organism including the important interactions between the separate parts. To be a scientist requires having a model in mind.

Simulation models also provide a multifaceted idea of causality; that is, of an effect having a number of possible causes and, conversely, of a single cause having many effects. A simple crop example of an 'effect with many causes' would be the reduction in leaf growth rate resulting from a shortage of water or nitrogen or both. A 'cause leading to more than one effect' would, for example, be an increase in growing temperature leading to an increase in respiration rate, a decrease or increase in life-cycle length and an increase or decrease in photosynthesis rate. The notion of a single limiting factor is thereby replaced by the idea of a sequence and network of different limitations operating through a crop's lifecycle, or of many causes having the same effect.

Quantitatively describing environmental effects on plant physiological processes is the basis of simulation models of crop production. However, the action of abiotic factors on developmental and growth processes differs markedly (Section 2.1). Growth is an irreversible increase in dry matter resulting from disequilibrium between the accumulation and loss of environmental resources. In plants, this balance is affected much more by the total amount of radiation received, than by for instance, temperature or the spectral composition of light. Temperature serves more to alter rates of metabolic change within a plant, by influencing enzymatic activity or membrane behaviour, rather than by making plants larger *per se* (Landsberg 1975). Temperature has a large effect on rates of respiration and the translocation of assimilates but much less an effect on the rate of photosynthesis at low light levels and even only a weak effect at light saturation (Grace 1989). Conversely, plant development is heavily dependent on both high and low temperatures that control the rate of development and the switch from a vegetative to a reproductive state (Chujo 1966). Similarly, light spectral composition, light quality, affects developmental rather than growth processes. At the whole-plant level, development can involve changes in the number of or dimensions of organs but not their weight (e.g. leaves, Harper 1977) or the time taken to morphological events, for example, flowering. Thus, cereal development is measured *via* the number of leaves produced (Section 3.2.2) as is similarly, plant senescence and the onset of dormancy as leaves die. In cereals we can also define stages of the development of the cereal apex (Tables 2.1, 2.2) thus enabling estimates of the rate of plant development to be made by measuring the time taken to move between consecutive apical stages. Models of crop growth need to describe both growth and development.

As seen in the earlier parts of this book, the interlinked processes that form the basis of crop growth and yield are phenological development, leaf canopy development, biomass production and its partitioning (Chapters 2 to 6). The degree to which the last three processes approach their potential rates is governed mainly by the availability of water and nitrogen (Chapter 7). Combined genetic, environmental and management factors drive such processes. Thus, crop cultivar, sowing date and seed population, fertiliser and irrigation levels are largely under management control. The factors listed under environment are largely incapable of being affected by the actions of a farmer, although temperature may be the exception to this general statement. Figure 9.1 shows which factors affect which processes in a qualitative sense.

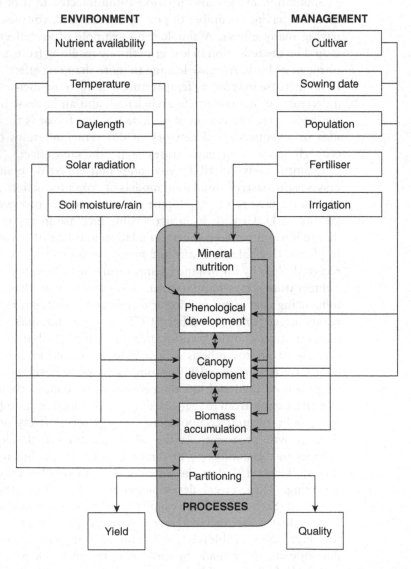

Figure 9.1 Environmental and management factors affecting the processes of growth and development of crops leading to yield and its quality.

The relative importance of environment and management was considered by Monteith (1981b) in relation to the question of climatic variation and the growth of crops. Monteith calculated the contribution from environmental and management factors to the between-year yield variation of winter- and spring-sown crops grown in eastern England. He concluded that the two largest climatic causes of variation in yield were temperature and rainfall and their independent effects were three to four times larger than those caused by variation in how much light was incident on crops. For winter cereals on heavy soils, 12% of their yield variation stemmed from variation in temperature, radiation and rainfall, with the value rising to 17% on more sandy soils. Therefore, other causes of yield variation, notably management, contribute much more to yield variability than environment – climate change not withstanding. It is the goal of simulation modelling to describe such genotype × environment × management (GEM) interactions quantitatively and thereby allow examination of their interactions.

9.2   Building a crop model

The concept of a rate, and a rate process, is a central concept in models and means how a quantity changes over time, such as weight or development stage. Therefore, rates are analogous to speed – that is the rate at which distance is traversed. The units of a rate, by definition, include a 'per unit time' element such as s^{-1} or d^{-1} and so forth. When rates are integrated over time they yield states, such as the total dry weight of a crop at a given time. The simplest model that could be constructed would be one in which it was assumed that the absolute rate of change of dry weight (dW/dt) of a crop was linear and constant and that the total weight was the integral of the rate over time. Representative values for a growth rate for crops growing under temperate conditions are in the range 10–$30 \text{ g m}^{-2} \text{ d}^{-1}$ (Evans 1998). The question then arises over the length of the time-period or integral for growth. The simplest value for a period of growth would be a number of days from sowing (s) to harvest (h). However, as has been seen earlier (Chapter 2), the time taken from one phenological stage to another is more similar between a range of sowing dates or sites when thermal time rather than number of days is used to express the interval. Thus in Equation 9.1, growth rate (dW) is integrated between the thermal time (Tt) from sowing (s) to harvest (h).

$$W = \int_{Tt=s}^{Tt=h} dW \cdot dt \qquad (9.1)$$

The dW term in Equation 9.1 can be disassembled into a term describing the amount of photosynthetically active radiation incident on the crop (Q; Chapter 3) and the efficiency of use of this radiation to produce dry matter (RUE; Chapters 3 and 4) as in Equation 9.2.

$$W = \int_{Tt=s}^{Tt=h} Q \cdot RUE \cdot dt \qquad (9.2)$$

A leaf canopy is needed first to intercept and then absorb the radiation and this is a function of the leaf area index of the crop (Chapter 3) and is designated $f(L)$, leading to:

$$W = \int_{Tt=s}^{Tt=h} Q \cdot RUE \cdot f(L) \cdot dt \qquad (9.3)$$

The dry matter produced is, at its most basic, partitioned between the harvested and non-harvested fractions and, if the harvested fraction is denoted as $g(W)$, then the harvested yield (Y) is given by:

$$Y = \int_{Tt=s}^{Tt=h} Q \cdot RUE \cdot f(L) \times g(W) \cdot dt \qquad (9.4)$$

Equation 9.4 contains the main elements of a crop simulation model to which many layers of detail can be added. For example, further sophistication and detail can be added by substituting RUE by mathematical descriptions of photosynthesis and respiration (Chapters 4 and 5) and by including constraints to production caused by suboptimal supplies and uptake of water and nutrients (Chapter 7) and the presence of pests, diseases and competing plants. Single crop models can also be combined to examine crops grown in important rotational systems, such as rice and wheat (Aggarwal *et al.* 2000) or in the presence of pollutants such as ozone (Ewert and Porter 2000).

A crucial notion in choosing either to use or to develop a model is balance. Balance means that the model should be sufficiently but not overly detailed for the question that is to be addressed. The different processes in the model, such as canopy and root development, should be described in broadly the same level of detail. A useful principle is that the fastest model elements should have rates that are between 10^2 and 10^3 times faster than the slowest model elements. Thus for a simulation model that simulates seasonal crop growth over a period of perhaps 300 days, a one-hour to one-day time-step for the most detailed processes is appropriate. For a soil carbon and nitrogen model, in which some of the organic matter pools have residence times of tens of years, monthly time-steps are more suitable. Such integration constraints are also seen within the biological hierarchies of cell ↔ tissue ↔ organ ↔ individual ↔ population ↔ community ↔ ecosystem ↔ landscape. It is almost impossible to make meaningful connections between the workings of a cell and that of an ecosystem, but it is possible to connect leaves as organs to crops as a population of plants. A second important concept in modelling is whether the availability of carbon and limiting factors such as nitrogen and water limit growth or whether the number and activity of sinks for assimilate is a restriction. The former is termed a source-driven and the latter a sink-driven limitation. As an example, leaf area development can be sink limited just after emergence because the initial leaves are small; later limitations to leaf area production can occur because the sinks that are formed cannot be supplied with sufficient C and/or N to satisfy their environmentally determined potential demand (Section 6.2 and Chapter 7).

Developing a crop or cropping system simulation model can be very intellectually demanding but also rewarding since, in the process, one is forced to state qualitative and quantitative hypotheses about how a crop develops and grows. Once the model is constructed, it must be tested against observations and experiments independent of those used to develop the model. The ability to predict novel observable outcomes based on the mechanistic understanding of phenomena is one of the most powerful means by which science advances knowledge. An example on the scientifically grand scale is the nineteenth-century Russian chemist Mendeleyev, who predicted the existence and the physico-chemical properties of as then undiscovered elements in his Periodic Table because he understood how such properties were linked to an element's position in the Table. At a much less grand level, simulation models enable predictions of crop behaviour to be made for new conditions of elevated CO_2, drought, nitrogen application and timings, weed competition and their interactions.

Constructing a crop simulation model requires a deep knowledge of the literature, both to understand processes and to derive numerical values for parameters that are used in the model's equations. The most important assets needed to build a simulation model are clarity of thought, an ability to synthesise a simple but not simplistic framework for the system of interest, knowledge of the literature and the essential engagement of the modeller in experimental work. There can be no simulation without experimentation. Mathematical and computing sophistication are of little importance in the development of models, although they may have a role in how models are presented to users. There is no substitute for constructing a model in order to appreciate the intellectual processes involved. However, it is possible to give guidance as to what requires consideration during the modelling process and, for this, the reader is referred to the paper by Wilkerson *et al.* (1983) that describes the development of a model for soybean growth. The next section will examine the workings of models for wheat, maize and soybean.

9.3   Crop models of wheat (AFRC2), soybean (CROPGRO) and maize

The crop models for wheat (*Triticum aestivum* L.), soybean (*Glycine max* L.) and maize (*Zea mays* L.) to be described below share some features. They each describe the growth, development and yield of monoculture field crops, have major time-steps of one day, and they include modifications to growth, development and yield caused by water and nitrogen shortage. These three models were chosen to illustrate the range of methods used to assemble crop models for three of the four most globally significant grain crops. The described models have been used extensively in studies of the effects of weather, climate and management on crop productivity per unit area. Many models can and have been used for the purposes of simulation and the reader is encouraged to use them with the goal of understanding the ways in which models describe the crop processes dealt with in the earlier chapters of this book.

The principal references for the three described models are:

- AFRC2: Porter 1984; Weir *et al.* 1984, 1985; Porter 1993; Jamieson *et al.* 1998b;
- CROPGRO: Wilkerson *et al.* 1983; Boote *et al.* 1998;
- Maize: Muchow *et al.* 1990; Muchow and Sinclair 1991; Sinclair and Muchow 1995.

9.3.1   The AFRC2 wheat model

AFRC2 is a wheat simulation model, the development of which was started as a joint research activity between four research institutes in the former Agricultural and Food Research Council of the UK. An initial version of the model (ARCWHEAT1) was described in Hay and Walker (1989) based on the work of Porter (1984) and (Weir *et al.* 1984, 1985). The model originally comprised five sub-models that described the phenology, leaf area development and growth of a wheat crop in which it was assumed that water and nitrogen were in optimal supply. Porter (1993) presented a version of the model in which the availability of water and nitrogen could affect simulated growth and development. Ewert and Porter (2000) extended the list of influences modifying growth and development to include elevated CO_2 and ozone (O_3) levels, both as separate and as interacting factors. The version below is based on that described by Porter (1993).

In addition to the original five sub-models described by Porter (1984) and Weir *et al.* (1984), two further sub-models were added (Porter 1993) to describe water and nitrogen movement in the soil, their uptake by a wheat crop and consequent effects on growth, leaf area development and yield. The weather data, sometimes referred to as driving variables, required to run the model are the daily maximum and minimum air temperature, daily wet- and dry-bulb temperatures or relative humidity, daily net short-wave radiation and daily precipitation. Starting conditions for the soil are defined in terms of the volumetric water content for 40 soil layers each 5 cm thick. Water content in each layer varies between a drained upper limit (−0.05 MPa), a middle limit (−2 MPa) and a lower limit (−15 MPa), below which roots cannot extract water (Chapter 7). Initial soil water values per layer have also to be given. Similarly, the amount of ammonium (NH_4) and nitrate (NO_3) nitrogen in the different layers and the date and amount of artificial fertiliser applications are input. The model is represented schematically in Figure 9.2.

Figure 9.2 shows that within the framework of the crop lifecycle, defined by the phenological development model, the model simulates the production of tillers and leaves and thus a canopy of photosynthesising leaves and a population of ear bearing shoots. The rate of photosynthesis is calculated from the amount of photosynthetically active radiation intercepted by the canopy, a photosynthetic light-response curve, with respiration subtracted to give a total daily increase in dry weight as carbohydrate. The resulting net dry matter is partitioned between roots, leaves, shoots and, later, ears, in percentage proportions that vary with stage of phenological development. The upward arrows in Figure 9.2 indicate that, in the model, shoot numbers, individual leaf areas, and thus crop canopy area, are reduced by suboptimal nitrogen levels (Chapter 7). At lower levels of

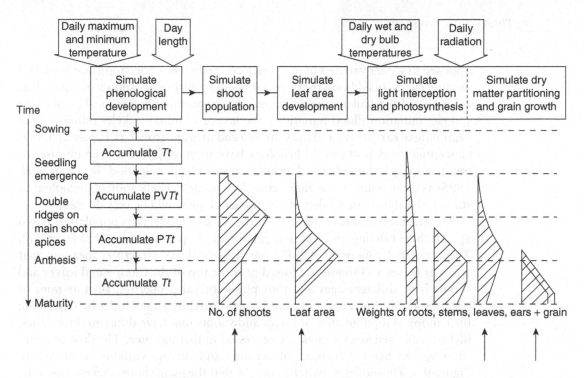

Figure 9.2 Diagrammatic representation of the AFRC2 wheat simulation model showing the interactions between the sub-models: phenological development, tiller and leaf growth, root growth, light interception and photosynthesis, dry-matter partitioning and grain growth. The arrowed boxes show the weather driving variables and the arrows at the bottom indicate the crop processes that are modelled as being affected by shortage of water and/or nitrogen (after Weir *et al.* 1984). Tt, thermal time; P V Tt, photo-vernal thermal time; P I t, photothermal time.

crop nitrogen status, maximum rates of photosynthesis are reduced (Chapter 4) and the same general pattern of suboptimal growth is induced by water shortage.

The timing of the major developmental events or phenological stages, and therefore the duration of the intervening phases, is determined mainly by accumulated temperature or thermal time (Sections 2.1.3 and 2.2.2). In AFRC2, thermal time is calculated as accumulated mean daily temperature above a base temperature of 1°C until anthesis and 9°C thereafter until crop maturity, with the condition that temperatures above 35°C or below −15°C can damage wheat development (Porter and Gawith 1999). Strictly, thermal time should be measured in Kelvin (K) since this is an absolute temperature scale and can be integrated, whereas °C measures the difference in enthalpy or heat content, expressed as temperature, between two objects (i.e. freezing and boiling water). However, this physical nuance has been sacrificed to the norm of using °Cday as the unit for time-integrated temperature. For the period between crop emergence and the appearance of double ridges (Sections 2.1.3 and 2.2.2), the rate of thermal time accumulation is reduced by lack of crop vernalisation and by photoperiods that are less than a defined optimal value. The accumulation of thermal time is thereby modified to give photo-vernal thermal or photo-thermal time (see Weir *et al.* 1984 for further details). In other words, the fastest rate of development, and thus the shortest time interval between emergence and double ridges, will occur in those varieties that have no vernalisation and photoperiod response (daylength neutral) or where photoperiods are longer than the optimal photoperiod of about 14 h for wheat. The faster development of spring as opposed to winter varieties of temperate cereals can be understood and modelled in these terms. In addition to the major phenological stages shown in Figure 9.2, the model also calculates the timing of other important phenological events: first spikelet initiation (floral initiation), the last or terminal spikelet initiation, the beginning of ear growth and the start and end of grain growth (Chapter 2). Other conceptual models of cereal phenology have been proposed that emphasise the importance of leaf number and its interaction with photo-period (Brooking *et al.* 1995) as the measure of vegetative crop development. Unification of phenological stage and leaf number models of cereal phenology would be very desirable.

A particularly distinctive feature of the AFRC2 model is its population-based approach to tillering and leaf and canopy development (Porter 1984, 1985b). The simulated canopy in AFRC2 has a vertical age structure; meaning that younger leaves and shoots are found near the top of the canopy and leaves and shoots have different ages and thus physiological properties, such as rates of photosynthesis. Such age-structure is introduced into the model by describing the canopy as a population of leaves and shoots that have different birth times, lifetimes and senescence rates, all expressed in thermal time. The flow diagram showing the main decisions, calculations and driving variables is shown in Figure 9.3. The model is 'switched on', so that the main shoots emerge when the development model has calculated that crop emergence has occurred. Leaves appear on the main shoot at intervals in thermal time, the length of which is calculated from the rate and direction of change of daylength at emergence. The correlation between rate of change of daylength and leaf emergence was first

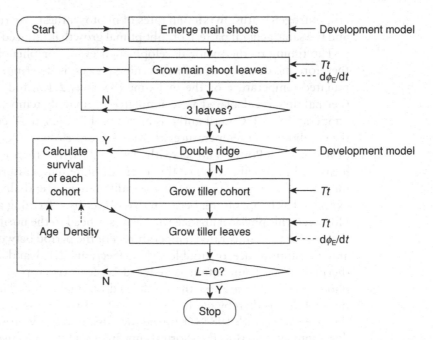

Figure 9.3 Flow diagram of the canopy development model in AFRC2. Solid lines indicate main driving variables in a calculation; broken lines indicate modifiers. Abbreviations: $d\phi_E/dt$, rate of change of daylength at crop emergence; L, green leaf area index; Y, yes; N, no; Tt, thermal time (Porter 1984).

shown by Baker *et al.* (1980) and has subsequently been found to be a robust if non-mechanistic way to predict leaf appearance rate and leaf number. New leaf appearance stops with the flag leaf, which is the last leaf to reach full size by the time of anthesis, once the phenology model has passed the information that this has occurred.

After the appearance of the third leaf on the main shoot, groups (cohorts) of tillers are produced each week. A tiller is a side shoot that grows from the axilliary bud of its parent leaf (Chapter 2). A cohort, a term borrowed from population biology, is any group of individuals, in this case tillers, born in the same time period (Harper 1977). Shoot number within a cohort increases if more leaves are initiated as there are more leaf axils. Leaf number increases with temperature and thus thermal time and, therefore, the number of shoots per cohort is calculated as number of shoots m^{-2} °Cday^{-1} and is set at 2.6 shoots m^{-2} °Cday^{-1}. This value was taken from observations of crops grown with high levels of nitrogen, adequate water and protection against disease (Thorne and Taylor 1980).

Tiller production, in the model, stops when the main shoot reaches the double-ridge stage and from then until anthesis, some tillers die. Those that survive are considered to develop ears. The surviving proportion of a cohort after double ridges is calculated from an equation that contains the following factors important in determining whether a shoot survives to produce an ear, or not. A tiller has a higher chance of dying if it is young at the time of flowering, that is to say it was born just before the double-ridge termination of shoot production, and also if it is in a high-density shoot population. The equation used in Porter (1984) is flexible enough to account for both the age and density effects, includes the most important population-regulating influences and produces proportional tiller-survival curves of the form seen in Figure 9.4.

The model allows for subtle interactions between phenology, tiller population dynamics and leaf and canopy leaf area production. For example, the timing of the double-ridge stage is influenced by the temperature, photoperiod and vernalisation sensitivities of a particular variety, characters that seem to be under strong genetic control. The number of cohorts and thus leaf area depends on the length of the period, between the three-leaf stage and double ridges, because one tiller cohort is produced per week. An alternative means of timing the appearance of shoots could be by leaf number, since a new leaf on the main shoot means the possibility to produce shoots from buds in the leaf axils. The maximum number of leaves produced by a shoot is linked to the environment through the interplay between the thermal time per leaf, which is set at emergence in the model and the thermal time to anthesis. Weather factors such as temperature between emergence and anthesis can influence the dynamics of leaf canopy construction and thereby radiation interception and dry matter production. Other factors affecting the size of the canopy leaf area are the thermal time taken for leaves to reach full expansion, the length of time from then until the start of their senescence and their rate of senescence. The modelled cycle of leaf production, growth and death continues until the expansion of the flag leaves, which occurs synchronously on all shoots. As no leaves are produced after this point, the green leaf canopy declines steadily throughout the grain-filling period and reaches zero

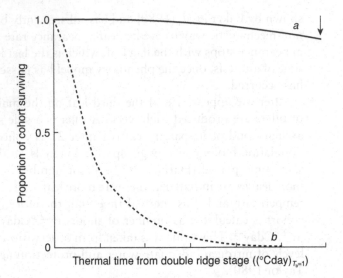

Figure 9.4 The simulated effect of the age of a cohort and the shoot population density per m²
on the proportion of shoots in the cohort surviving from the double ridge stage to anthesis.
a, cohort produced early in the life of a crop with a crop density of 200 shoots per m²;
b, a later produced cohort for a crop density of 1600 shoots per m² (Porter 1984).
T_b is the base temperature (°C) above which thermal time is accumulated.

green area at about crop maturity. The dynamics of leaf canopy expansion for various management and environmental factors are laid out in Section 3.3.

The canopy model developed in AFRC2 is detailed and well suited to studying the effect of combinations of temperature, photoperiod and other environmental influences on leaf area. Later, it will be shown how lack of water and nitrogen influence the population dynamics and growth of shoots and leaves and how a leaf canopy can be reduced in size and duration. A simpler version of the AFRC2 canopy model has been proposed recently (Lawless *et al.* 2005); this version models leaf area development in terms of layers of leaf area added with the emergence of a new leaf.

In AFRC2, as in other models, the sub-models of phenological and canopy development come together to estimate the duration and size of the absorbing leaf surface that intercepts radiant energy throughout the growing season. Radiant energy is used to drive the initial stages of photosynthesis (Chapter 4) and the evapotranspiration of water from a crop. The diurnal variation in incident PAR (Chapter 1) at an hourly scale is calculated from the daily values of net short-wave radiation using an equation that describes a sine-wave distribution of radiation level during daylight. The mean PAR incident on leaf surfaces within a horizontal layer of the canopy is calculated for each canopy layer and hour during daylight using Equation 9.5:

$$I(z) = k/(1-m)\, I_0 e^{kL(z)} \tag{9.5}$$

where $I(z)$ is the PAR incident on the leaf surfaces at level z, the mid-point of a canopy layer, with $z = 0$ at the top of the canopy. I_0 is the amount of incident PAR at the top of the canopy and $L(z)$ is the cumulative green leaf area index from the top of the canopy to level z; k is an extinction coefficient and m is a leaf transmission coefficient for PAR, with a typical value of 0.1, meaning that 90% of light incident on a leaf is either absorbed or reflected and 10% (i.e. 0.1) is transmitted through. In the AFRC2 model, L is divided into layers of thickness of 1 m^2 leaf m^{-2} ground. Equation 9.5 is clearly the Monsi–Saeki equation (Section 3.4.4) modified to include leaf transmission. This is necessary because the amount of radiation incident on a leaf is greater than the amount absorbed and photosynthesis is calculated in the AFRC2 model based on absorbed PAR. It should be noted that the amount of available energy to drive evapotranspiration is the total short-wave and not PAR.

The response of photosynthetic rate, P (net or gross) to irradiance $I(z)$ has often been modelled by an equation of the form:

$$P = \alpha I(z) P_{max}/(\alpha I(z) + P_{max}) \qquad (9.6)$$

where α is the quantum efficiency and P_{max} is the value of P at saturating PAR (Chapter 4). Equation 9.6 describes a rectangular hyperbola and its general shape is shown in Figure 9.5.

Leaves of crops grown in the field have a different response to PAR from those in controlled environmental conditions. The main difference is that the light response curve for field-grown crops has a sharper 'shoulder' than that shown in Figure 9.5. This is included in the AFRC2 model by using the quadratic Equation 9.7:

$$\theta P_g^2 - P_g[P_{max} + \alpha I(z)] + \alpha I(z)P_{max} = 0 \qquad (9.7)$$

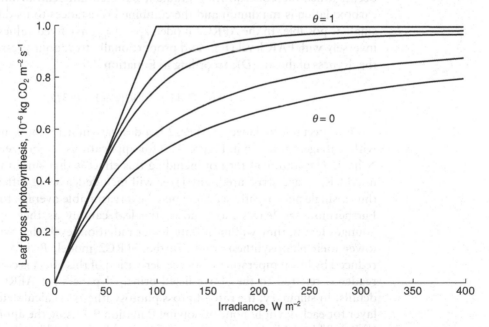

Figure 9.5 Photosynthetic light response curve of a leaf according to Equation 9.6 (after Thornley and Johnson 1990). The meaning of θ is discussed in relation to Equation 9.7.

where P_g is the rate of photosynthesis net of photorespiration and θ is the ratio of the physical resistances to CO_2 diffusion to the total resistance, that is the sum of the physical and biochemical resistances (Chapter 4). As θ approaches zero, Equation 9.7 reduces to the rectangular hyperbola with a smooth 'shoulder' described by Equation 9.6. This describes a situation in which the biochemical resistances are much larger than the physical ones as might be found in the regulated conditions of a glasshouse. As θ increases towards unity, the physical resistances become predominant as might be expected in a crop growing in the field and subjected to still air within the canopy, warm dry conditions and higher radiation levels than can be achieved in growth chambers. The sharp transition from a linear light-limited to a linear light-saturated photosynthetic rate that results when θ is close to unity is known as the Blackman response, after the physiologist G E Blackman who first identified it. From Figure 9.6c, it can be seen that the two straight lines or Blackman-type response describe field measurements (Figure 9.6a) of winter wheat photosynthesis more closely than the rectangular hyperbola (Figure 9.6b) and field measurements of θ were found to be from 0.85 to 1.0, showing no trend with leaf age (Marshall and Biscoe 1980). The dimensionless value used for θ in AFRC2 is 0.995 and α is given a value of 0.009 mol CO_2 mol^{-1} photons.

The maximum photosynthesis rate, P_{max}, is calculated as:

$$P_{max} = \theta c_a / (r_a + r_s + r_m) \qquad (9.8)$$

where c_a is the ambient CO_2 concentration and the resistance terms are as defined in Chapter 4 and, as explained therein, the maximum rate of CO_2 influx occurs when the concentration gradient between ambient air and the sites of carboxylation is maximum and the combined resistances to its diffusion are as small as possible. In the AFRC2 model, r_a and r_m have fixed values and r_s varies inversely with PAR level ($I(z)$) and proportionally to vapour pressure deficit, or the dryness of the air (D), according to Equation 9.9:

$$r_s = 1.56 \times 75[1 + 100/I(z)](1 - 0.3D) \qquad (9.9)$$

One effect not included in AFRC2 is a decrease in net photosynthesis or P_{max} with leaf age, as shown in Figure 4.20 for applications of both zero and 151 kg N ha^{-1}. The argument for not including this effect is that since the leaf canopy in AFCR2 is age-structured, any layer will include leaves of different ages and thus a single photosynthesis value may be a reasonable average for a leaf layer. Furthermore, as leaves move down the leaf canopy as they age relative to younger leaves, they will move into lower radiation levels that will also clearly lower their photosynthesis rates. In the AFRC2 model, P_g in Equation 9.7 is reduced by low temperatures, but the derivation of this effect is complex and the reader is referred to the above listed principal papers for AFRC2 for further details. In summary, the rate of gross photosynthesis is calculated for each leaf layer for each daylight hour by solving Equation 9.7 using the appropriate value of $I(z)$. This yields P_g in units of mg CO_2 m^{-2} leaf s^{-1}, which can be converted to μmol m^{-2} leaf s^{-1}, using information from Chapter 4. The hourly canopy

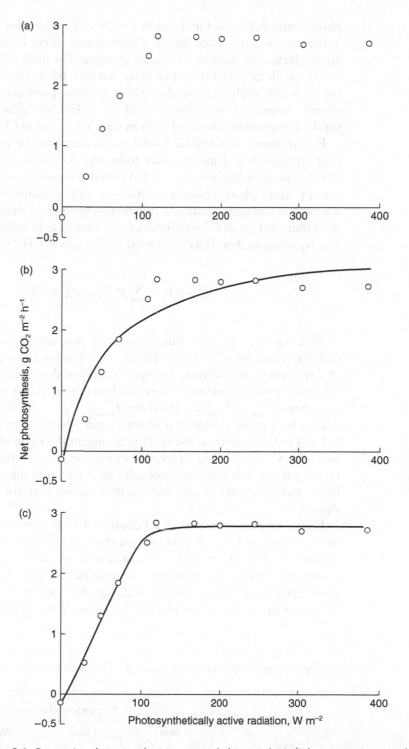

Figure 9.6 Comparison between the (a) measured photosynthetic light response curve of the fully expanded flag leaf of wheat in the field and either (b) a rectangular hyperbola model (Equation 9.6) or (c) a non-rectangular hyperbola (Equation 9.7 with θ equal to 0.995). (Marshall and Biscoe 1980).

photosynthesis is scaled up from this value and the number of whole leaf layers in the canopy. This yields the gross photosynthesis per hour and per m^2 ground area, which is summed to give the daily value. The daily CO_2 fixed is converted to carbohydrate (CH_2O) by multiplication by 0.65, as the approximate ratio of the respective molecular weights. Other models, particularly those for oil and protein crops, such as soybean, need more detailed calculations of metabolic product conversion ratios and costs as seen below and in Chapter 5.

Respiration in the AFRC2 model is calculated as the sum of losses used to support growth and maintenance following the classical analysis of McCree (1974), as described in Chapter 5. Growth respiration is a function of daily canopy gross photosynthesis production and is unaffected by temperature, whereas maintenance respiration increases with crop weight and temperature such that the rate of C loss doubles for a 10°C rise in mean daily temperature. The equation used in AFRC2 for total respiration as CH_2O (R_m) is:

$$R_m = 0.65 \, a \sum_{h=0}^{h=H} P_g(h) + b \, W \, 2^{0.1.T_{mean}} \tag{9.10}$$

This equation assumes that a constant proportion (a) of currently fixed CO_2 fixed each hour (h) during daylight ($h = 0$ to $h = H$) by gross photosynthesis (P_g) is respired. In addition, a proportion (b) of dry weight is lost but this loss increases with temperature. The power function in Equation 9.10 is a short way of writing $[(T_{max} + T_{min})/2 \times 10]$ since $(T_{max} + T_{min})/2$ equals T_{mean}. Experimental studies have given a value of 0.34 for a, and b has a value of 0.002 before, but reduces to 0.001 after, anthesis. This change in the value of b reflects the lower maintenance cost associated with an increasing proportion of the crop's weight represented by relatively metabolically inert starch in the grains as opposed to lower molecular weight metabolic carbohydrates that are found in leaves and shoots.

Daily net photosynthate (P_n; Equation 4.2) is partitioned between leaves, roots, shoots and ears in proportions that change with stage of development (Table 9.1). Initially, most dry weight is allocated to leaves and roots until the double ridge stage when shoots start to take 40% of daily P_n. Following the start of ear growth, 30% of dry matter is allocated to ears until anthesis. At anthesis, 10 mg of ear weight is assumed to be equivalent to one grain, although the

Table 9.1 The partitioning of dry matter between plant parts in the AFRC2 model (after Weir *et al.* 1984).

	Proportions of current assimilate			
	Roots	Leaves	Stems	Ears
Emergence to double ridge	0.35	0.55	0.10	0
Double ridge to beginning ear growth	0.20	0.40	0.40	0
Beginning ear growth to anthesis	0.10	0.30	0.30	0.30

number of grains can be reduced by high threshold temperatures in the period before anthesis to take account of the damaging effects of temperatures above 30°C on anther meiosis (Ferris *et al.* 1998). The simple relationship used to calculate grain number proposes that grain number is positively correlated with total radiation receipts during the ear formation phase but negatively correlated with mean temperature. This is partly because high temperatures accelerate development and thus shorten the number of days for ear formation and also because of higher maintenance respiration losses associated with warmer days. Grain growth, after anthesis, is supported by a pool of assimilate that comprises all current net photosynthate plus a potential 30% of the shoot and leaf weight at anthesis. The period from anthesis to crop maturity is divided into three phases: initiation, linear growth and maturity as an approximation of an S-shaped curve. There is no grain growth during the initiation phase and assimilate is accumulated in the pool. During the linear phase, each grain has the potential to grow at the following temperature-dependent rate (dG_{max}/dt; mg grain^{-1} d^{-1}):

$$dG_{max}/dt = 0.045(T_{mean}) + 0.4 \tag{9.11}$$

On a daily basis, if the production of assimilate is insufficient to meet total grain carbon demand, the product of the maximum grain growth rate and the number of grains, assimilate can be taken from the pool to cover the shortfall. Should the pool empty then grain growth rate will be restricted to that coming from daily assimilation, an amount that will clearly decline as the leaf canopy dies during this period. Thus, grain growth and yield in AFRC2 is modelled as being regulated by either lack of source assimilates or by lack of grain sink demand or by both constraints operating sequentially. A more extensive analysis of grain set, population density and their importance for grain yield is given in Chapter 6.

A necessary simplification when constructing models is to assume, in the first instance, that growth, development and yield are limited only by the amount of radiation received and daily temperature. Although correlated, as days with high radiation levels are generally warm, radiation and temperature can be thought of as having antagonistic effects on growth and yield. Clearly, high radiation receipts are required for high yields but associated high temperatures generally shorten the time during which a field crop is able to absorb light since the crop life cycle is completed more quickly as temperature increases and thus yield is limited. The other two main limitations to dry matter accumulation in field crops come from inadequate supplies of water and nitrogen, the modelling of whose effects will be described next.

It is likely that there will be intense competition for water supplies among crop irrigation, household use, industry and tourism in the future. The potential disruption of regional rainfall patterns by global climate change could also have important implications for the world's staple crops, of which wheat is the most important. Since the 1960s efforts have been made by environmental physicists and crop physiologists to understand and predict, using models, the processes by which water flow through the soil–crop–atmosphere pathway is regulated, and associated effects on growth. Central to these studies have been the notions of water supply to roots and atmospherically driven demand by canopies for water. The balance between these two processes determines the degree to which 'water limited' yield falls below 'potential' yield, as defined above. The most important processes in defining demand for water are its evaporation from soil and its transpiration from leaves, termed evapotranspiration in combination. Conversion of water from a liquid to an evaporative gas at the same temperature requires latent energy that is supplied by solar radiation. Air dryness and the level of resistance to air flow between the canopy and the atmosphere influence the basic energy-driven process. Evaporation from the soil surface in temperate conditions contributes much less to the flux of water vapour of a cropping system because it is usually wet with relatively low radiation receipts when the soil is bare, i.e. in winter. In addition, only a small amount of radiant energy reaches the soil to drive evaporation when the ground is covered by a crop. However in tropical systems the soil is often bare in the dry season with high radiation receipts when the farmers are waiting for the rains to start before they plant, and this leads to high soil-water evaporation rates. In hot dry conditions, irrigated cereal crops can evapotranspire 5–10 mm of water per day, although in cool temperate areas, maximum values of 3–4 mm per day are more typical.

There is a strong association between plant growth and transpiration because the inflow of CO_2 into a leaf (Chapter 4) and the outflow of water (Chapter 7) share much of the same pathway. Crucial steps in the calculation of crop evapotranspiration were the formulation of models of the unrestricted or potential evaporation of water from bare soil and short grass (Penman 1948) and their later extension to conditions where evapotranspiration was reduced by, amongst other factors, drying soils (Monteith 1965; Tanner and Jury 1976). The next stage was to link models of water vapour exchange with the atmosphere to the soil water balance and thence to dry matter production and yield *via* descriptions of plant water status and its effect on canopy expansion and photosynthesis. A fully coupled crop–environment water cycle is then made possible since reduced canopy expansion or photosynthesis will in turn reduce evapotranspiration, and so on.

To simulate the effects of drought on crop production (Section 7.1) requires calculation of both evapotranspiration and the depletion of soil water (Jamieson *et al.* 1998b) and involves four steps: (1) determination of the demand for water imposed on the crop by available energy from solar radiation and the overlying atmosphere; (2) definition of the amount of soil water available to plants from the soil reservoir as the amount held in the soil and the ability of the roots to extract it; (3) calculation of evapotranspiration when the soil water reservoir limits uptake; and (4) calculation of the effects of water shortage on physiological processes such as canopy expansion and photosynthesis.

In a comparison of the performance of five globally used wheat models for drought conditions Jamieson *et al.* (1998b) found that each of the models calculated crop evapotranspiration (E_t) as potential evapotranspiration (E_p: Penman 1948), multiplied by the current solar radiation intercepted by L:

$$E_t = E_p[1 - e^{(-kL)}] \qquad (9.12)$$

Readers will remember that the bracketed part of Equation 9.12 is of a similar form to Equation 9.5 that describes the radiation amount at levels in the crop canopy with k as an extinction coefficient for solar radiation.

The most widely used equation to calculate E_t is that derived by Monteith (1965) in his extension of the Penman equation to account for the aerodynamic resistance of crops. In its full version it can be written as follows:

$$E_t = \frac{sR_n[1 - e^{(-kL)} + \rho C_p/r_a(e^* - e)]}{\lambda[s + \gamma(1 + r_s/r_a)]}$$ (9.13)

If Equation 9.13 is broken down into its separate parts, it states that E_t is increased by the interception of radiation (R_n; J m^{-2} s^{-1}) as described by the first term in the numerator and where s (Pa °C^{-1}) is the slope of the saturated vapour-pressure temperature curve, that measures how saturated vapour pressure changes per unit change in temperature. The second term in the numerator states that E_t will decrease as the difference between saturated vapour pressure (e^*; Pa) and actual vapour pressure at the same temperature (e; Pa), that is the dryness of the air, increases since this increases the aerodynamic resistance (r_a). The air density (ρ; g m^{-3}) and the air heat capacity (C_p; J g^{-1} °C^{-1}) terms convert the resistance term into the same units (Pa °C^{-1} J m^{-2} s^{-1}) as the first expression in the numerator. The denominator in Equation 9.13 states that E_t decreases as the ratio between the within canopy (r_s; s m^{-1}) to the overall aerodynamic resistance (r_a, s m^{-1}) increases. This means, for example, in still conditions, where r_a is large compared with r_s, that E_t will be less than in windy conditions where r_a would have a lower value. The latent heat of vaporisation of water (λ; J g^{-1}) and the psychometric constant (γ; Pa °C^{-1}) are required in the denominator to keep the overall units consistent. It is recommended that readers attempt to understand the Penman–Monteith equation by programming it into a spreadsheet and making trial simulations with it. The Penman–Monteith equation is a stunning example of the application of the principle of conservation of energy to cropping systems in order to calculate the latent heat water flux or, as it is better known, crop transpiration.

In common with other wheat models, AFRC2 reduces E_t if the availability of water from the soil reservoir is depleted because roots are not in soil profile(s) containing sufficient water for unlimited evapotranspiration (Section 7.1). This can occur if the soil profile has dried out by previous cropping or if the roots have not penetrated into wet layers, and both effects are modelled in AFRC2. As a soil dries out so water becomes more difficult to extract and the force per unit area, or pressure, required to overcome the potential of the soil moisture increases. In AFRC2, water in the soil is described as plant available or unavailable and the available fraction is further divided into water that is downwardly mobile in the soil and that which is retained within a layer. Mobile water is that which is held at suctions or water potentials between the saturated moisture content at zero suction and the drained upper limit of −5 kPa; all other water is assumed non-mobile. Plants can extract water when it is held at suctions between zero and −1.5 MPa, known as the lower limit. High levels of soil moisture content would be between 0.35 and 0.40, meaning that 35–40% of the layer volume was water and the relationship between a soil's water content and the suction needed to extract the water depends partly on its relative proportions of clay, sand and silt.

Total water available for extraction within the root zone is calculated as the product of the moisture content of all the layers that have roots in them between the saturated upper limit and the lower limit, an upper limit for root transport of water and the depth of the water in the profile. Thus in the model, water is most difficult to extract from dry and deep layers that contain low root densities. All mobile water and 25% of the retained water in a layer are available for uptake on any day. The actual daily amount of water transpired is calculated as the lower of that set by atmospheric demand and calculated by the Penman–Monteith equation and the profile available water. The ratio between these two amounts is used as an index of crop water stress and is used to restrict canopy expansion and net photosynthesis, described later. The actual water extracted from an individual soil layer is the proportion of the water available within the whole root zone that is available within the layer. This has the effect that water is taken preferentially from wetter layers.

Water may be extracted only from that volume in the soil in which there are roots. Thus, the size of the soil water reservoir is limited by the bottom of the rooting zone and the rate of root vertical extension partially determines the exploration of the reservoir. The root front in AFRC2 increases downward through the soil by 1.8 mm °Cday^{-1} from crop emergence until the start of grain filling, typically reaching depths of 1.5–1.8 m. In the AFRC2 model roots are divided into seminal and lateral with an absolute upper water uptake rate of 0.3 mm H_2O km^{-1} root m^{-2} d^{-1} (Weir and Barraclough 1986) or 0.3 cm^3 water (m root)$^{-1}$ d^{-1} and is independent of demand for water (Jamieson et al. 1998b). Other models, such as CERES-Wheat (Ritchie and Otter 1985) incorporate a true root resistance to water uptake by making root water flow depend on crop water demand.

As stated above, AFRC2 calculates a ratio of root and soil water supply to water demand that forms the basis of a 'stress factor' (F_w) that reduces the potential rates of L expansion and biomass accumulation. AFRC2 calculates L as a population of leaves and shoots and the factor used to reduce leaf area (F_l) is calculated as:

$$F_l = 0.5 F_w \qquad (9.14)$$

Equation 9.14 states that reduction in the daily increment of the expansion of a leaf and therefore its final size as well as leaf lifespan, and thus its rate of senescence, start to occur when there is less than twice as much root extractable water as is demanded by E_t. In other words, the model hypothesises that the processes most sensitive to lack of water are leaf expansion and senescence (Jamieson et al. 1998b). Other effects, such as the rate of crop assimilation and partitioning, are also reduced directly by water shortage but these effects start at more severe levels of drought.

In AFRC2, the effects of nitrogen shortage are included as soon as crop nitrogen concentration per unit dry matter falls below a specified upper limit, the level

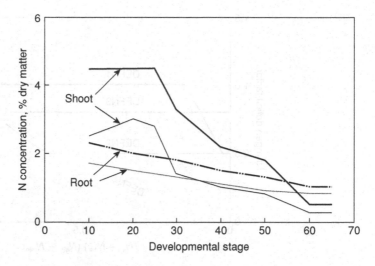

Figure 9.7 Upper and lower nitrogen concentrations (as % dry matter) for shoots and roots as a function of developmental stage in the AFRC2 wheat model. 10, emergence; 20, double ridges; 25, terminal spikelet; 30, begin ear-growth; 40, anthesis; 50, start-of-grain filling; 60, end-of-grain filling (Porter 1993).

of which declines with crop development stage (Figure 9.7). In the model, shoots have a wider concentration range than roots and have higher maximum and minimum nitrogen concentrations for most stages. Maximum shoot nitrogen concentration is set at 4.5% of shoot dry matter from emergence to the terminal spikelet stage, from which point it falls to 0.5% by the end of grain filling. The minimum value for shoot nitrogen concentration starts at 2.5%, falling to 0.25% by the end of growth. Equivalent values for root nitrogen are 2.3% falling to 1.0% for the maximum value and 1.7% falling to 0.8% for the minimum.

In calculating crop uptake of nitrogen, demand is calculated from the difference between current nitrogen concentration for shoots and roots at the current developmental stage and their maximum value. This difference multiplied by the change in shoot or root dry weight, gives the total daily crop nitrogen demand, which can or cannot be met by current nitrogen uptake from the soil. The amount of nitrogen taken up permits a nitrogen concentration to be calculated for the shoots and roots. Values of crop nitrogen concentration below the maximum designated for the phenological stage can affect physiological processes, as detailed below. At the start of grain filling, a pool of nitrogen is created jointly by shoots and roots for use by the grain; the pool size is calculated as the difference in nitrogen concentration of shoots and roots minus their minimum values and multiplied by their dry mass at anthesis. The maximum demand by grain for nitrogen in the model is 1.7 mg N grain^{-1} °Cday^{-1} (Vos 1981).

The effect of nitrogen shortage in the crop is modelled in a manner analogous to that of water. A calculation is made of the current relative crop nitrogen deficit (N_{fac}) as:

$$N_{fac} = (N_{act} - N_{min})/(N_{max} - N_{min}) \qquad (9.15)$$

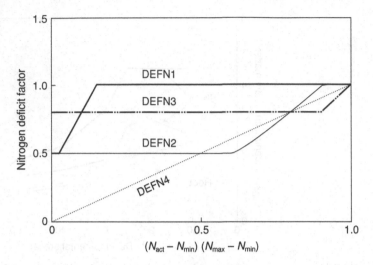

Figure 9.8 Proposed relationship between the ratio of current shoot nitrogen concentration (N_{act}) and the minimum (N_{min}) and maximum (N_{max}) nitrogen concentration for different phenological stages (see Figure 9.7) and factors used to increase tiller death rate (DEFN1), reduce leaf expansion rate (DEFN2), reduce tiller production rate (DEFN3) and increase leaf senescence rate (DEFN4) (Porter 1993).

where, N_{act} is the current shoot nitrogen concentration and N_{min} and N_{max} are the minimum and maximum concentrations for the current developmental stage. Equation 9.15 estimates the nitrogen deficit on the basis of current nitrogen concentration relative to the maximum and minimum concentrations. The N_{fac} comparator controls the value of four factors that operate with differential sensitivities and on different components of crop growth (Figure 9.8; Marshall and Porter 1991; Porter 1993).

The first factor (DEFN1) reduces the time during which tiller production occurs and increases the death rate of tillers from double ridge to anthesis. The effect of DEFN1 becomes apparent only at moderately severe levels of nitrogen shortage (Figure 9.8; Pearman *et al.* 1978). A more severe effect at equivalent levels of nitrogen is included in the model for leaf expansion (DEFN2), where up to 50% of the daily increment in lamina and sheath length can be lost when N_{fac} falls to 60% of its maximum value for a developmental stage. DEFN3 reduces the tiller production rate as soon as the nitrogen concentration falls below its maximum level. However, the effect is constrained such that the minimum tillering rate is 80% of the maximum (Pearman *et al.* 1978). The final factor (DEFN4) hastens leaf ageing and thus canopy senescence. In summary AFRC2, in common with other models, hypothesises that there are differentially severe effects of nitrogen shortage on different crop growth processes, with leaf growth and tillering being the most sensitive. Other concepts included in the model but not described here (Porter 1993) are that there is a time lag between the nitrogen concentration falling below its maximum level and an effect on a process. This is to take account of the capacity of plants to translocate nitrogen internally, from older senescing to younger growing organs, thus delaying a crop response to a lowered nitrogen concentration. It should be noted that neither shortage of water nor nitrogen is thought to affect the rate of phenological or stage development as indicated in Chapter 2 and found in many experiments.

In considering leaf and canopy growth in the AFRC2 and other models, two processes, leaf ageing and leaf area expansion, are modelled as being sensitive to nitrogen and water. Leaf canopy expansion rate is reduced and leaf senescence is enhanced by shortages of either factor. In the AFRC2 model, the more severe shortage is assumed to operate singly on any day and this is referred to as 'Law of the Minimum Factor' of von Leibig. This notion has been challenged (Sinclair 1992) but it is important to emphasise the time aspect of von Leibig's axiom. As a hypothesis, the controlling minimum factor is an instantaneous and not a cumulative concept. This means that, viewed over the course of a growing season, crop growth might be seen as being limited by both water and nitrogen but with them operating *in succession* as the most limiting factor. An analysis of variance might well show a significant interaction between these factors in terms of their significance for yield, but over a shorter timescale only one factor predominates. In other words, interpretation of Leibig's minimum factor law is timescale dependent with a limiting factor depending on the time over which it operates.

In conclusion, AFRC2 is a detailed model of the effects of water and nitrogen on the growth and development of a wheat crop. It contains many feedbacks that, for instance, go from root growth to soil water extraction to canopy expansion to evapotranspiration to dry matter production to partitioning and back to root growth again. It has proved most useful in aiding our understanding of wheat crop growth and has formed the basis of a simpler and more management-oriented wheat model, Sirius (Jamieson *et al.* 1998c). Many of the ideas used in AFRC2 are also to be found in the crop models for soybean and maize to be described next.

9.3.2 The CROPGRO soybean model

CROPGRO started in 1980 at the University of Florida, USA with the release of the soybean simulation model SOYGRO (Wilkerson *et al.* 1983). The other grain legume crops, peanut and common bean (*Vicia faba*) were added later to the generic CROPGRO and soil N balances and N fixation routines were added between 1990 and 1994. The CROPGRO model is generic for grain legumes with the different species and their varieties described in external input files rather than within the model computer code. The following describes CROPGRO V3.1 (Boote *et al.* 1998).

As with the AFRC2 model for wheat, CROPGRO is a process descriptive model that considers crop phenology and canopy development and crop carbon, nitrogen and water balances. Crop phenology includes the rates of vegetative and reproductive development that govern the partitioning of C and N to plant organs over time. Crop N balance includes daily soil N uptake, N_2 fixation, N mobilisation from vegetative to storage tissues and N loss in abscised parts. Soil water balance includes infiltration of rain and irrigation, crop and soil evapotranspiration, root uptake and water drainage and distribution within the soil profile.

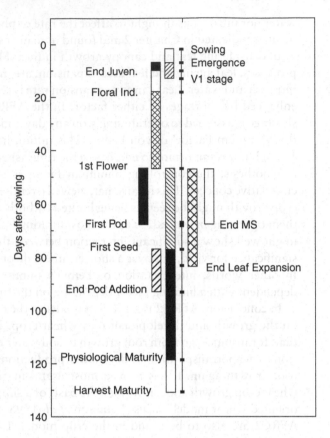

Figure 9.9 Ontogeny of vegetative and reproductive stages of grain legume crops. Timing of occurrences are shown for soybean, cultivar MG7 grown in Florida, USA. MS, main shoot (Boote *et al.* 1998).

Crop development in CROPGRO during the various growth phases is differentially sensitive to temperature and photoperiod. In CROPGRO there are 13 possible life-cycle phases from sowing to maturity, each with its own unique development rate and a developmental phase change occurs when the integrated development rate reaches a cultivar-dependent threshold and, for example, seed growth starts. This is very similar to the method used in the AFRC2 model, except that the life cycle is divided into fewer phases in that model. The full range of the ontogeny of vegetative and reproductive stages of the grain is shown in Figure 9.9 (Boote *et al.* 1998).

The physiological development rate, expressed as physiological days per calendar day, is modelled as a function of temperature, photoperiod and water deficit. It may be the case that the effects of temperature and water deficit are confounded, since drought leads to the closing of stomata, an increase in the net solar energy load on the crop and an increase in its temperature and rate of development. If conditions are optimal, one physiological day is accumulated per calendar day and the number of physiological days equals the number of calendar days for a developmental phase. If conditions are not optimal for development, the number of physiological days for a developmental phase will be

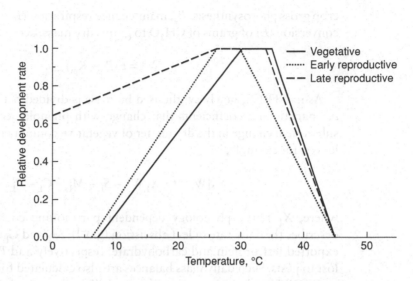

Figure 9.10 Effect of temperature on the rate of vegetative development, early reproductive development from first flower to first seed and late reproductive development from first seed to maturity. Note that the optimal temperature range, when relative development rate equals one, differs for the different developmental phases and that the response space is defined by a minimum temperature, a lower and a higher optimal temperature and a maximum temperature (Boote *et al.* 1998).

higher than the number of calendar days. Optimality of conditions is defined in terms of how the relative rate of development is affected by, for example, different temperature. Figure 9.10 shows a temperature curve for three developmental phases in which the area enclosed by each polygon represents the possible combinations of temperatures and development rate for a given phase. From this it is clear that vegetative development has a higher base and optimal temperature than progress towards flowering. The later developmental stage has a derived base temperature of −48°C, as this is the extrapolated temperature at which the relative rate of development, plotted on the vertical axis, equals zero. This postulates that rate of progress towards physiological maturity changes little with change in temperature during the later phases of soybean development.

Total dry matter growth of the crop is based on a carbon balance that includes photosynthesis, respiration, partitioning of carbohydrate and N, remobilisation of protein and carbohydrate from vegetative tissues and loss of plant parts *via* abscission. Carbohydrate, as glucose, is the molecular building block used in CROPGRO for the calculation of C gains and losses and the costs of synthesis of lipids, proteins and higher molecular weight plant components. The rate of change in total crop dry matter, W (g m^{-2}), is described as:

$$dW/dt = W^* - S - C - P \qquad (9.16)$$

where, W^* is currently synthesised biomass, S represents abscised plant parts such as leaves, C is the carbohydrate loss to symbiotic bacteria and P is the loss of dry matter to pests, which is not considered further. W^* is a function of crop gross photosynthesis, P_g, maintenance respiration (R_m) and the efficiency of conversion (E) of grams of CH_2O to grams dry mass, according to:

$$W^* = E(P_g - R_m) \qquad (9.17)$$

As in AFRC2, newly synthesised biomass is divided between the plant parts *via* partitioning coefficients that change with phenological development. The subsequent change in the dry matter of vegetative component plant parts such as leaves, for example, is;

$$dW_L/dt = X_L W^* - S_L - M_L - C_L - P_L \qquad (9.18)$$

where, X_L is the phenology dependent partitioning coefficient to leaves; S_L describes the daily rate of leaf abscission (g d^{-1}); M_L and C_L are the daily mass of exported leaf protein and carbohydrate, respectively and P_L is the mass of leaf lost to pests. Such daily mass balances are also calculated for shoots and roots.

CROPGRO has two options for modelling photosynthesis: (1) daily canopy photosynthesis or (2) hourly leaf-level photosynthesis with row-based light interception. As the hourly version describes a more mechanistic response of photosynthesis to CO_2, temperature and irradiance, it will be described. Radiation is most commonly recorded as the daily sum of incident light energy falling on an area of ground, and this integral needs to be divided to give a diurnal cycle of radiation that starts with low values in the morning, builds to a maximum intensity at solar noon and then declines towards sunset. This requires calculation of the length of the photoperiod that changes with both day of the year and latitude (Spitters 1986; Spitters *et al.* 1986). Equivalent calculations are made in AFRC2.

Hourly temperatures are calculated from daily maximum and minimum air temperatures using a sinusoidal curve linking air temperature to time of day such that temperature increases from its initial low value to a maximum, thereafter falling again. Each hour, light interception and absorption, which equals the light intercepted minus the light transmitted, are calculated for both direct and diffuse radiation as a function of canopy height, L, and its orientation relative to the passage of the sun. As soybean is a row crop, meaning that radiation can penetrate the canopy horizontally as well as vertically (see Equation 9.5), a more complicated radiation interception and transmission model is needed than the one-dimensional one used for wheat. Hourly canopy photosynthesis is computed as the sum of leaf photosynthetic rates of sunlit and shaded leaves, measured as their leaf area indices. Leaf photosynthesis rates are calculated using the asymptotic exponential light response equation that yields a response curve to irradiance, similar to that in Figure 9.5. In the CROPGRO light response curve, quantum efficiency (α) and the light saturated photosynthesis rate (P_{max}) are modelled as being dependent on CO_2 level, O_2 concentration and temperature. The Farquhar and von Caemmerer (1982) equations for the RuBP regeneration limited region (Chapter 4) are used to model the kinetics of Rubisco and to compute the initial slope or the quantum efficiency. Included in the calculation is the effect of temperature on the enzymatic specificity of Rubisco for its substrate and on the CO_2 compensation point (Chapter 4). Single leaf P_{max} also depends on the amount of leaf nitrogen per unit leaf dry weight and as this decreases down the modelled canopy, so does P_{max}. Several responses of the CROPGRO photosynthesis model to temperature, leaf nitrogen concentration and L are shown in Figures 9.11–9.13.

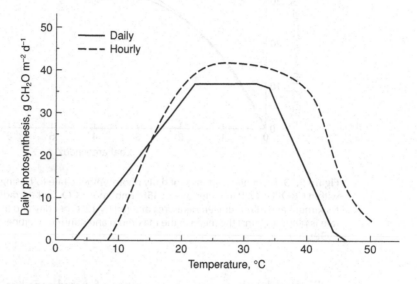

Figure 9.11 Modelled responses of daily photosynthetic rate to temperature using the hourly model within CROPGRO. Parameters were: 350 μmol mol^{-1} CO_2; solar irradiance 22 MJ m^{-2} d^{-1}; $L = 5.85$. Maximum and minimum temperatures differed by 12°C per day and the photosynthetic rate is plotted against the mean of the maximum and minimum temperatures (Boote *et al.* 1998).

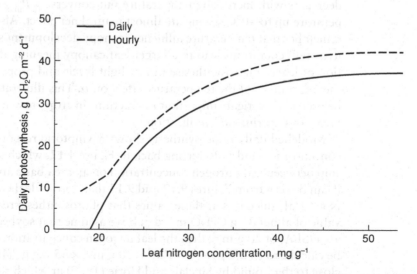

Figure 9.12 Modelled responses of daily photosynthetic rate to canopy mean leaf nitrogen concentration using the hourly model within CROPGRO. Parameters were: 350 μmol mol^{-1} CO_2; solar irradiance, 22 MJ m^{-2} d^{-1}; maximum and minimum temperatures, 32°C and 20°C (Boote *et al.* 1998).

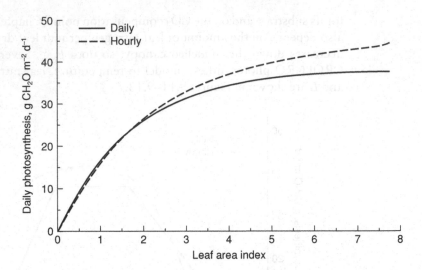

Figure 9.13 Modelled responses of daily photosynthetic rate to L using the hourly model within CROPGRO. Parameters were: 350 μmol mol^{-1} CO_2, solar irradiance 22 MJ m^{-2} d^{-1}. Maximum and minimum temperatures differed by 12°C per day and the photosynthetic rate is plotted against the mean of the maximum and minimum temperatures (Boote *et al.* 1998).

Simulated daily canopy photosynthesis for soybeans has a broad temperature optimum within the range of mean temperature between 22°C and 36°C. This is a wider range, particularly at the high temperature end, than has been typically recorded for C_3 species (cf. Figure 4.16) but is explained in the model by two temperature functions operating in opposite directions. Quantum efficiency decreases with increasing temperature but conversely P_{max} increases with temperature up to 40°C, giving an almost neutral net effect. Also, it is important to remember that temperature influences canopy development strongly; warm (and irrigated) conditions lead to a larger leaf canopy meaning that a larger proportion of leaves photosynthesise in low light levels and proportionally increasing the importance of the temperature effect on α. This illustrates how a model can be used to investigate important interactions in crop growth that suggest further areas for experimental verification.

Modelled daily photosynthesis shows asymptotic responses to leaf nitrogen concentration and L. Referring back to Figure 4.12, which showed the relationship between leaf nitrogen concentration on an area basis and RUE for soybean, it can be seen from Figures 9.12 and 9.13 that 4 m^2 of leaf contains about 35 mg N g^{-1} leaf, since it is at these values that photosynthesis reaches its asymptotic value of about 35 g CH_2O m^{-2} d^{-1}. If we assume that soybean has a specific leaf area (SLA) of 20 g m^{-2}, then the leaf nitrogen concentration on an area basis can be calculated as 2.8 g m^{-2} (= 4 m$^2 \times$ 20 g m$^{-2} \times$ 35 mg g^{-1}/1000), a value that is close to that found by Sinclair and Horie (1989) at which soybean has its maximum RUE (Figure 4.12). The point is that it is both possible and necessary to verify predictions from a model by cross-checking them with independent values from the literature.

Table 9.2 Percentage composition by weight of plant components of soybean as protein, lipids, carbohydrate and lipids (after Wilkerson *et al.* 1983). Minor fractions such as organic acids and minerals are excluded.

Component	Protein	Lipid	Carbohydrate	Lignin
Leaves	29.4	2.5	58.7	*c.* 8.0
Stems	18.8	0.4	76.2	*c.* 5.0
Seed cases	25.0	1.5	65.6	*c.* 7.0
Seeds	39.8	19.7	35.7	*c.* 5.0
Roots	9.2	1.0	84.1	*c.* 5.0

Up to this point many of the ways to describe crop physiological processes in the AFRC2 and CROPGRO models have been similar: thermal time and photoperiod controlling phenological development; single leaf photosynthesis modelled at an hourly time step within an age-structured canopy. However, in partitioning the primary carbohydrate product of photosynthesis into plant tissues for growth CROPGRO needs to be and is more sophisticated than AFRC2. The reason is that the diversity of plant compounds produced by soybeans, and legumes in general, is much larger than that found in the graminaceous cereals. In addition to higher molecular weight carbohydrates such as starch, soybean contains large amounts of protein and lipids. Wilkerson *et al.* (1983) give fractions of proteins, lipids and carbohydrate in the final tissue products of soybean (Table 9.2).

The proportions of protein and lipid are about two and three times, respectively, the quantities found in wheat. Carbohydrates make up about 85% of the dry matter of wheat and maize. Lipids and proteins are complex molecules and, compared with starch, their synthesis from glucose requires more energy. The metabolic conversion costs can be expressed as the number of grams of glucose needed to produce one gram of product (Section 5.4). Costs can be broken down into the glucose-C cost of biosynthesis involving the respiratory energy source (ATP) plus the reduction compounds (NADH and NADPH) and the stoichiometric amount of glucose-C in the end product, either lipid, cellulose or protein. These glucose-C costs are given in Table 9.3 and where the conversion efficiency (i.e. g product g glucose^{-1}) is the reciprocal of the metabolic conversion cost. It is clear that the synthesis of lipids is very expensive in terms of the energy required and the energy stored in the lipid as C is very concentrated. Protein conversion costs are computed from the rate of uptake of NO_3 uptake, because most of the NH_4 mineralised from soil organic matter is converted rapidly to NO_3 (Chapter 5; Section 7.2.1). Protein synthesis from NH_4 as a substrate has lower costs, even though the C in the final product is the same (Table 9.3).

Tables 9.2 and 9.3 demonstrate two important features about soybean and legumes in general. First, because they expend more C to produce their energy-rich plant compounds they have lower RUE than C_3 and C_4 cereals (Figure 4.12),

Table 9.3 Glucose equivalent costs for respiration and biosynthesis of the main plant compounds (after Boote *et al.* 1998).

Plant compound	Glucose cost for respiration and biosynthesis (g glucose g product^{-1})		
	Energy cost	C in product	Total cost
Protein			
from NH_4	0.36	1.34	1.70
from NO_3	1.22	1.34	2.56
from N_2 fixation	1.49	1.34	2.83
Cellulose (starch)	0.11	1.13	1.24
Lipid	1.17	1.94	3.11
Lignin	0.62	1.55	2.17

since this is calculated on a dry weight basis. If RUE were expressed based on the energy content of the products, then these differences would be much less. On the other hand, the high carbon, nitrogen and energy concentration of their products makes, particularly, legumes and, generally, oil crops, nutritious feed for herbivores. With the above glucose equivalent costs of plant compounds, CROPGRO calculates the amounts of each of them produced daily from the formula:

$$E = 1/(2.56F_P + 1.24F_C + 3.11F_F + 2.17F_L) \tag{9.19}$$

in which F is the plant fraction composition of protein (P), starch carbohydrates (C), lipids, or fat (F) and lignin (L), respectively. E is the overall growth conversion efficiency, defined as g product g glucose^{-1} and is calculated daily in the CROPGRO model. Maintenance respiration depends on temperature and current crop biomass, as in AFRC2, but does not include oil and protein in the seeds. Therefore the Q_{10} value is 1.85, a value lower than that for wheat (2.0) but still following the general approach of McCree (1974). During vegetative growth before pod formation, all primary C assimilate is allocated to vegetative tissues, with partitioning between leaf, stem and root dependent on the phenological development stage (Figure 9.14), as in AFRC2 (Table 9.1) and as discussed in Section 6.2.

As reproductive development progresses following the vegetative phases, new sinks are formed as pod walls and seeds, and assimilate is increasingly partitioned to reproductive structures. Thereafter, the vegetative tissues share the fraction of assimilates that remains after supplying reproductive structures. Partitioning to the reproductive structures of seed cases and seeds is more complex and, as in AFRC2, depends on the balance between a temperature-driven demand and the supply of assimilate. In contrast to the determinate ear of wheat in which all the grains are initiated within a short space of time, soybean produces a temporal succession of fruits that changes the overall size of the reproductive sink, in addition to the temperature effects on fruit growth rate. Thus, in soybean, new age-classes or cohorts of fruits are added daily (Chapter 2). Fruits

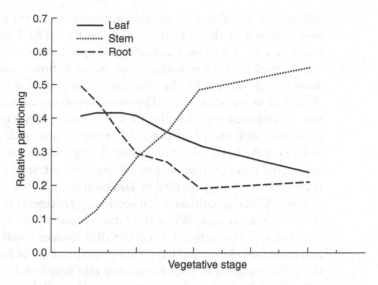

Figure 9.14 Partitioning of dry matter among leaf, stem and root components as a function of the vegetative growth stage of soybeans (Boote *et al.* 1998).

of each cohort age in thermal time and pass through a seed case-growth phase, during which seeds in each fruit are set and their rapid growth phase starts. Thus, the total reproductive sink in soybean comprises many age-defined cohorts of different reproductive organs, each having a potential assimilate demand determined by temperature. The priority order for assimilated carbon use is seeds followed by seed cases, new pods and then vegetative tissues, although the carbon required for N_2 fixation has the absolute highest priority if N uptake is inadequate. It is possible qualitatively to see that this reproductive sink–source model will have the consequence of diverting an increasing proportion of carbon to soybean reproductive structures *via* the combined effect of their ontogeny and their growth, leaving little assimilate for vegetative growth. It is also important to appreciate the difference between the monopodial, coordinated reproductive architecture of the graminaceous cereals and the sympodial, sequential form of the legume in defining their respective components of yield (Chapter 2). In summary CROPGRO is a source-driven model, meaning that limitations to growth are the result of lack of assimilates rather than lack of sinks for assimilates, except under two circumstances. The first is during early vegetative development when potential leaf area expansion can be too low, because of low temperature, to allow full light interception and crop photosynthesis. Second, severe N deficits that limit the potential growth of, particularly, seed components and other factors such as temperatures above 30–35°C, can limit the number of seedpods. This will be manifested as sink limitations to growth and yield of soybean and arable crops in general.

CROPGRO allows several methods for calculating climatically driven potential evapotranspiration. These include the Priestley–Taylor method (Priestley and Taylor 1972) that requires only temperature and total short-wave solar radiation to calculate the equilibrium evapotranspiration, and the Penman equation, referred to above in the section on the AFRC2 model, that additionally requires wind speed and humidity as inputs. As with the AFRC2 model, the energy used to drive potential plant transpiration is calculated from the total solar energy receipts by the crop and its extinction down the crop canopy, as described in Equation 9.12. The water-supplying capacity of the soil–root system is calculated from root length and soil water content in each layer and then compared with the plant potential transpiration, with actual plant transpiration being the lower of the two rates. A fuller discussion of the methods used by wheat and other crop models to calculate evapotranspiration and water uptake from roots has been provided by Jamieson *et al.* (1998b).

CROPGRO hypothesises that some crop processes are more sensitive to water shortage than others. When the ratio of root water uptake to potential crop transpiration is less than 1.5, CROPGRO reduces specific leaf area (m^2 leaf g^{-1}) and internode elongation but increases partitioning of assimilate to roots. When the ratio reaches 1.0, root elongation rate is reduced and N_2 fixation declines when the volumetric soil water content in the nodule zone (5–40 cm) falls below 40%. Leaf senescence is accelerated when potential crop transpiration is larger than the supply of water. In summary, these constraints in CROPGRO hypothesise that under water shortage soybean will initially respond by producing thicker leaves, a shorter stem and more root growth. More severe water shortage leads to shallower roots, decreased nitrogen fixation and faster leaf canopy senescence. Many of these hypothesised responses resemble those seen above in the AFRC2 model.

The soil nitrogen balance and root nitrogen uptake processes in CROPGRO follow those used in the CERES-Wheat model (Godwin *et al.* 1989). Uptake of NO_3 and NH_4 are functions of their concentrations in the soil layers, soil water availability and the root length density in the layer, an approach shared with AFRC2. Should total N supply exceed N demand by the crop, soil N can be leached since uptake cannot be higher than demand. As in AFRC2, total crop demand for N is equal to the phenology-dependent maximum N concentration, which falls as the crop progresses through its life cycle, multiplied by the increment in crop dry weight. In practice, CROPGRO differentiates the nitrogen demands for different organs (shoots, roots, leaves and seeds) and is more comprehensive in its crop nitrogen balance calculations than is AFRC2.

As a legume, soybean can also receive nitrogen *via* its symbiosis with nitrogen-fixing bacteria present in root nodules and this N_2 fixation has to be accounted for in modelling the crop's N balance. CROPGRO approaches this by initiating nodule production at a given thermal time after crop emergence. N_2 fixation is metabolically expensive for legumes since C, fixed by the crop, is exchanged with the bacteria as part of the symbiosis. Thus, when soil N uptake is sufficient to meet demand, nodule growth is slow and vice versa. Should uptake fall below demand for the growth of new tissues, plant carbohydrates are used to stimulate nodule growth at a rate that depends on soil temperature and on a crop-species-defined nodule growth rate. Soil dryness, aeration and crop phenology are also modelled to affect the rate of N_2 fixation. N_2 fixation is modelled as a reserve system for cases in which the preferred mineral soil nitrogen is in short supply. A reciprocal and dynamic balance exists between atmospheric and terrestrial sources of nitrogen for soybean that maintains production under low-input conditions. However, it remains the case that high yields of carbon and protein from legumes require non-symbiotic sources of nitrogen.

When the sum of N uptake and N_2 fixation is less than N demand, vegetative parts continue to photosynthesise at the non-limited rate (Figure 9.12). As specific leaf nitrogen (g N g^{-1} leaf) continues to fall, leaf and canopy photosynthesis are reduced following the non-linear curve shown in Figure 9.12. Should tissue N concentration approach a predefined lower limit, carbohydrate starts to accumulate and is directed to nodule formation and an enhancement of N_2 fixation. Such responses will alter the plant C:N ratio and the supply of these elements to the seeds. Recent modifications to CROPGRO permit seed protein and lipid composition to change with C and N supply, thus moving in the direction of simulating yield quality as well as quantity. Protein, lipids and carbohydrate for seeds are mobilised from shoots and leaves at a rate that increases with the speed of reproductive development, making the process demand-driven and therefore supply-limited. During this process, the plant is involved in a delicate balancing act; protein mobilisation and withdrawal from leaves is less metabolically costly and thus more efficient than *de novo* synthesis from reduced sources of N (NH_4). However, as leaf N declines so does leaf area and leaf photosynthesis leading to declines in leaf and stem mass and protein. Mobilisation can also occur from the shoots and they play an important role as a protein and carbohydrate buffer. Selection for high yields and protein content in legumes and other crops should consider the genetic variation in shoot and leaf storage properties more seriously than is done at present.

9.3.3 The maize model

The final described model was developed for maize (*Zea mays* L.) over a period of several years by R C Muchow and T R Sinclair and their co-workers. As with the other two crop models, the Muchow–Sinclair model is designed to simulate the growth and development of a maize crop under conditions of optimal and restricted supplies of water and nitrogen.

Phenological development is modelled *via* the accumulation of thermal time, above a base of 8°C, and, thereby, calculation of the final leaf number (Dwyer and Stewart 1986; Muchow and Carberry 1990). This allows calculation of the developing and senescing canopy leaf area per plant based on leaf sizes. The green and photosynthetically active leaf area is the difference between the total expanded leaf area and the fraction senesced. Crop leaf area is green leaf area per plant multiplied by plant population per m^2 and this marks a difference between this and the other two models. The maize model describes a single plant that is multiplied up to provide a description of a crop, whereas the wheat and soybean models describe crops. The concept of RUE (Chapters 3 and 4) is used in the maize model to calculate dry matter production from daily incident solar radiation and its extinction down the canopy. An extinction coefficient of 0.4 is used and the RUE has an experimentally determined value of 1.6 g MJ^{-1}, that declines to 1.2 g MJ^{-1} during grain growth to account for the declining N level in the leaves. Silking (Section 2.1.4) occurs 67°Cdays after the final leaf is fully expanded and grain growth starts 3 d after silking. Grain growth is simulated as biomass accumulation multiplied by a linear increase in harvest index with time (Muchow 1989; Moot *et al.* 1996) with a rate of 0.015 d^{-1}. The maximum harvest index is set to 0.5 to reflect the genetic potential of commercial maize hybrids. The thermal time from silking to physiological maturity is 1150°Cdays, although this total can vary between varieties (McGarrahan and Dale 1984). Simulations with the original model (Muchow *et al.* 1990) showed how changes in temperature and radiation level could affect maize growth and yield across a region, under the assumption of adequate irrigation and nitrogen applications. High maize yields were associated in the model with low temperatures and high solar radiation. Such a combination of meteorological conditions is not common in mid-latitude zones since sunny days tend to be warm and *vice versa*. More northerly sites with longer periods of daylight, where radiation will be less intense but have a longer duration, can provide lower temperatures, longer life cycles and grain filling, lower respiration but high radiation levels that will result in high yields.

As the availability of water imposes one of the major limitations to rain-fed maize production, Muchow and Sinclair developed their potential production model to include the effects of limited water on maize growth, development and yield (Muchow and Sinclair 1991). A soil-water budget was incorporated into the model by accounting for inputs from rainfall and irrigation and water loss *via* soil and crop evapotranspiration. Responses linking the soil-water budget to leaf-area development and RUE were derived from experimental studies to derive two logistic functions that describe the proportional reduction in RUE (Figure 9.15) and leaf area development (Figure 9.16) as a function of the fraction of transpirable soil water (FTSW). FTSW is the fraction of soil water remaining out of the total potential store of soil water that can be transpired. FTSW, the remaining fraction of the available soil water, is the index of soil drying used in the model and is adjusted daily using the addition and removal of water to the soil profile. It has a value of unity in the model when the soil water content equals the maximum soil water content. Soil evaporation is calculated using the Penman energy balance equation (Equation 9.13) but the soil is divided in the model such that upper layers dry out before lower layers (Muchow and Sinclair 1991). Daily transpiration is calculated using the method of Tanner and Sinclair (1983). Crop biomass accumulation declines as FTSW decreases, since leaf area development is reduced and less radiation intercepted. In addition, both transpiration and RUE are reduced as the ratio of crop biomass accumulation to transpiration is inversely proportional to vapour pressure deficit (Tanner and Sinclair 1983).

The above two formulations of water stress effects on maize (Figure 9.15 and 9.16) show that relative leaf area development is reduced more strongly than relative transpiration for the same FTSW. Thus in the model, relative transpiration has values of 0.92, 0.70 and 0.34 for FTSW values of 0.3, 0.2 and 0.1, respectively; relative leaf area development has values of 0.98, 0.70 and 0.08 for the same FTSW values. Such modelling of the effects of water availability on

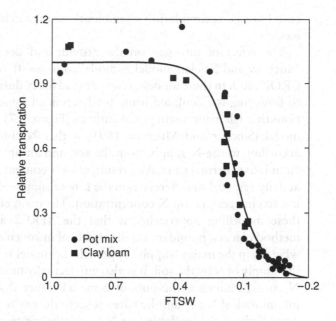

Figure 9.15 Relationship between relative transpiration (1 = no effect) and the fraction of transpirable soil water (FTSW). The solid line is the logistic function used in the maize model to constrain transpiration on the basis of soil water availability (Muchow and Sinclair, 1991).

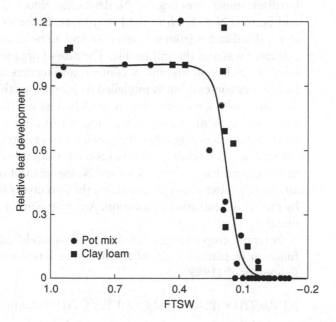

Figure 9.16 Relationship between relative leaf area development (1 = no effect) and the fraction of transpirable water (FTSW). The solid line is the logistic function used in the maize model to constrain leaf area development on the basis of soil water availability (Muchow and Sinclair 1991).

crop biomass accumulation resembles those used in the AFRC2 and CROPGRO models.

The effect of nitrogen on the growth and development of maize in the Muchow and Sinclair model is modelled in a different way from AFRC2 and CROPGRO. In these models, crop demand for N during the vegetative period up to flowering is calculated using predetermined upper and lower N concentrations that fall from sowing until anthesis (Figure 9.7). The approach in the maize model (Sinclair and Muchow 1995) is that N enters the crop from the soil according to the N supply from the soil and an uptake function dependent on cumulative thermal time. As a result, crop N content and resulting physiological activity respond to differences in the rate of supply of N in the maize model and not to changes in crop N concentration. The main effect of differences between these modelling approaches is that the AFRC2 and CROPGRO 'demand' method places a boundary on the effects of nitrogen shortage on crop processes, whereas in the maize 'supply' approach the model is freer to vary in response to the supply of N by the soil. It is also intellectually more satisfactory to have tissue N concentrations as outputs of a model rather than as specified inputs. The minimum leaf N per unit leaf area is set in the maize model at 0.55 g N m^{-2} leaf area. If the daily available leaf N is insufficient to maintain this general level, daily leaf area growth is zero and leaf area is lost so that the minimum N level is maintained for the remaining leaf area. This approach has been utilised in later wheat models (Jamieson and Semenov 2000).

The soil N model of Sinclair and Amir (1992) was used to simulate the supply of soil N. The top layer of two in the modelled soil profile receives inorganic fertiliser, mineralises organic N, denitrifies N to N$_2$O under flooded conditions and permits N to leach to the lower layer. In the lower soil layer, organic N is mineralised and N from whatever source can be leached from this layer if there is sufficient water in the soil profile. The rate of organic N mineralisation is calculated from the soil organic N content, and increases with soil temperature and high water content, but is modelled to decrease as the mineralised N concentration increases. Denitrification is modelled as increasing with temperature and soil water content. Nitrogen leaching out of the lower of the two soil layers is modelled as occurring when the potential store of transpirable water of 135 mm (Muchow and Sinclair 1991) is exceeded. Under this circumstance, water drains out of the profile carrying dissolved N, the amount of which is calculated as the amount of water in the profile above the potentially transpirable limit multiplied by the N concentration in solution. Any nitrogen leaching is assumed to occur in one day.

Potential crop nitrogen uptake (NUP) is modelled as a non-linear hyperbolic function (Equation 9.20) of thermal time based on a regression from data of Bennett *et al.* (1989):

$$\text{NUP} = TtU \times (\text{PNU} \times (5.24 \times 10^{-8}) \times CTtU^{1.6} \times \exp(-(CTtU/958)^{2.6})) \qquad (9.20)$$

TtU is the thermal time for the day (above 8°C), PNU the potential N uptake per day and $CTtU$ is the cumulative thermal time since sowing. If insufficient soil mineral N is available to meet the N demand, then actual uptake (NU) is made equal to N supply from the soil, assuming that N in soil solution above 1.0 mg N l^{-1} of water is available to the crop. At low soil water levels, as indicated by the FTSW value, NU is restricted by the reduced ability of a dry soil to transport soluble N.

Biomass accumulation in the maize model uses the simple radiation use efficiency approach, but with the value of RUE linked to leaf N concentration per unit leaf area (L_N). The empirical equation (Equation 9.21) is:

$$RUE = 0.12 + 1.09 \times L_N \qquad (9.21)$$

For the minimum leaf nitrogen concentration in the model of 0.55 g N m^{-2} leaf, Equation 9.21 returns an RUE value of 0.72 g MJ^{-1} and its maximum value is set at 1.6 g MJ^{-1} solar radiation or 3.2 g MJ^{-1} PAR.

The translocation of N to the grain follows the same philosophy as for the uptake of N from the soil. For each day, and based on the thermal time for the day, N is transferred to the grain. The supply of this N comes from stem and leaves that are drained of N from their levels at the beginning of seed growth down to a minimum N level of 0.4 g N m^{-2} leaf and 2.5 g N kg^{-1} for stems (Muchow 1994). The available N is transferred to the grain pool over the thermal time (1150°Cday) that defines the duration of the grain-filling period. Nitrogen uptake from the soil during grain filling can also contribute to grain N, thus the total amount of N in the grain is not fixed at the start of grain filling. As described above, biomass accumulation into the grain is modelled as the harvest index of total biomass production during grain filling, meaning that grain biomass and N accumulation are modelled independently of each other. In these circumstances, grain N concentration would be predicted to rise for maize crops that do not experience decreased rates of grain biomass accumulation. In contrast, wheat responds to low N availability during grain growth by hastening the translocation of N from leaves, thus killing leaf area and reducing biomass accumulation and thereby raising grain N concentration (Muchow 1994).

9.4 Modelling variety differences and traits

Simulation models distinguish between crop varieties mainly on the basis of development rather than of growth. Thus models reduce the thermal time totals used to pass between development stages for rapidly developing varieties (i.e. Kiniry 1991, maize; Jones *et al.* 1991, soybean), or reduce the responsiveness of final leaf number to photoperiod and temperature (Jamieson *et al.* 1998c). Leaf size and extinction coefficient can also change with variety and Moot *et al.* (1996) showed that old and new varieties of wheat could be distinguished based on the duration of the lag phase between anthesis and the start of grain filling. Once grain filling commenced it did so at statistically similar rates so that the high harvest indices of modern varieties (*c.* 50%) *vis-à-vis* older varieties (*c.* 35%) could be partly explained but their shorter duration of active grain filling. Differences in photosynthetic or respiratory parameters are not thought to offer significant variation in crop performance.

An example of how crop models can be used to identify plant traits that may prove useful in particular growth conditions, such as increased temperature or reduced water supply is given by Sinclair and Muchow (2001) for maize. Using the maize simulation model described above, they analysed the possible effect of eight putative crop traits on the mean simulated yield of maize grown over 20 years at Columbus, Missouri in the USA. The site was chosen because its high inter-annual variation in rainfall (a 20-year mean of 411 mm year^{-1} with a range from 169 mm year^{-1} to 772 mm year^{-1}) could be expected to ensure seasons of contrasting crop water relations. Traits for adaptation to drought conditions fall into three broad categories:

- a reduction in the length of pre-flowering and/or post-flowering phases to escape drought (e.g. Bolaños *et al.* 1993);
- expansion of the soil water extracted volume to increase water supply by, for example, deeper roots (Salih *et al.* 1999);
- decreases in the rate of soil water extraction by reduction in canopy size *via* smaller leaves (Salih *et al.* 1999) or a slower rate of leaf appearance (Muchow and Carberry 1989, 1990) and/or stomatal responses (Ludlow and Muchow 1990; Ray and Sinclair 1997).

Such individual phenotypic traits can be combined but can also have negative implications for other plant processes. For example, smaller leaves may reduce evapotranspiration but a smaller canopy will intercept less radiation and produce less dry matter, leaving the biomass yield to transpiration ratio largely unchanged. Such trade-offs become most readily apparent within a logically and quantitatively portrayed cropping system, which is the definition of a simulation model.

Sinclair and Muchow (2001) examined the effects on grain yield and its variation of eight plant traits that are known to vary between cultivars and are presumably heritable to a degree. The traits were chosen to resemble those of sorghum, a crop adapted to drier conditions than maize:

- two increased depths (100 cm and 120 cm versus 80 cm) of soil water extraction;
- decreased leaf size;
- slower rate of leaf appearance;
- decreased CO_2 assimilation;
- increased grain growth rate but decreased duration;
- early stomatal closure;
- delayed stomatal closure;
- a combination of those of the above eight traits related to leaves, CO_2 and grain growth.

The yearly simulated yields over 20 years for the site in Columbus, Missouri, USA formed the baseline for comparison for yearly yields with and without the 'sorghum' traits. Table 9.4 shows the mean yields, evapotranspiration and their ratio for the changed crop traits. For the baseline over the 20 years, the simulated grain yields ranged from 87 g m^{-2} to 976 g m^{-2} with a mean of 409 g m^{-2}.

The three traits that produced the largest yield enhancement relative to the baseline were increased rooting depth and thereby the soil volume explored for

Table 9.4 Summary of mean values and coefficient of variation (CV) for simulated grain yield, evapotranspiration (ET) and their ratio over 20 years of simulations with changes in traits described in the text (after Sinclair and Muchow 2001).

Trait	Yield (g m^{-2})	CV (%)	ET (mm m^{-2})	Yield/ET (g mm^{-1})
Baseline maize	409	54	387	0.99
100 cm rooting depth	597	47	419	1.16
120 cm rooting depth	563	43	442	1.23
Smaller leaves	370	52	389	0.90
Slower leaf appearance	540	57	408	1.23
Lower RUE	421	47	386	1.04
Higher grain growth rate	368	50	366	0.96
Early stomatal closure	445	60	391	1.08
Delayed stomatal closure	381	55	383	0.93
Combined traits	474	36	402	1.14

RUE, radiation use efficiency.

water, followed by a slower rate of leaf emergence, canopy development, and thus evapotranspiration and their combination. The transpiration efficiency (g biomass (g water)$^{-1}$) was simulated to increase by between 14% and 23% with these trait changes. The effect of slowing the rate of leaf emergence (Figure 9.17) reveals an interesting pattern. In the two years with the largest change in yield, the baseline simulation predicted very low yields. Generally, all years benefited a little from slower canopy development as evapotranspiration was thereby spread over the growing season more equitably. In a few years, the yield was predicted to be lower than the baseline and this occurred when the crop ran into

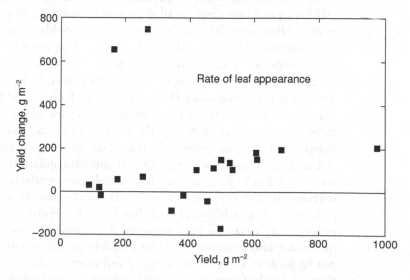

Figure 9.17 Yield change in each year from a maize crop with a relatively rapid leaf appearance rate (baseline) to one with a slower rate, plotted against yield simulated for the baseline (Sinclair and Muchow 2001).

an early frost and died prematurely. Larger positive yield changes were seen for the larger baseline yields, indicating that extending the length of the growing season is advantageous both in conditions where drought limitations are small to begin with and in extreme drought conditions, where resources are used more sparingly (Section 7.1.3). The general point is that there seem to be traits that can benefit both ends of the continuum of high and low resource inputs into cropping.

The coefficient of variation (standard deviation as a percentage of the mean) of yields measures their inter-annual variation and is an indication of yield stability. Table 9.4 shows that in the variable rainfall of Missouri the CV of the modelled baseline yields was 54%. Plant traits, such as increased rooting depth and a lower RUE that increased predicted mean yield, also reduced inter-annual yield variation as shown by the lowered CV. Early stomatal closure was predicted both to reduce yields and increase their variation, and thus would not seem to be a desirable trait for adapting to drought. The combined traits offered an increase in mean yield and a substantial reduction in their CV, thus illustrating and emphasising the conclusion that adapting crops to drought needs to use a range of phenotypic strategies and that single traits are unlikely to succeed.

Changing the response of stomata to drought was predicted to have little general effect on improving yield, although in extremely dry years early closure did ensure that the crop avoided a drought that reduced the baseline yield substantially. This trait was predicted to have a neutral effect for most years and may be useful in conditions where extremely severe droughts occur. Allowing stomata to remain open even at very low leaf water potentials *via* leaf osmotic adjustment resulted in continued evapotranspiration and increased the likelihood that the crop experienced a lethal stress before reaching maturity. Reliance on a single physiological factor such as the maintenance of leaf turgor to preserve yield is challenged by such a prediction. Combining traits led to stabilisation of yields in poor years but small decreases in good baseline years, although the overall effect over 20 years was positive for yields. A similar conclusion was also reached by Porter *et al.* (1995) in modelling the effects of climatic change and genetic modification on nitrogen use by wheat. Such predictions of the modelled effects of traits on crop production need to be tested experimentally. The general conclusion of the Sinclair and Muchow (2001) exercise was that breeding to increase the below-ground exploration of the soil for water is likely to be very beneficial. Reducing the efficiency of radiation 'conversion' into dry matter had a lower beneficial effect than slowing canopy development that reduced the demand for water. Altered internal regulation of stomatal sensitivity was predicted to have a neutral or a deleterious effect. Such results sound a warning to the relevance of molecular and other detailed studies of single processes of physiological and biochemical regulation for improving crop production in extreme environments. It is important to initiate a dialogue with molecular and conventional plant breeders to utilise both molecular and computing tools to identify, incorporate and examine the $G \times E \times M$ combinations that are likely to prove successful in breeding and cultivating the future crop varieties that will be required to enable humans to feed and clothe themselves (Chapter 10).

9.5   Conclusions

The described models illustrate the range of approaches used in simulating crop growth and development of three important crops. However, modelling forces one to make choices both about the processes to be included in a crop model and the level of detail to be used in describing processes. In summary, the choices of approach to the main crop physiological processes can be narrowed down to the following. Crop models need to model phenology, since this provides the temporal framework into which the growth processes have to fit. Decisions to be taken in modelling the life cycle of crops include the phases into which the crop life cycle is divided. As has been seen, the models described above for wheat, maize and soybean include juvenile, floral inductive and reproductive stages. Other models (Kropff and van Laar 1993; Lindquist, 2001) distinguish only two phases: the vegetative phase from emergence to anthesis and the reproductive phase from anthesis to maturity. A second major decision in modelling phenology concerns the environmental factors that determine the rate of development, with temperature, photoperiod and exposure to low temperatures being the important ones. The main considerations to be borne in mind are that maize and soybean are quantitative short-day plants, meaning that long photoperiods slow their development rate, whereas wheat is a long-day plant. Some wheat varieties require exposure to low temperatures before they can become reproductive and produce ears and grain; this is not required in maize or soybean but they both have juvenile phases during which their development is unresponsive to photoperiod but responds positively to temperature. An alternative approach to modelling phenology in wheat, described Jamieson *et al.* (1998a), links developmental stage to the number of leaves produced on the main shoot. This approach models development as a continuous process rather than as a series of distinct phases, each with their unique responses to temperature, photoperiod and vernalising temperatures. It thus attempts to link the external appearance of a wheat plant to its internal developmental processes as it moves from the initiation of vegetative to reproductive structures. Crop models view the post-flowering phases, during which grains, cobs and pods are being filled, either as a single phase or else divided into an initiation phase, followed by a grain-filling phase, followed by a maturing phase before physiological maturity is reached and the crop can be harvested.

As seen above, there are three main approaches to simulating the development of a leaf canopy. The AFRC2 wheat model has a canopy model driven mainly by temperature that includes age-specific details of the leaf, shoot and tiller populations. The canopy thus simulated is age structured, responds to crop water and nitrogen status and models photosynthesis on a leaf layer-by-layer basis. Most importantly, this type of model asserts that canopy development is a sink-regulated process, meaning that the rate of supply of carbon to the leaves is not the main factor limiting canopy expansion. An alternative approach (Kropff and van Laar 1993) has been to partition weight to the leaf pool and then convert weight into area *via* a value for specific leaf area (SLA, area per unit weight of leaves). This is a source-limited model and it has been found necessary to restrict the value used for SLA in order to prevent overestimation of canopy leaf area.

The reason for this is the possibility in such a source-driven model that a positive feedback loop is created where leaf weight drives leaf area, which drives leaf weight and so on. The final general approach to canopy modelling has been simply to make total green area index (GAI) development depend positively on temperature, up to a maximum GAI value and then decline or senesce during grain filling (Jamieson *et al.* 1998c). This extreme sink-limited approach obviously avoids the problems associated with the source-regulated idea but describes the properties of the canopy in terms of light interception, photosynthesis and the responses to water and nitrogen as being the canopy average.

Model descriptions of the process of light interception and dry matter production have generally used one of three methods. Important considerations in modelling light interception are whether the crop is planted in widely spaced rows as in soybean, or at low (maize) or high (wheat) densities per m^2 and how much of the radiation is diffuse as opposed to direct sunlight. It is generally the case that a form of the Monsi–Saeki equation is used to compute PAR levels within the leaf canopy and the amount of PAR intercepted and absorbed and thereby available for photosynthesis. Models of different crops (wheat: Jamieson *et al.* 1998c; maize: Jones and Kiniry 1986; Kiniry *et al.* 1997; soybean: Jones *et al.* 1987) use the notion of RUE (Chapters 3 and 4) to compute the amount of dry matter produced per unit of PAR intercepted by the leaf canopy. Such an approach leads to robust estimates of dry matter accumulation, since as shown in Chapter 4, the numerical value of RUE is conservative over quite a wide range of growing conditions. Respiration is included in RUE since the dry matter calculated by this method is net of CO_2 losses, but this can be less than satisfactory if one wishes to understand and quantify the relative magnitudes of CO_2 gains and losses that result in net dry matter accumulation. As shown in Chapters 4 and 5, models have been developed that attempt to describe, in varying levels of detail, CO_2 influx and efflux from leaves as functions of PAR, CO_2, O_x, water and nitrogen status and references are found therein. The latest versions of the maize, wheat and soybean models described above contain examples of all three approaches to modelling dry matter production as illustrations of the above comments. It is generally the case that models are not static and are continually being updated and released both to remove errors and to introduce new ideas. For example, the generic CROPGRO model (Boote *et al.* 1998) now contains options for photosynthesis calculations as daily canopy photosynthesis or hourly leaf-level photosynthesis.

The partitioning of dry matter to plant organs such as leaves, stem and roots is almost completely descriptive in crop models. It has not proved possible to incorporate mechanistic descriptions of the generation of concentration gradients between the many sources and sinks for carbohydrates in models and how these change with time. Allocation to yield components has proved to be more tractable from a modelling viewpoint. For example, in the AFRC2 wheat model, grain number, a major determinant of grain-sink strength, is calculated using the accumulation of ear dry matter and a function that reduces grain number at temperatures above 30°C immediately before and after anthesis (Ferris *et al.* 1998). Similarly in soybean, with its semi-determinate growth habit, change in the timing of important phenological stages by only a few days can have important effects on crop yields. Such insights have been fostered by crop models and have acted as a counterbalance to the overemphasis on photosynthesis as the process that determines crop production.

What capability and architecture is required for crop modelling to be valued in genetic applications? It seems clear that current capability in crop modelling will need to be enhanced if it is to play an integral role. Gene to phenotype modelling requires prediction based on scientific understanding of plant function and control. Models that integrate understanding of processes across levels of biological organisation, providing insight into key phenomena and responses that emerge, at higher levels, will probably be valuable. This does not necessarily imply, however, any increase in detail or complexity in crop modelling. Models that quantify the functional controls driving crop responses to environmental conditions will also be valuable. The rules required to quantify the functional controls may be quite simple but should enable a complicated array of crop responses to emerge given differing combinations of conditions. The rules most likely reflect the basis of metabolic signalling in plants and thus provide focal points for linking crop models to genetic analysis and exploration of gene function. Modular model architecture will be needed in crop modelling to assist advance *via* enhanced scientific understanding at component level. Such architecture provides a means to capture unifying principles, test new insights, and compare approaches to component modelling, while maintaining a focus on predictive capability at the whole plant and crop levels. This will be essential to generate the dialectic between 'bottom-up' and 'top-down' approaches required for advance.

To date crop simulation modelling has operated mostly at the organ–plant–crop level and has been associated primarily with understanding and predicting plant–environment interactions for applications to crop and cropping systems management. From a current perspective, it is likely that crop simulation modelling will broaden in two directions in the future. There is ample evidence for a move from the cropping level to the wider system level for studies and education in land use associated with environmental and ecological aspects of farming in the landscape. This will mainly require better use of existing crop modelling capability and a focus on connections to other aspects of the broader system. The other end of the spectrum sees a move from the crop level to the plant and genetic level for studies connecting to plant breeding and functional genomics. This will require enhanced crop models based on improved scientific understanding of plant function and control. This frontier provides a unique opportunity for crop modelling to play a significant role in enhancing the integration of molecular genetics with crop improvement, while offering new intellectual challenges to those who assemble logically constructed frameworks of how plant systems work (Hammer *et al.* 2002). Simply concentrating on identification and description of putative molecular mechanisms of isolated biochemical or physiological processes will not result in effective breeding strategies for crops in extreme conditions. Any advances need to be put into the context of whole-crop systems and models, that permit trade-offs between traits to be identified and the $G \times E \times M$ relationship to be optimised.

Chapter 10

Crop physiology: the future

The discipline of crop physiology was named by W L Balls in 1917 and given the aim of understanding the dynamics of yield development in crops. Balls' expectations were rather modest – it would elucidate the reasons for conventional farm practice, hence the tag often given to it of the 'retrospective science' and 'might occasionally indicate the path' to further advances in yield . . . and like the Roman god Janus, our discipline has two faces.

(Evans 1994)

10.1 Introduction

The heyday of crop physiology was in the years between 1960 and 1990, since when it has not played such a prominent part in plant and crop science, except in the development and use of crop simulation models. The crop physiological focus on whole plants, their populations, and their interactions, with the abiotic environment was superseded by developments in molecular biology, molecular genetics and a general 'omics-ology' that continues to this day. Science cannot stand still, priorities should change, and crop physiologists probably paid insufficient attention to the consequences of genetic differences within species. Today's molecular biologists tend to assume that the presence and/or expression of a gene means that it automatically makes a difference to crop yield and quality. Nevertheless, as molecular biology matures, it is becoming generally recognised that the emerging generation of crop scientists must be equipped to operate at both molecular and whole plant levels.

Much more is now known about the genetics of crop function, both in terms of the inheritance of genetic traits, and the genetic control of physiological and biochemical processes, than was the case 16 years ago, when the predecessor of this book (Hay and Walker 1989) appeared. This knowledge is about single genes, groups of genes or, more likely, less defined quantitative trait loci (qtl), that are involved in disease resistance, the control of flowering time *via* photoperiod and/or vernalisation (e.g. Sung and Amasino 2004) and plant architecture.

What we are, as yet, unable to assess is the importance of single plant traits and qtls for the productivity and quality of crops in the field. A full range of disciplines need to be pursued actively to ensure the integration of knowledge at scales from the genome to the crop in the field. Crop physiology can play a role, by identifying those plant processes that are important in determining yield and quality in the context of other plant processes (Sinclair and Muchow 2001; Chapter 9), integrating them in conceptual models (Chapter 9), and making predictions for different combinations of environment and management that can then be tested – the classical scientific procedure. Molecular scientists can take part in this process through strengthening the basic scientific background to crop breeding but with equal focus on the phenotype and on the genotype. Such integration will be needed in the development of new crop varieties for the twenty-first century, when altered patterns of temperature and precipitation are likely to increase the importance of the abiotic and biotic effects of weather on crop yield and quality. A goal of this book has been to provide a synopsis of the quality- and yield-determining processes for scientists and students who study plant physiological processes at the sub-plant and sub-organ levels and who look for a context for their work at the level of the whole crop. The fundamental issues are how to make the necessary connections, and the extent to which processes that seem important at the genomic level have impact at higher levels of organisation. In this context, integrative models have an important role to play in the future of crop physiology as tools to bridge the gap between reductionism and holism.

Crop physiology can play a role in four main areas in the future in dealing with: the consequences of lowering inputs to improve nutrient and other resource use efficiencies; the effects of climatic change and climatic variability on crop yields; the nutritional, technological and environmental quality of products; and the development of new crops for non-food products such as biomass energy and biopolymers. The remainder of this chapter considers each of these in turn.

10.2   Lowering inputs

One if the most familiar icons in agronomy is the nitrogen response curve (Figure 10.1), which expresses yield as a function of applied nitrogen fertiliser. Such asymptotic curves, which portray many processes in biology, from leaf photosynthesis to crop–weed competition (Landsberg 1977), have an initial slope that rises to an asymptote beyond which the rate of change of the dependent variable (such as yield) is close to zero or may become negative. The most efficient use of the resource, such as N, in terms of the kg yield kg^{-1} N applied, occurs during the initial linear part of the curve and efficiency decreases as the maximum is reached. It thus seems impossible simultaneously to maximise output and efficiency, as there is a fundamental trade-off between them, and this introduces an important dilemma into the discussion of the future needs of global food production: whether it is more environmentally sustainable to attempt to maximise resource use efficiency or total production. An argument for the former is that losses and damage to the wider environment are likely to be

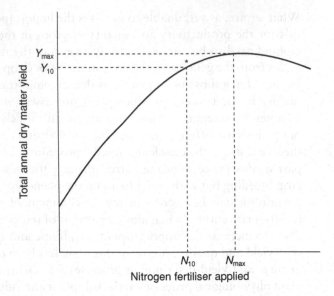

Figure 10.1 Curve to illustrate the influence of the amount of nitrogen fertiliser applied on the total annual dry matter yield of a sward of perennial ryegrass. N_{max}, the quantity of fertiliser N required to give the maximum yield, Y_{max}; N_{10}, the quantity of fertiliser N that gives a return of 10 kg dry matter kg^{-1} nitrogen applied; Y_{10}, the dry matter yield for N_{10} (from Hay and Walker 1989).

lower but, conversely, more land area may have to be used to meet the demand for food since production would be below its potential maximum level. In addition, it has been shown (Sinclair 1992) that the efficiency of use of one nutrient, such as N, depends on the presence of adequate levels of others such as P. Such synergies need to be appreciated when attempting to assess the overall resource use efficiency of crop production. Of course, many non-biological factors, such as the provision of production subsidies, and the structure of global food markets, influence the use of resources for agricultural production and, in reality, socio-economic drivers and agricultural policies are the dominant influences on yields in farmers' fields. Therefore there is debate about whether the optimal use of all resources occurs only at high input levels (de Wit 1992; Zoebel 1996) and whether such conclusions are specific to small spatial scales and well-monitored experiments. Resource use efficiency in the real world of crop production and agriculture seems a more complicated concept. For example, at a regional scale it can be that resources such as fertilisers are overused leading to wastage, whereas other areas experience deficit. Can a nitrogen response curve be applied at a regional spatial scale and what would it mean? In the future, agronomists will be required to work at spatial scales larger than the field and at temporal scales longer than one year. This will require the adoption of new thinking and skills in addition to those of understanding and predicting the yield development in crops, and these could be in the directions exemplified by van Ittersum and Rabbinge (1997), Evans and Fischer (1999) and Lu *et al.* (2004). Crop physiological concepts described in this book will be applied at scales larger than the field plot and, at the same time, crop physiologists will recognise the limits to their insights in the processes determining crop growth, yield and quality.

10.3   Climate change

For most of the time that humans have cultivated the land, the atmospheric CO_2 concentration has been between 260 and 280 μmol mol^{-1} (Indermühle *et al.* 1999) and the northern hemisphere temperature variation, in relation to its long-term mean, has been about 1°C, with a decreased amplitude since about 1500 (Mann *et al.* 1999). However, since 1900, both the mean and variance of the temperature anomaly have been increasing, as has the atmospheric CO_2 level to its current value of about 370 μmol mol^{-1}. Issues for crop production under these circumstances are: the potential of climate change to reduce the productivity of cropping systems; the possible adaptations of systems to change; and the potential of farming activities to slow the build-up of greenhouse gases. Crop physiology has had and will continue to have an important role to play in each of these areas. The impacts of climate change on crops have received much attention from crop physiologists, indicating the potential for adaptation by breeding and agronomy (Evans and Fischer 1999). As there has been a substantial lowering of soil organic content under modern farming practice throughout the world, the agricultural soil sink for carbon is large, and agricultural management offers great potential for carbon mitigation (Royal Society of London 2001). A very useful overview of these issues for European farming is given by Olesen and Bindi (2002).

From the start of the debate about the impacts of changing CO_2 levels and enhanced radiative warming of the global atmosphere, in the mid-1980s, many studies of agricultural and non-agricultural species concentrated on responses of physiological processes involved in C fixation to CO_2 concentrations above pre-industrial values (Drake *et al.* 1997; Kimball *et al.* 2002; Ewert 2004; Long *et al.* 2004). Until the mid-1990s, modelling studies of the impacts of climate change on agriculture (Rosenzweig and Parry 1994) focused on the effects of increases in CO_2 level and average climate conditions, such as a rise in mean global temperature or in the amount of rainfall, on crop production. However, it was soon realised that these analyses were conceptually incomplete because: (1) crops, and plants in general, respond non-linearly to changes in growing conditions, exhibiting discontinuous threshold responses; and (2) crops are often subject to combinations of stress factors that affect their growth and development. Therefore, the effect of climatic variability, the frequency of extreme events, and the effects of combinations of factors have since assumed greater importance (Semenov and Porter 1995). The latest version of the Intergovernmental Panel on Climate Change report on climate change (IPCC 2001) has emphasised issues of climatic variability, mitigation, and adaptation to climate change for cultivated and natural ecosystems, as well as likely impacts on the global biosphere.

Climatic variability, or more accurately the variability of the weather, since crop growth and development respond to local weather and not general climate, is seen in IPCC (2001) as stemming from three sources:

- changes in the mean weather such as a change, and most likely an increase, in annual mean temperature and/or precipitation;

- a change in the distribution of weather, such that there are more frequent extreme weather events such as physiologically damaging temperatures or longer periods of drought;
- a combination of changes to the mean and its variability.

These three scenarios are illustrated in Figure 10.2 (Porter and Gawith 1999; IPCC 2001). All three would reveal themselves in higher maximum temperatures, hotter days and heat waves over nearly all land areas; and more intense precipitation and drought events over many areas associated with, for example, El-Niño events and a likely increase in the variability of Asian summer monsoon precipitation. Variability at a range of spatial and temporal scales is now a key concern in studies of impacts on ecosystems in general and agroecosystems in particular. The question for agriculture is how any changes in the variability of weather are mirrored in changes in the variability of crop production. Increased

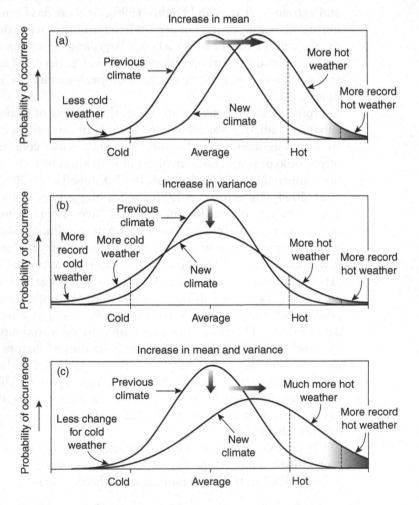

Figure 10.2 Illustrations of the effects of increases in (a) mean temperature, (b) increase in temperature variance and (c) increase in both on the frequency of extreme hot weather.

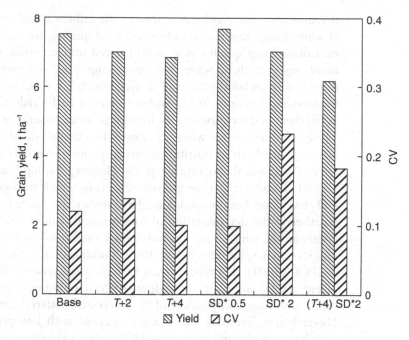

Figure 10.3 Modelling of the importance of the variation in temperature on crop yields and its variation (as CV) for wheat. Base, baseline conditions; *T*+2, mean annual temperature increased by 2°C; *T*+4, mean annual temperature increased by 4°C; SD*0.5, standard deviation of annual temperature decreased by 50% without change in its mean value; SD*2, standard deviation of annual temperature doubled without change in its mean value; (*T*+4) SD*2, combination of raised mean and standard deviation of temperature (Semenov and Porter 1995).

spatial or temporal variability in crop production, measured by the coefficient of variation (CV) of yield, implies lower security of the food supply in terms of amount and quality.

Simulation modelling of the effects of climatic variability has pointed to the general conclusion that increased annual variability in weather causes increased variation in yields. For wheat (Figure 10.3), it was found that doubling the standard variation of annual temperature, while holding its mean value unchanged (i.e. the scenario in Figure 10.2a compared with that in Figure 10.2b), gave the same decrease in yield as a 4°C increase in mean temperature (Figure 10.3) but a more than doubled coefficient of yield variation. It remains a challenge for experimental studies to test these model predictions but there are sound reasons to expect verification of the predictions. The mechanisms that lie behind such responses are likely to be complex but will involve the non-linear relationship of respiration with temperature, and temperature threshold effects on reproductive fertility and phenology (e.g. Porter and Gawith 1999).

10.4   Quality

The majority of analyses of the effects of climate change on food and forage crops have been concerned with production, either per unit area or for a region.

A major omission has been studies of the influence and mechanisms by which climate change might affect crop and food quality, for either human or animal nutrition. Crop quality is a multi-faceted and complex subject (Chapter 8) involving growth, storage and processing pre- and post-harvest, including nutritional, technological and environmental facets. For example, in both humans and livestock, it is possible to have a carbohydrate sufficiency but still suffer from malnutrition – in the form of protein, mineral or vitamin deficiencies. The general picture for wheat (Section 8.1) is that elevated CO_2 is detrimental to flour quality both in nutritional terms (protein content) and in technological terms (rheological properties, e.g. the Hagberg Falling Number) (IPCC 2001). Closed chamber experiments with wheat, in which nitrogen, temperature and CO_2 level have been manipulated, have shown that there is a strong influence of weather in the determination of total protein content and, as importantly, its composition, which affects its nutritional and breadmaking rheological properties (Section 8.1; Kettlewell *et al.* 1999; Gooding *et al.* 2003; Martre *et al.* 2003).

IPCC (2001) also reports that, in high-quality grass species for ruminants, elevated CO_2 and temperature increase have only minor impacts on the digestibility and fibre composition of the harvested material (Soussana *et al.* 1997). Nevertheless, livestock that graze rangeland with low protein-content forage may be more affected by increased C:N ratios than energy-limited livestock that graze protein-rich pastures (Gregory *et al.* 1999). Lowering the ratio of protein to energy in forage could reduce the availability of microbial protein to ruminants for growth and production, leading to less efficient utilisation of feed and more waste, including emissions of methane. The effect of climate change on food quality will be an important issue for future research in the development of a healthy and sustainable diet. To play an active role in this area, crop physiologists will need to take more account of the interests of breeders and processors by studying, quantifying and modelling the differences in quality among crop varieties and species.

10.5   New crops

Although the production and quality of food and forage for human and animal consumption have to remain the central issues for crop physiologists, there is growing interest in the production of non-food crops for energy or for specialised products such as biopharmaceuticals. Energy crops (Figure 10.4) are expected to make a significant contribution to the global 'energy mixture', designed to reduce emissions of CO_2 (IPCC 2001; Figure 10.5). This scenario has been questioned on the grounds that there will be intense competition for land between biomass and food production particularly in the developing world and, second, that biomass is a low energy-density fuel that requires input of energy to process it into a high-quality and useful form (Smil 2003).

Crop physiologists could play a role in the *design* of suitable biomass crops. One of the seminal papers in crop physiology, formulated for the guidance of plant breeders (Donald 1968) defined a crop ideotype for a high-yielding cereal, identifying the architectural, anatomical and physiological traits of the

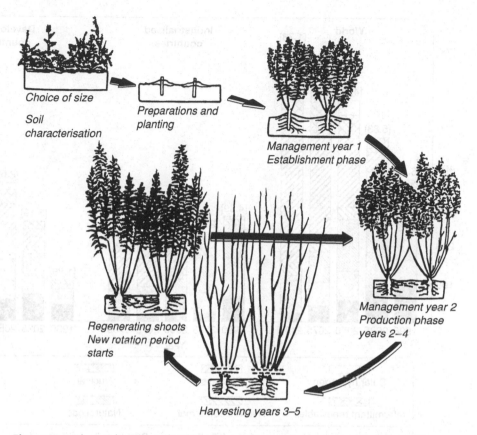

Figure 10.4 The production cycle of an energy crop based on a perennial such as willow (*Salix* spp.). After planting as vegetative shoots, plants develop roots and shoot systems and, in temperate zones, grow for up to 5 years before harvesting. After harvesting the wood for fuel that is either burned for heat or fermented for alcohol, the remaining cut shoots in the soil re-sprout and the harvesting cycle is repeated but without replanting.

phenotype that promote high yield. Since then, characters such as leaf shape, leaf angle, tillering, reduced respiration and increased photosynthetic rate have been investigated (Evans 1994). Donald's approach was deductive, in moving from the theory of crop physiology to the practice of plant breeding, although yield is the combined outcome of complex genotype × environment × management interactions and thus uncertain. In the spirit of the ideotype, it is possible to postulate some characters that would make for high-yielding biomass crops.

Biomass is harvested for its carbon and not for its nitrogen content, and, in fact, low plant N uptake is an advantage in reducing N emissions on combustion. In contrast to cereal crops, where grain N concentrations can be between 1.5% and 2.5% on a dry weight basis, biomass crops such as willow and *Miscanthus* have around one-tenth of these values. The agronomy of biomass crops does not, therefore, involve large additions of N fertiliser. Quality in biomass crops intended for fermentation rather than combustion means a high level of low molecular weight carbohydrates, in order to reduce energy inputs. A phenology that permits a long growing season is clearly necessary and biomass crops should have low harvest indices when defined in terms of

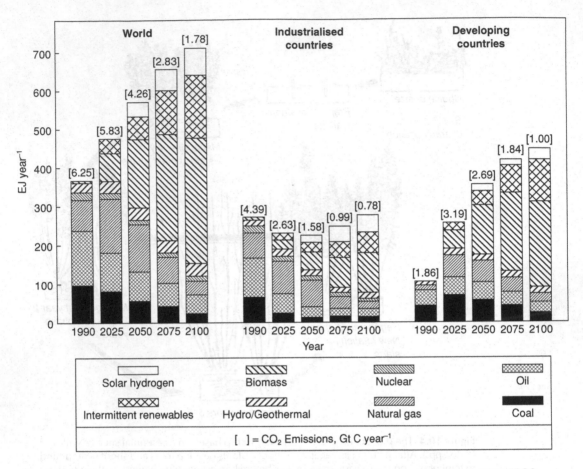

Figure 10.5 The Low Emissions Supply Scenario (LESS) for global primary energy production to 2100 (IPCC 2001). The proportion of global energy derived from biomass is predicted to increase sharply over the next century with a consequent reduction in CO_2 emissions (taken from IPCC 2001).

reproductive structures, in contrast to cereals. Water use efficiency should be high and leaf size and shape should be effective in competition against weeds. Fungal disease resistance should be multi-genic, and mixed cultivar plantings should be encouraged, as spraying against diseases is technically difficult in stands 4–5 m high. Furthermore, since the perenniality of biomass crops means that cultivars cannot be regularly substituted in a plantation, field resistance against diseases and pests should be given high priority.

10.6   The potential for increasing crop photosynthesis and yield

The projected global population within the next 30–40 years is 9–10 billion people. A significant proportion of these will be in Asia where it has been calculated that a population equal to the current total of 6 billion will live in the future. The world demand for fine grain cereals, such as wheat, is predicted to increase by 30% by 2020 and the demand for animal feed crops to rise by up to 50% (Rosegrant *et al.* 2001) including high-protein animal feeds such as soybean. Most of this increased demand must be met from increased productivity per unit area and per unit resource since there is little unused land of agricultural potential left in the world. Although it is true that the main reasons for current malnutrition are socio-political and economic in a world in which the majority has sufficient food, the addition of another 50% to the current population binds humans to increasing crop yields once more. Appeals for global vegetarianism and a fairer distribution of food can help moderate but not deny the clinical nutritional arithmetic of feeding 9–10 billion people.

Past experience suggests that the necessary increases in yield cannot come from increasing the harvest index of grain crops much beyond the current highest values of about 0.55. As there are theoretical reasons for this (Section 6.3), the search for higher yields has to focus on raising the biomass production of the major crops. Following this general guideline, two major routes can be followed. Dry matter can be raised either by raising RUE or by increasing the amount of PAR that a crop intercepts during its lifecycle at unchanged RUE. Raising RUE means either increasing the rates of photosynthesis or reducing the rates of respiration (Equation 4.3). Increasing the interception of PAR interception is more complex as it involves the architectural arrangement of leaves and shoots in the crop canopy, as well as the efficiency with which resources such as water and nitrogen are used, but some improvement can be achieved by practices such as double cropping (Timsina and Connor 2001). Despite huge research efforts aimed at raising the level of dry matter production by improving knowledge of the details of leaf photosynthesis, there have been no examples, for the major crops, where higher rates of photosynthesis have resulted in harvested yield increases. In fact maximal rates of photosynthesis per unit leaf area correlate poorly or even negatively with crop yield; this is because it is canopy behaviour that determines yield and, within a canopy, leaves are in a variety of micrometeorological and radiation environments (Chapters 3 and 4).

As shown specifically for photosynthesis and respiration in Chapters 4 and 5, plant processes are multi-scaled and complex and, therefore, resemble analogue processes more than the digital switches of flowering and other developmental events, upon which molecular biology has concentrated hitherto. Different approaches are needed: for example, an empirically based 'thought experiment' (Sinclair *et al.* 2004) on soybean, where high leaf photosynthetic rate is a quantitatively inherited trait (Wiebold *et al.* 1981). The authors calculate the impact that enhanced photosynthetic activity at one level has on the activity at the next level of complexity, ending with grain yield (Figure 10.6). In other words, they determine how much of an enhancement at the molecular effect remains after cascading from the genome to the grain yield of a crop. The approach is similar to the calculation of the proportion of sunlight that can be used by a crop in Chapter 4.2.

If the cells in soybean leaves can be transformed to produce 50% more mRNA for synthesis of the small and large subunits of Rubisco, there should be 37% more Rubisco, which, in turn, is estimated to give a 33% increase in light-saturated leaf photosynthetic rate (Jiang *et al.* 1993; Sinclair *et al.* 2004). Assuming that an isolated plant has as much as 60% of its leaf area exposed

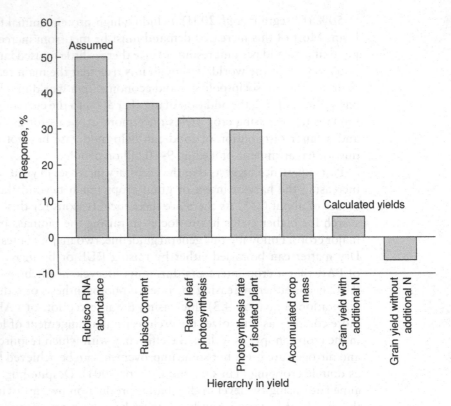

Figure 10.6 Carbon assimilation capacity and grain yield calculated at each increasing scale of crop hierarchy beginning with an assumed 50% increase in mRNA concentration (from Sinclair *et al.* 2004).

to saturating levels of PAR, it can be estimated that photosynthetic carbon assimilation would increase by 30% because of the initial increase in mRNA. However, allowing for competition for light in the stand of plants, the increase in carbon assimilation falls from the original potential of 30% to 18%. As the soybean crop needs to produce harvestable grain containing protein, derived mainly from the senescing leaves, yield is dependent on continuing uptake of nitrogen during grain growth and remobilisation from Rubisco, which is the primary store of nitrogen during *vegetative* development. If nitrogen is readily available in the soil to meet the requirement for grain nitrogen, then sufficient nitrogen will have been stored in Rubisco to allow a 6% increase in soybean grain yield. However, any possible yield increase from genetic modification can be reversed if nitrogen accumulation by the transformed plants does not also increase. Perversely, larger vegetative plants that result from stimulated photosynthesis require more nitrogen to be incorporated into structural components of the vegetative tissue and this nitrogen is therefore not available for subsequent transfer to the grain. Hence, without additional nitrogen accumulation there will be a net decrease in nitrogen available for the grain and a calculated grain yield *decrease* of 6%. Similar conclusions were reached by Porter *et al.* (1995)

in a modelling exercise where the maximum rate of uptake of N into the crop was increased. Many factors play a role in the damping and 'averaging out' of responses on moving from the genome to the whole plant. These include issues covered in this book such as the non-linear relationships between light level and rate of photosynthesis, plant architecture and the interception of radiation. Taken together these mean that a 'plant response' is always an average of a series of optimal, supra- and sub-optimal responses by different plant parts. Such considerations go some way to explain the difficulty of increasing yield by improving leaf photosynthetic capacity only. Current or future targets for molecular modification of photosynthesis to introduce the precursor organic acid C4 pathway into C3 species may yield understanding but are very unlikely to raise C3 crop yields.

If it is unlikely that improvements in photosynthesis and its associated processes will lead to improved yields and higher food production, where are they likely to come from? It is a matter of speculation but improved crop management, in which existing yield potential is allowed to express itself more widely, is likely to be very beneficial. Much more is now known about the responses of crops to their environment and this can be combined with farmers' knowledge to allow a better and more effective use of, especially, water and fertilisers. Techniques such as predictive modelling (Chapter 9), used in cooperation with farmers as a platform for their decision making, can give them an idea of what yields are possible for their site and season and suggest management options. Thus, the second Green Revolution is likely to be as much based on improved communication and knowledge of cropping systems at the farm level as the first Green Revolution was on improving the basic biology of food production. Techniques that could play a useful role include the precision application of water and nutrients with the goal of raising the poorest yields, and thereby, the average.

10.7 The last words

Interest in the physiology of crop yield started during that part of the technical revolution in agriculture characterised as the mechanisation phase (1900 onwards); progressed through the periods of intensification of use of chemicals and energy in the rich world (1940s and 1950s); was formative to the background thinking and interpretation of the Green Revolution (1960s and 1970s); was a strong driver for the introduction of information technologies and modelling in agriculture (1980s and 1990s); but has since played a small part in the bio(techno)logical revolution (1990 onwards). If crop physiology has a role, it has to be in the middle ground between the extreme reductionism of genetic engineering and the wholesale integration of efforts to predict yields and to breed better plants. Two conditions must be met before this fusion of knowledge from different disciplines can occur. The first is that molecular geneticists and crop physiologists learn each other's 'language'. The second is that funding agencies actively support joint interdisciplinary research activities, rather than merely paying lip service as in the past. Molecular genetics risks running into an intellectual dead-end if the issue of scaling to the crop level is not grasped; cooperation with physiologists and modellers can provide the means to overcome this. Central to these developments must be the training of students in crop physiology. This book has been written to synthesise existing knowledge for the benefit of students, teachers and professionals.

References

Aggarwal P K, Talukdar K K, Mall R K (2000) Potential yields of rice–wheat system in the Indo-Gangetic Plains of India. *Rice–Wheat Consortium Paper Series 10.* Rice–Wheat Consortium for the Indo-Gangetic Plains, New Delhi, India, 16 pp

Allen E J (1979) Effects of cutting seed tubers on number of stems and tubers and tuber yields of several potato varieties. *Journal of Agricultural Science, Cambridge* 93: 121–8

Allen E J, Scott R K (1980) An analysis of growth of the potato crop. *Journal of Agricultural Science, Cambridge* 94: 583–606

Allen E J, Scott R K (1992) Principles of agronomy and their application in the potato industry. In Harris P M (ed) *The Potato Crop.* 2nd edition. Chapman and Hall, pp 816–81

Allen E J, Wurr D C E (1992) Plant density. In Harris P M (ed) *The Potato Crop.* 2nd edition. Chapman and Hall, pp 292–333

Alves B J R, Boddey R M, Segundo U (2003) The success of BNF in soybean in Brazil. *Plant and Soil* 252: 1–9

Amthor J S (1989) *Respiration and Crop Productivity.* Springer Verlag, USA

Amthor J S (1991) Respiration in a future, higher CO_2 world. *Plant, Cell and Environment* 14: 13–20

Amthor J S (1994) Plant respiratory responses to the environment and their effects on the carbon balance. In Wilkinson R E (ed) *Plant–Environment Interactions.* Marcel Dekker, pp 501–4

Amthor J S (2000) The McCree–de Wit–Penning de Vries–Thornley respiration paradigm: 30 years later. *Annals of Botany* 86: 1–20

Andrade F H, Vega C, Uhart S, Cirlo A, Cantarero M, Valentinuz O (1999) Kernel number determination in maize. *Crop Science* 39: 453–9

Andrade F H, Cirilo A G, Echarte L (2000a) Factors affecting kernel number in maize. In Otegui M E, Slafer G A (eds) *Physiological Bases for Maize Improvement.* Food Products Press, pp 59–74

Andrade F H, Otegui M E, Vega C (2000b) Intercepted radiation at flowering and kernel number in maize. *Agronomy Journal* 92: 92–7

Angus J F, Hasegawa S, Hsiao T C, Liboon S P, Zandstra H G (1983) The water balance of post-monsoonal dryland crops. *Journal of Agricultural Science, Cambridge* 101: 699–710

Appeldoorn N J G, de Bruijn S M, Koot-Gronsveld E A M, Visser R G F, Vreugdenhil D, van der Plas L H W (1999) Developmental changes in enzymes involved in the conversion of hexose phosphate and its subsequent metabolites during early tuberization of potato. *Plant, Cell and Environment* 22: 1085–96

Appleford N E J, Lenton J R (1991) Gibberellins and leaf expansion in near-isogenic wheat lines containing *Rht1* and *Rht2* dwarfing alleles. *Planta, Berlin* 183: 229–36

Araus J L, Slafer G A, Reynolds M P, Royo C (2002) Plant breeding and drought in C_3 cereals: what should we breed for? *Annals of Botany* 89: 925–40

Armstrong M J, Milford G F M, Pocock T O, Last P J, Day W (1986) The dynamics of nitrogen uptake and its remobilization during growth of sugar beet. *Journal of Agricultural Science, Cambridge* 107: 145–54

Austin R B, Morgan C L, Ford M A, Blackwell R D (1980a) Contributions to grain yield from pre-anthesis assimilation in tall and dwarf barley phenotypes in two contrasting seasons. *Annals of Botany* 45: 309–19

Austin R B, Bingham J, Blackwell R D, Evans L T, Ford M A, Morgan C L, Taylor M (1980b) Genetic improvements in winter wheat since 1900 and associated physiological changes. *Journal of Agricultural Science, Cambridge* 94: 675–89

Austin R B, Ford M A, Morgan C L (1989) Genetic improvement in the yield of winter wheat: a further evaluation. *Journal of Agricultural Science, Cambridge* 112: 295–301

Austin R B, Ford M A, Morgan C L, Yeoman D (1993) Old and modern wheat cultivars compared on the Broadbalk wheat experiment. *European Journal of Agronomy* 2: 141–7

Azcón-Bieto J, Osmond C B (1983) Relationship between photosynthesis and respiration. The effect of carbohydrate status on the rate of CO_2 production by respiration in darkened and illuminated wheat leaves. *Plant Physiology* 71: 574–81

Bachem C, van der Hoeven R, Lucker J, Oomen R, Casarini E, Jacobsen E, Visser R (2000) Functional genomic analysis of potato tuber life-cycle. *Potato Research* 43: 297–312

Bainbridge G, Madgwick P, Parmar S, Mitchell R A, Paul M, Pitts J, Keys A J, Parry M A J (1995) Engineering rubisco to change catalytic properties. *Journal of Experimental Botany* 46: 1269–76

Baker C K, Gallagher J N, Monteith J L (1980) Daylength change and leaf appearance in winter wheat. *Plant, Cell and Environment* 3: 285–7

Baker D N, Hesketh J D, Duncan W G (1972) Simulation of the growth and yield in cotton. 1. Gross photosynthesis, respiration and growth. *Crop Science* 12: 431–5

Bange M P, Hammer G L, Rickert K G (1997) Effect of radiation environment on radiation use efficiency and growth of sunflower. *Crop Science* 37: 1208–14

Bänziger M, Edmeades G O (1997) Genotypic variation for transpiration efficiency in a lowland tropical maize population. In Edmeades G O, Bänziger M, Mickelson H R, Peña-Valdivia C B (eds) *Developing Drought- and Low N-Tolerant Maize*. CIMMYT, Mexico D.F., pp 189–91

Bänziger M, Edmeades G O, Lafitte H R (1999) Selection for drought tolerance increases maize yields across a range of nitrogen levels. *Crop Science* 39: 1035–40

Bänziger M, Edmeades G O, Beck D, Bellon M (2000) *Breeding for Drought and Nitrogen Stress Tolerance in Maize: From Theory to Practice*. CIMMYT, Mexico D.F.

Barneix A J, Cooper H D, Stulen I, Lambers H (1988) Metabolism and translocation of nitrogen in two *Lolium perenne* populations with contrasting rates of mature leaf respiration and yield. *Physiologia Plantarum* 61: 357–62

Beever D E, Offer N, Gill M (2000) The feeding value of grass and grass products. In Hopkins A (ed) *Grass, its Production and Utilization*. Blackwell Science, pp 140–95

Below F E (1995) Nitrogen metabolism and crop productivity. In Pessarakli M (ed) *Handbook of Plant and Crop Physiology*. Marcel Dekker, pp 275–301

Bennett J M, Mutti L S M, Rao P S C, Jones J W (1989) Interactive effects of nitrogen and water stresses on biomass accumulation, nitrogen uptake, and seed yield of maize. *Field Crops Research* **19**: 297–311

Bernacchi C J, Singaas E L, Pimental C, Portis A R, Long S P (2001) Improved temperature response functions for models of Rubisco-limited photosynthesis. *Plant, Cell and Environment* **24**: 253–60

Bindi M, Porter J R, Miglietta F (1995) Comparison of models to simulate leaf appearance in wheat. *European Journal of Agronomy* **4**: 15–25

Bindi M, Sinclair T R, Harrison J (1999) Analysis of seed growth by linear increase in harvest index. *Crop Science* **39**: 486–93

Bindraban P S, Sayre K D, Solis-Moya E (1998) Identifying factors that determine kernel number in wheat. *Field Crops Research* **58**: 223–34

Birch C J, Vos J, Kiniry J, Bos H J, Elings A (1998) Phyllochron responds to acclimation to temperature and irradiance in maize. *Field Crops Research* **59**: 187–200

Bird I F, Cornelius M J, Keys A J (1982) Affinity of RuBP carboxylases for carbon dioxide and inhibition of the enzymes by oxygen. *Journal of Experimental Botany* **33**: 1004–13

Biscoe P V, Gallagher J N (1977) Weather, dry matter production and yield. In Landsberg J J, Cutting C V (eds) *Environmental Effects on Crop Physiology*. Academic Press, pp 75–100

Biscoe P V, Willington V B A (1984) Cereal crop physiology – a key to accurate nitrogen timing. In *Marketable Yield of Cereals, Course Papers December 1984*. Arable Unit, National Agricultural Centre, Coventry, UK, pp 67–74

Biscoe P V, Scott R K, Monteith J L (1975a) Barley and its environment. 3. Carbon budget of the stand. *Journal of Applied Ecology* **12**: 269–93

Biscoe P V, Gallagher J N, Littleton E J, Monteith J L, Scott R K (1975b) Barley and its environment. 4. Sources of assimilate for the grain. *Journal of Applied Ecology* **12**: 295–318

Blad B L, Baker D G (1972) Orientation and distribution of leaves within soybean canopies. *Agronomy Journal* **64**: 26–9

Blum A (1998) Improving wheat grain filling under stress by stem reserve mobilisation. *Euphytica* **100**: 77–83

Bly A G, Woodward H J (2003) Foliar nitrogen application timing influence on grain yield and protein concentration of hard red winter and spring wheat. *Agronomy Journal* **95**: 335–8

Boedhram N, Arkebauer T J, Batchelor W D (2001) Season-long characterization of vertical distribution of leaf area in corn. *Agronomy Journal* **93**: 1235–42

Bolaños J, Edmeades G O (1993) Eight cycles of selection for drought tolerance in lowland tropical maize. I. Responses in grain yield, biomass, and radiation utilization. II. Responses in reproductive behavior. *Field Crops Research* **31**: 233–52, 253–68

Bolaños J, Edmeades G O (1996) The importance of the anthesis-silking interval in breeding for drought tolerance in tropical maize. *Field Crops Research* **48**: 65–80

Bolaños J, Edmeades G O, Martinez L (1983) Eight cycles of selection for drought tolerance in lowland tropical maize. 3. Responses to drought-adaptive physiological and morphological traits. *Field Crops Research* **31**: 269–86

Bonhomme R (2000) Bases and limits to using 'degree.day' units. *European Journal of Agronomy* **13**: 1–10

Boote K J, Jones J W, Hoogenboom G (1998) Simulation of crop growth: CROPGRO model. In Peart R M, Curry R B (eds) *Agricultural Systems Modelling and Simulation*. Marcel Dekker Inc, pp 651–92

Borrell A K, Incoll L D, Dalling M J (1993) The influence of the Rht_1 and Rht_2 alleles on the deposition and use of stem reserves in wheat. *Annals of Botany* 71: 317–26

Bouma T J, de Visser R (1993) Energy requirements for maintenance of ion concentrations in roots. *Physiologia Plantarum* 89: 133–42

Bouma T J, de Visser R, Janssen J H J A, de Kock M J, van Leeuwen P H, Lambers H (1994) Respiratory energy requirements and rate of protein turnover *in vivo* determined by the use of an inhibitor of protein synthesis and a probe to assess its effect. *Physiologia Plantarum* 92: 585–94

Brancourt-Hulmel M, Doussinault G, Lecomte C, Berard P, Le Buanec B, Trottet M (2003) Genetic improvement of agronomic traits of winter wheat cultivars released in France from 1946–1992. *Crop Science* 43: 37–45

Bray E A (2002) Classification of genes differentially expressed during water-deficit stress in *Arabidopsis thaliana*: an analysis using microarray and differential expression data. *Annals of Botany* 89: 803–11

Brim C A, Burton J W (1979) Recurrent selection in soybeans. 2. Selection for increased percent protein in seeds. *Crop Science* 19: 494–8

Brooking I R (1996) Temperature response of vernalization in wheat: a developmental analysis. *Annals of Botany* 78: 507–12

Brooking I R, Jamieson P D (2002) Temperature and photoperiod response of vernalization in near-isogenic lines of wheat. *Field Crops Research* 79: 21–38

Brooking I R, Jamieson P D, Porter J R (1995) The influence of daylength on final leaf number in spring wheat. *Field Crops Research* 41: 155–65

Brooks A, Farquhar G D (1985) Effect of temperature on the CO_2/O_2 specificity of ribulose-1,5-bisphosphate carboxylase/oxygenase and the rate of respiration in the light. *Planta* 165: 397–406

Brown R H, Blaser R E (1968) Leaf area index in pasture growth. *Herbage Abstracts* 38: 1–9

Bruhn C M (2003) Meeting demand for food safety, quality and environmental protection. *Proceedings of the BCPC International Congress – Crop Science and Technology*, pp 27–34

Brumm T J, Hurburgh C R (2003) United States Soybean Crop Quality Survey. http://www.soygrowers.com/international/quality/US-SB-Quality-03.pdf

Burton J W (1997) Soyabean (*Glycine max* (L.) Merr.). *Field Crops Research* 53: 171–86

Burton J W, Brim C A (1981) Recurrent selection in soybeans. 3. Selection for increased percent oil in seeds. *Crop Science* 21: 31–4

Burton W G (1966) *The Potato*. Veenman and Zonen NV

Byrd G T, Sage R T, Brown R H (1992) A comparison of dark respiration between C_3 and C_4 plants. *Plant Physiology* 100: 191–8

Calderini D F, Torres-Léon S, Slafer G A (1995) Consequences of wheat breeding on nitrogen and phosphorus yield, grain nitrogen and phosphorus concentration and associated traits. *Annals of Botany* 76: 315–22

Calderini D F, Miralles D J, Sadras V O (1996) Appearance and growth of individual leaves as affected by semidwarfism in isogenic lines of wheat. *Annals of Botany* 77: 583–9

Caligari P D S (1992) Breeding new varieties. In Harris P M (ed) *The Potato Crop* 2nd edition. Chapman and Hall, pp 334–72

Camp P J, Huber S C, Burke J J, Moreland D E (1982) Biochemical changes that occur during the senescence of wheat leaves. I Basis for the reduction of photosynthesis. *Plant Physiology* 70: 1641–6

Campbell G S, van Evert F K (1994) Light interception by plant canopies: efficiency and architecture. In Monteith J L, Scott R K, Unsworth M H (eds) *Resource Capture by Crops*. Nottingham University Press, pp 35–52

Cannell M G R, Thornley J H M (2000) Modelling the components of plant respiration and some guiding principles. *Annals of Botany* 85: 45–54

Cao W, Tibbitts J W (1995) Leaf emergence on potato stems in relation to thermal time. *Agronomy Journal* 87: 474–7

Carlson J B, Lersten N R (1987) Reproductive morphology. In Wilcox J R (ed) *Soybeans: Improvement, Production and Uses*. Agronomy Society of America, Madison, Wisconsin, pp 95–134

Chapman S C, Edmeades G O (1999) Selection improves drought tolerance in tropical maize populations 2. Direct and correlated responses among secondary traits. *Crop Science* 39: 1315–24

Chimenti C A, Cantagallo J, Guevara E (1997) Osmotic adjustment in maize: genetic variation and association with water uptake. In Edmeades G O, Bänziger M, Mickelson H R, Peña-Valdivia C B (eds) *Developing Drought- and Low N-tolerant Maize*. CIMMYT, Mexico, pp 200–3

Chujo H (1966) Differences in vernalisation effect in wheat under various temperatures. *Proceedings of the Crop Science Society of Japan* 35: 177–86

Clutterbuck B J, Simpson K (1978) The interactions of water and fertiliser nitrogen in effects on growth pattern and yield of potatoes. *Journal of Agricultural Science, Cambridge* 91: 161–72

Cober E R, Voldeng H D (2000) Developing high-protein, high-yield soybean populations and lines. *Crop Science* 40: 39–42

Cober E R, Stewart D W, Voldeng H D (2001) Photoperiod and temperature responses in early-maturing, near-isogenic soybean lines. *Crop Science* 41: 721–7

Cochrane M P, Duffus C M (1981) Endosperm cell number in barley. *Nature* 289: 399–401

Cochrane M P, Duffus C M (1983) Endosperm cell number in cultivars of barley differing in grain weight. *Annals of Applied Biology* 102: 177–81

Collins RP, Rhodes I (1995) Stolon characteristics related to winter survival in white clover. *Journal of Agricultural Science, Cambridge* 124: 11–16

Collins RP, Glendining MJ, Rhodes I (1991) The relationships between stolon characteristics, winter survival and annual yields in white clover (*Trifolium repens* L.) *Grass and Forage Science* 46: 51–61

Colnenne C, Meynard J M, Reau R, Justes E, Merrien A (1998) Determination of a critical nitrogen dilution curve for winter oilseed rape. *Annals of Botany* 81: 311–7

Condon A G, Richards R A (1993) Exploiting genetic variation in transpiration efficiency in wheat: an agronomic view. In Ehleringer J R, Hall A E, Farquhar G D (eds) *Stable Isotopes and Plant Carbon-Water Relations*. Academic Press, pp 435–62

Connor D J, Jones T R (1985) Responses of sunflower to strategies of irrigation. 2. Morphological and physiological responses to water stress. *Field Crops Research* 12: 91–103

Cook M G, Evans L T (1978) Effect of relative size and distance of competing sinks on the distribution of photosynthetic assimilates in wheat. *Australian Journal of Plant Physiology* 5: 495–509

Corbel G, Robin C, Frankow-Lindberg BE, Ourry A, Guckert A (1999) Regrowth of white clover after chilling: assimilate partitioning and vegetative storage proteins. *Crop Science* 39: 1756–61

Cosgrove D J (2000) Loosening of plant cell walls by expansins. *Nature* 407: 321–6

Cramer G R, Bowman D C (1991) Kinetics of maize leaf elongation. 1. Increased yield threshold limits short-term, steady-state elongation rates after exposure to salinity. *Journal of Experimental Botany* 42: 1417–26

Curran P J, Foody G M, van Gardingen P R (1997) Scaling up. In van Gardingen P R, Foody G M, Curran P J (eds) *Scaling Up from Cell to Landscape*. Cambridge University Press, pp 1–6

Dale J E (1982) *The Growth of Leaves*. Edward Arnold

Dale J E (1985) The carbon relations of the developing leaf. In Baker N R, Davies W J, Ong C K (eds) *Control of Leaf Growth*. Cambridge University Press, pp 135–53

Dale J E, Milthorpe F L (1983) General features of the production and growth of leaves. In Dale J E, Milthorpe F L (eds) *The Growth and Functioning of Leaves*. Cambridge University Press, pp 151–78

Darby H M, Lauer J G (2002) Harvest date and hybrid influence on corn forage yield, quality, and preservation. *Agronomy Journal* 94: 559–66

Darwinkel A (1978) Patterns of tillering and grain production of winter wheat at a wide range of plant densities. *Netherlands Journal of Agricultural Science* 26: 283–98

Davies W J, Gowing D J G (1999) Plant responses to small perturbations in soil water status. In Press M C, Scholes J D, Barker M G (eds) *Physiological Plant Ecology*. Blackwell Science, pp 67–89

Davies W J, Tardieu F, Trejo C L (1994) How do chemical signals work in plants that grow in drying soil? *Plant Physiology* 104: 309–14

Davies W J, Wilkinson S, Loveys B (2002) Stomatal control by chemical signalling and the exploitation of this mechanism to increase water use efficiency in agriculture. *New Phytologist* 153: 449–60

de Pury D G G, Farquhar G D (1997) Simple scaling of photosynthesis from leaves to simple canopies without the errors of big-leaf models. *Plant, Cell and Environment* 20: 537–57

de Visser R, Spitters C J T, Bouma T J (1992) Energy cost of protein turnover: theoretical calculation and experimental estimation from regression of respiration on protein concentration of full-grown leaves. In Lambers H, van der Plas L H W (eds) *Molecular, Biochemical and Physiological Aspects of Plant Respiration*. SPB Academic Publishing, the Netherlands, pp 493–508

de Wit C T (1958) Transpiration and crop yields. *Verslagen van Lanbouwkundige Onderzoekingen* 64.6. Wageningen, the Netherlands

de Wit C T (1992) Resource use efficiency in agriculture. *Agricultural Systems* 40: 125–51

Delécolle R, Hay R K M, Guérif M, Pluchard P, Varlet-Grancher C (1989) A method of describing the progress of apical development in wheat, based on the time-course of organogenesis. *Field Crops Research* 21: 147–60

Dennett M D, Auld B A, Elston J (1978) A description of leaf growth in *Vicia faba* L. *Annals of Botany* 42: 223–32

Devienne-Barret F, Justes E, Machet J M, Mary B (2000) Integrated control of nitrate uptake by crop growth rate and soil nitrate availability under field conditions. *Annals of Botany* 86: 995–1005

Dewar R C, Medlyn B E, McMurtrie R E (1997) A mechanistic analysis of light and carbon use efficiencies. *Plant, Cell and Environment* 21: 573–88

Dofing S M (1999) Inheritance of phyllochron in barley. *Crop Science* 39: 334–7

Donaghy D J, Fulkerson W J (1998) Priority for allocation of water-soluble carbohydrate reserves during regrowth of *Lolium perenne*. *Grass and Forage Science* 53: 211–8

Donald C M (1968) The breeding of crop ideotypes. *Euphytica* 17: 385–403

Donatelli M, Hammer G L, Vanderlip R L (1992) Genotype and water limitation effects on phenology, growth and transpiration efficiency in grain sorghum. *Crop Science* 32: 781–6

Drake B G, Gonzàlez-Meier M A, Long S P (1997) More efficient plants: a consequence of rising atmospheric CO_2? *Annual Review of Plant Physiology and Plant Molecular Biology* 48: 609–39

Drake B G, Azcón-Bieto J, Berry J, Bunce J, Dikstra P, Farrar J, Gifford R M, Gonzàlez-Meier M A, Koch G, Lambers H, Siedow J, Wullschleger S (1999) Does elevated atmospheric CO_2 inhibit mitochondrial respiration in green plants? *Plant, Cell and Environment* 22: 649–57

Dreccer M F, van Oijen M, Schapendonk A H C M, Pot C S, Rabbinge R (2000) Dynamics of vertical leaf nitrogen distribution in a vegetative wheat canopy. Impact on canopy photosynthesis. *Annals of Botany* 86: 821–31

Drouet J-L, Bonhomme R (1999) Do variations in local leaf irradiance explain changes to leaf nitrogen within row maize canopies? *Annals of Botany* 84: 61–9

Drouet J-L, Moulia B, Bonhomme R (1999) Do changes in the azimuthal distribution of maize leaves over time affect canopy light absorption? *Agronomie* 19: 281–94

Duru M, Ducrocq H (2000) Growth and senescence of the successive grass leaves on a tiller. Ontogenic development and effect of temperature. *Annals of Botany* 85: 635–43

Duvick D N (1992) Genetic contributions to advances in yield of U.S. maize. *Maydica* 37: 69–79

Duvick L M, Cassman K G (1999) Post-green revolution trends in yield potential of temperate maize in the North-Central United States. *Crop Science* 39: 1622–30

Dwyer L M, Stewart D W (1986) Leaf area development in field grown maize. *Agronomy Journal* 78: 334–43

Dwyer L M, Stewart D W, Hamilton R I, Houwing L (1992) Ear position and vertical distribution of leaf area in corn. *Agronomy Journal* 84: 430–8

Earl H J, Tollenaar M (1998) Differences among commercial maize (*Zea mays* L.) hybrids in respiration rates of mature leaves. *Field Crops Research* 59: 9–19

Echarte L, Luque S, Andrade F H, Sadras V O, Cirilo A, Otegui M E, Vega C R C (2000) Responses of maize kernel number to plant density in Argentinian hybrids released between 1965 and 1993. *Field Crops Research* 68: 1–8

Edmeades G O, Daynard T B (1979a) The development of plant-to-plant variability in maize at different planting densities. *Canadian Journal of Plant Science* 59: 561–76

Edmeades G O, Daynard T B (1979b) The relationship between final yield and photosynthesis at flowering in individual maize plants. *Canadian Journal of Plant Science* 59: 585–601

Edmeades G O, Bolaños J, Chapman S C, Lafitte H R, Bänziger M (1999) Selection improves drought tolerance in tropical maize populations: 1. Gains in biomass, grain yield, and harvest index. *Crop Science* 39: 1306–15

Edmeades G O, Bolaños J, Elings A, Ribaut J-M, Bänziger M, Westgate M E (2000) The role and regulation of the anthesis-silking interval in maize. In *Physiology and Modelling Kernel Set in Maize*. Crop Science Society of America Special Publication 29: 43–73

Edwards G, Walker D A (1983) C_3, C_4: *Mechanisms and Cellular Regulation of Photosynthesis*. Blackwell Scientific Publications

Egli D B (1998) *Seed Biology and the Yield of Grain Crops*. CAB International

Egli D B, Zhen-wen Y (1991) Crop growth rate and seeds per unit area in soybean. *Crop Science* 31: 439–42

Egli D B, Ramseur E L, Zhen-wen Y, Sullivan C H (1989) Source–sink alterations affect the number of cells in soybean cotyledons. *Crop Science* **29**: 732–5

Ehleringer J R, Björkman O (1977) Quantum yields for CO_2 uptake in C_3 and C_4 plants. Dependence on temperature, CO_2 and O_2 concentrations. *Plant Physiology* **59**: 86–90

Ehleringer J R, Pearcy R W (1983) Variations in quantum yield for CO_2 uptake among C_3 and C_4 plants. *Plant Physiology* **73**: 555–9

Ehleringer J R, Hall A E, Farquhar G D (eds) (1993) *Stable Isotopes and Plant Carbon–Water Relations*. Academic Press

Engels C, Marschner H (1986) Allocation of photosynthate to individual tubers of *Solanum tuberosum* L. *Journal of Experimental Botany* **37**: 1813–22

Engels C, Marschner H (1987) Effects of reducing leaf area and tuber number on the growth rates of tubers on individual potato plants. *Potato Research* **30**: 177–86

Evans J R (1989) Photosynthesis and nitrogen relationships in leaves of C_3 plants. *Oecologia* **78**: 9–19

Evans J R, Farquhar G D (1992) Modeling canopy photosynthesis from the biochemistry of the C_3 chloroplast. In Boote K J, Loomis R S (eds) *Modeling Crop Photosynthesis – from Biochemistry to Canopy*. Crop Science Society of America, pp 1–15

Evans J R, von Caemmerer S (1996) Carbon dioxide diffusion inside leaves. *Plant Physiology* **110**: 339–46

Evans J R, Sharkey T D, Berry J A, Farquhar G D (1986) Carbon isotope discrimination measured concurrently with gas exchange to investigate CO_2 diffusion in leaves of higher plants. *Australian Journal of Plant Physiology* **13**: 281–92

Evans L T (1975) The physiological basis of crop yield. In Evans L T (ed) *Crop Physiology*. Cambridge University Press, pp 327–335

Evans L T (1993) *Crop Evolution, Adaptation and Yield*. Cambridge University Press

Evans L T (1994) Crop physiology: prospects for the retrospective science. In Boote K J, Bennett J M, Sinclair T R, Paulsen G M (eds) *Physiology and Determination of Crop Yield*. Crop Science Society of America, pp 19–35

Evans L T (1998) *Feeding the Ten Billion*. Cambridge University Press

Evans L T, Dunstone R L (1970) Some physiological aspects of evolution in wheat. *Australian Journal of Biological Sciences* **23**: 725–41

Evans L T, Fischer R A (1999) Yield potential: its definition, measurement, and significance. *Crop Science* **39**: 1544–51

Evans L T, Wardlaw I F, Fischer R A (1975) Wheat. In Evans L T (ed) *Crop Physiology*. Cambridge University Press, pp 101–49

Evans M S, Poethig R S (1995) Gibberellins promote vegetative phase change and reproductive maturity in maize. *Plant Physiology, Lancaster* **108**: 475–87

Ewert F (2004) Modelling plant response to elevated CO_2: how important is leaf area index? *Annals of Botany* **93**: 619–27

Ewert F, Porter J R (2000) Ozone effects on wheat in relation to CO_2: modeling short-term and long-term responses of leaf photosynthesis and leaf duration. *Global Change Biology* **6**: 735–50

Farage P K, Long S P (1995) An *in vivo* analysis of photosynthesis during short-term O_3 exposure in three contrasting species. *Photosynthesis Research* **43**: 11–18

Farquhar G D, von Caemmerer S (1982) Modelling of photosynthetic response to environment. *Encyclopedia of Plant Physiology* **12B**: 549–87

Farquhar G D, von Caemmerer S, Berry J A (1980) A biochemical model of photosynthetic CO_2 assimilation in leaves of C_3 species. *Planta* **149**: 78–90

Farquhar G D, Ehleringer J R, Hubick K T (1989) Carbon isotope discrimination and photosynthesis. *Annual Review of Plant Physiology and Plant Molecular Biology* **40**: 503–37

Farrar J F (1980) The pattern of respiration in the vegetative barley plant. *Annals of Botany* 46: 71–6

Farrar J F (1992) The whole plant: carbon partitioning during development. In Pollock C J, Farrar J F, Gordon A J (eds) *Carbon Partitioning within and between Organisms.* Bios Scientific Publishers, pp 163–79

Farrar J F (1999) Carbohydrate: where does it come from, where does it go? In Bryant J A, Burrell M M, Kruger N J (eds) *Plant Carbohydrate Biochemistry.* Bios Scientific Publishers, pp 29–46

Farrar J F, Lewis D H (1987) Nutrient relations in biotrophic infections. In Pegg G F, Ayres P G (eds) *Fungal Infection of Plants.* Cambridge University Press, pp 92–132

Faurie O, Soussana J F, Sinoquet H (1996) Radiation interception, partitioning and use in grass-clover mixtures. *Annals of Botany* 77: 35–45

Fehr W R, Caviness C E (1977) *Stages of Soybean Development. Special Report* 80. Cooperative Extension Service, Iowa State University

Feil B (1997) The inverse yield-protein relationship in cereals: possibilities and limitations for genetically improving the grain protein yield. *Trends in Agronomy* 1: 103–119

Ferris R, Ellis R H, Wheeler T R, Hadley P (1998) Effect of high temperature stress at anthesis on grain yield and biomass of field-grown crops of wheat. *Annals of Botany* 82: 631–9

Field C (1983) Allocating leaf nitrogen for the maximization of carbon gain: leaf age as a control on the allocation program. *Oecologia* 56: 341–7

Firman D M, O'Brien P J, Allen E J (1992) Predicting the emergence of potato sprouts. *Journal of Agricultural Science, Cambridge* 118: 55–61

Firman D M, O'Brien P J, Allen E J (1995) Appearance and growth of individual leaves in the canopies of several potato cultivars. *Journal of Agricultural Science, Cambridge* 125: 379–94

Fischer R A (1985) Number of kernels in wheat crops and the influence of solar radiation and temperature. *Journal of Agricultural Science, Cambridge* 105: 447–61

Fischer R A, Stockman Y M (1986) Increased kernel number in Norin 10-derived dwarf wheat: evaluation of the cause. *Australian Journal of Plant Physiology* 13: 767–84

Fitter A H, Hay R K M (2002) *Environmental Physiology of Plants.* 3rd edition. Academic Press

Fleming A J, Caderas D, Wehrli E, McQueen-Mason S, Kuhlmeier C (1999) Analysis of expansin-induced morphogenesis on the apical meristem of tomato. *Planta* 208: 166–74

Flénet F, Kiniry J R, Board J E, Westgate M E, Reicosky D C (1996) Row spacing effects on light extinction coefficient of corn, sorghum, soybean and sunflower. *Agronomy Journal* 88: 185–90

Flinn A M, Pate J S (1970) A quantitative study of carbon transfer from pod and subtending leaf to the ripening seeds of the field pea (*Pisum arvense* L.). *Journal of Experimental Botany* 21: 71–82

Flintham J E, Börner A, Worland A J, Gale M D (1997) Optimizing wheat grain yield: effect of *Rht* (gibberellin-insensitive) dwarfing genes. *Journal of Agricultural Science, Cambridge* 128: 11–25

Forde B G, Clarkson D T (1998) Nitrate and ammonium nutrition of plants: physiological and molecular perspectives. *Advances in Botanical Research* 30: 1–90

Foroutan-pour K, Dutilleul P, Smith D L (1999) Soybean canopy development as affected by population density and intercropping with corn: fractal analysis in comparison with other quantitative approaches. *Crop Science* 39: 1785–91

Foulkes M J, Scott R K, Sylvester-Bradley R (2001) The ability of wheat varieties to withstand drought in UK conditions: resource capture. *Journal of Agricultural Science, Cambridge* **137**: 1–16

Frankow-Lindberg BE, von Firks HA (1998) Population fluctuations in three contrasting white clover cultivars under cutting, with particular reference to overwintering properties. *Journal of Agricultural Science, Cambridge* **131**: 143–53

Frankow-Lindberg B E, Svanäng K, Höglind M (1997) Effects of an autumn defoliation on overwintering, spring growth and yield of a white clover/grass sward. *Grass and Forage Science* **52**: 360–9

Fray M J, Evans E J, Lydiate D J, Arthur A E (1996) Physiological assessment of apetalous flowers and erectophile pods in oilseed rape (*Brassica napus*). *Journal of Agricultural Science, Cambridge* **127**: 193–200

Freeling M (1992) A conceptual framework for maize leaf development. *Developmental Biology* **153**: 44–58

Gales K (1983) Yield variation of wheat and barley in Britain in relation to crop growth and soil conditions – a review. *Journal of the Science of Food and Agriculture* **34**: 1085–104

Gallagher J N, Biscoe P V (1978) Radiation absorption, growth and yield of cereals. *Journal of Agricultural Science, Cambridge* **91**: 47–60

Gallo K P, Daughtry C S T (1986) Techniques for measuring intercepted and absorbed photosynthetically active radiation in corn canopies. *Agronomy Journal* **78**: 752–6

Garnier E, Gobin O, Poorter H (1995) Nitrogen productivity depends on photosynthetic nitrogen use efficiency and on nitrogen allocation within the plant. *Annals of Botany* **76**: 667–72

Geiger R (1965) *The Climate near the Ground*. Harvard University Press

Gentry B, Briantais J M, Baker N R (1989) The relationship between the quantum yield of photosynthetic electron transport and quenching of chlorophyll fluorescence. *Biochimica Biophysica Acta* **990**: 87–92

Giamoustaris A, Mithen R (1995) The effect of modifying the glucosinolate content of leaves of oilseed rape (*Brassica napus* ssp. *oleifera*) on its interactions with specialist and generalist pests. *Annals of Applied Biology* **126**: 347–63

Girardin P (1992) Leaf azimuth in maize canopies. *European Journal of Agronomy* **1**: 91–7

Godwin D C, Ritchie R T, Singh U, Hunt L A (1989) *A User's Guide to CERES-Wheat V2.10*. International Fertiliser Development Center, USA

Golz J F, Hudson A (2002) Signalling in plant lateral organ development. *The Plant Cell* **Supplement 2002**: S277–88

Gonzalez B, Boucaud J, Salette J, Langlois J, Duyme M (1989) Changes in stubble carbohydrate content during regrowth of defoliated perennial ryegrass (*Lolium perenne* L.) on two nitrogen levels. *Grass and Forage Science* **44**: 411–5

Gooding M J, Dimmock J P R E, France J, Jones S A (2000) Green leaf area decline of wheat leaves: the influence of fungicides and relationships with mean grain weight and grain yield. *Annals of Applied Biology* **136**: 77–84

Gooding M J, Pinyosinwat A, Ellis R H (2002) Responses of wheat grain and quality to seed rate. *Journal of Agricultural Science, Cambridge* **138**: 317–31

Gooding M J, Ellis R H, Shewry P R, Schofield J D (2003) Effects of restricted water availability and increased temperature on the grain filling, drying and quality of winter wheat. *Journal of Cereal Science* **37**: 295–309

Goulding K W T (1990) Nitrogen deposition to land from the atmosphere. *Soil Use and Management* **6**: 61–3

Grace J (1989) Temperature as a determinant of plant productivity. In Long S P, Woodward F I (eds) *Plants and Temperature*. Cambridge University Press, pp 91–107

Gray D (1973) The growth of individual tubers. *Potato Research* 16: 80–4

Green C F (1987) Nitrogen nutrition and wheat growth in relation to absorbed solar radiation. *Agricultural and Forest Meteorology* 41: 207–48

Greenwood D J, Lemaire G, Gosse G, Cruz P, Draycott A, Neeteson J J (1990) Decline in percentage N of C3 and C4 crops with increasing plant mass. *Annals of Botany* 66: 425–36

Gregory P J, Marshall B, Biscoe P V (1981) Nutrient relations of winter wheat. 3. Nitrogen uptake, photosynthesis of the flag leaves and translocation of nitrogen to the grain. *Journal of Agricultural Science, Cambridge* 96: 539–47

Gregory P J, Ingram J, Campbell B, Goudriaan J, Hunt T, Landsberg J J, Linder S, Stafford-Smith M, Sutherst B, Valentin C (1999) Managed production systems. In Walker B, Steffen W, Canadell J, Ingram J (eds) *The Terrestrial Biosphere and Global Change. Implications for Natural and Managed Ecosystems. Synthesis Volume.* International Geosphere-Biosphere Program Book Series 4, Cambridge, UK

Griffin K L, Sims D A, Seerman J R (1999) Altered night-time CO_2 concentration affects the growth, physiology and biochemistry of soybean. *Plant, Cell and Environment* 22: 91–9

Grindlay D J C (1997) Towards an explanation of crop nitrogen demand based on the optimization of leaf nitrogen per unit leaf area. *Journal of Agricultural Science, Cambridge* 128: 377–96

Hall A E, Richards R A, Condon A G, Wright G C, Farquhar G D (1994) Carbon isotope discrimination and plant breeding. *Plant Breeding Reviews* 4: 81–113

Hammer G L, Wright G C (1994) A theoretical analysis of nitrogen and radiation effects on radiation use efficiency in peanut. *Australian Journal of Agricultural Research* 45: 575–89

Hammer G L, Farquhar G D, Broad I J (1997) On the extent of genetic variation for transpiration efficiency in sorghum. *Australian Journal of Agricultural Research* 48: 649–55

Hammer G L, Kropff M J, Sinclair T R, Porter J R (2002) Future contributions of crop modeling – from heuristics and supporting decision making to understanding genetic regulation and aiding crop improvement. *European Journal of Agronomy* 18: 15–31

Hansen S (2002) DAISY, a flexible soil–plant–atmosphere system model. http://www.dina.kvl.dk/~daisy/ftp/DaisyDescription.pdf

Harley P C, Loreto F, Di Marco G, Sharkey T D (1992) Theoretical considerations when estimating the mesophyll conductance to CO_2 flux by analysis of the response of photosynthesis to CO_2. *Plant Physiology* 98: 1429–36

Harper J L (1977) *The Population Biology of Plants.* Academic Press

Harper J L (1986) Modules, branches and the capture of resources. In Jackson J B C, Buss L E, Cook R E (eds) *Population Biology and the Evolution of Clonal Organisms.* Yale University Press, pp 1–34

Harris P M (1992) Mineral nutrition. In Harris P M (ed) *The Potato Crop* 2nd edition. Chapman and Hall, pp 162–213

Harris W, Rhodes I, Mee S S (1983) Observations on environmental and genotypic influences on the overwintering of white clover. *Journal of Applied Ecology* 20: 609–24

Hatfield J L, Carlson R E (1979) Light quality distributions and spectral albedo of three maize canopies. *Agricultural Meteorology* 20: 215–26

Haverkort A J, van de Waart M, Bodlaender K B A (1990) Interrelationships of the number of initial sprouts, stems, stolons and tubers per potato plant. *Potato Research* 33: 269–74

Hay R K M (1981) Timely planting of maize – a case history from the Lilongwe Plain. *Tropical Agriculture, Trinidad* 58: 147–55

Hay R K M (1985) The microclimate of an upland grassland. *Grass and Forage Science* **40**: 201–212

Hay R K M (1986) Sowing date and the relationships between plant and apex development in winter cereals. *Field Crops Research* **14**: 321–37

Hay R K M (1990) The influence of photoperiod on the dry-matter production of grasses and cereals. *The New Phytologist* **116**: 233–54

Hay R K M (1995) Harvest index: a review of its use in plant breeding and crop physiology. *Annals of Applied Biology* **126**: 197–216

Hay R K M (1999) Physiological control of growth and yield in wheat: analysis and synthesis. In Smith D L, Hamel C (eds) *Crop Yield. Physiology and Processes*. Springer Verlag, pp 1–38

Hay R K M, Delécolle R (1989) The setting of rates of development of wheat plants at crop emergence: influence of the environment on rates of leaf appearance. *Annals of Applied Biology* **115**: 333–41

Hay R K M, Ellis R P (1998) The control of flowering in wheat and barley: what recent advances in molecular genetics can reveal. *Annals of Botany* **82**: 541–54

Hay R K M, Gilbert R A (2001) Variation in the harvest index of tropical maize: evaluation of recent evidence from Mexico and Malawi. *Annals of Applied Biology* **138**: 103–9

Hay R K M, Hampson J (1991) Sprout and stem development from potato tubers of differing physiological age: the role of apical dominance. *Field Crops Research* **27**: 1–16

Hay R K M, Kemp D R (1992) The prediction of leaf canopy expansion in the leek from a simple model dependent on primordial development. *Annals of Applied Biology* **120**: 537–45

Hay R K M, Kirby E J M (1991) Convergence and synchrony: a review of the co-ordination of development in wheat. *Australian Journal of Agricultural Research* **42**: 661–700

Hay R K M, Tunnicliffe Wilson G (1982) Leaf appearance and extension in field- grown winter wheat plants: the importance of soil temperature during vegetative growth. *Journal of Agricultural Science, Cambridge* **99**: 403–10

Hay R K M, Walker A J (1989) *An Introduction to the Physiology of Crop Yield.* Longman Scientific and Technical

Hay R K M, Galashan S, Russell G (1986) The yields of arable crops in Scotland 1978–82: actual and potential yields of cereals. *Research and Development in Agriculture* **3**: 59–64

Heath M C, Hebblethwaite P D (1985) Solar radiation interception by leafless, semileafless and leafed peas (*Pisum sativum*) under contrasting field conditions. *Annals of Applied Biology* **107**: 309–18

Heldt H W (1997) *Plant Biochemistry and Molecular Biology*. Oxford University Press

Hellmann H, Barker L, Funck D, Frommer W B (2000) The regulation of assimilate allocation and transport. *Australian Journal of Plant Physiology* **27**: 583–94

Henderson S, von Caemmerer S, Farquhar G D, Wade L, Hammer G (1998) Correlation between carbon isotope discrimination and transpiration efficiency in lines of the C_4 species *Sorghum bicolor* in the glasshouse and the field. *Australian Journal of Plant Physiology* **25**: 111–23

Hipps L E, Asrar G, Kanemasu E T (1983) Assessing the interception of photosynthetically-active radiation in winter wheat. *Agricultural Meteorology* **28**: 253–9

Hirel B, Bertin P, Quilleré I, Bourdoncle W, Attagnant C, Dellay C, Guoy A, Cadiou S, Retailliau C, Falque M, Gallais A (2001) Towards a better understanding of the genetic and physiological basis for nitrogen use efficiency in maize. *Plant Physiology, Lancaster* 125: 1258–70

Hodgson J, Bircham J S, Grant S A, King J (1981) The influence of cutting and grazing management on herbage growth and utilisation. In Wright C E (ed) *Plant Physiology and Herbage Production. Occasional Symposia of the British Grassland Society* 13: 51–62

Hofstra G, Hesketh J D, Myhre D L (1977) A plastochron model for soybean leaf and stem growth. *Canadian Journal of Plant Science* 57: 167–75

Holliday R (1960) Plant population and crop yield. *Field Crops Abstracts* 13: 159–67

Hopkins A (ed.) (2000) *Grass, its Production and Utilization.* Blackwell Science

Hotsonyame G K, Hunt L A (1997) Sowing date and photoperiod effects on leaf appearance in field-grown wheat. *Canadian Journal of Plant Science* 77: 23–31

Hotsonyame G K, Hunt L A (1998) Effects of sowing date, photoperiod and nitrogen on variation in main culm leaf dimensions in field-grown wheat. *Canadian Journal of Plant Science* 78: 35–49

Hsiao T C (1973) Plant responses to water stress. *Annual Review of Plant Physiology* 24: 519–70

Hsiao T C, Acevedo E, Fereres E, Henderson D W (1976) Water stress, growth and osmotic adjustment. *Philosophical Transactions of the Royal Society, London* B273: 479–500

Huber S C, Huber J L A, McMichael R W (1992) The regulation of sucrose synthesis in leaves. In Pollock C J, Farrar J F, Gordon A J (eds) *Carbon Partitioning within and between Organisms.* Bios Scientific Publishers, pp 1–26

Huffaker R C, Peterson L W (1974) Protein transport in plants and possible means of its regulation. *Annual Review of Plant Physiology* 25: 363–92

Hurrell R F (2001) Modifying the composition of plant foods for better human health. In Nösberger J, Geiger H H, Struik P C (eds) *Crop Science: Progress and Prospects.* CABI Publishing, pp 53–64

Hutmacher R B, Kreig D R (1983) Photosynthetic rate control in cotton. Stomatal and non-stomatal factors. *Plant Physiology* 73: 658–61

Ikeda T, Matsuda R (2002) Effects of soyabean leaflet inclination on some factors related to photosynthesis. *Journal of Agricultural Science, Cambridge* 138: 367–73

Indermühle A, Stocker T F, Joos F, Fischer H, Smith H J, Wahlen M, Deck B, Mastroianni D, Tschumi J, Blunier T, Meyer R, Stauffer B (1999) Holocene carbon-cycle dynamics based on CO_2 traped in ice at Taylor Dome. *Nature* 398: 121–6

Ingestad T, Lund A B (1986) Theory and techniques for steady state mineral nutrition and growth of plants. *Scandinavian Journal of Forest Research* 1: 439–53

IPCC (2001) *The Third Assessment Report of the Intergovernmental Panel on Climate Change.* Cambridge University Press

Jackson S D, Pratt S (1996) Control of tuberisation in potato by gibberellins and phytochrome B. *Physiologia Plantarum* 98: 407–12

Jame Y W, Cutforth H W, Ritchie J T (1998) Interaction of temperature and daylength on leaf appearance rate in wheat and barley. *Agricultural and Forest Meteorology* 92: 241–9

Jame Y W, Cutforth H W, Ritchie J T (1999) Temperature response function for leaf appearance rate in wheat and corn. *Canadian Journal of Plant Science* 79: 1–10

Jamieson P D, Semenov M A (2000) Modelling nitrogen uptake and redidtribution in wheat. *Field Crops Research* 68: 21–9

Jamieson P D, Brooking I R, Porter J R, Wilson D R (1995a) Prediction of leaf appearance in wheat: a question of temperature. *Field Crops Research* 41: 35–44

Jamieson P D, Martin R J, Francis G S, Wilson D R (1995b) Drought effects on biomass production and radiation-use efficiency in barley. *Field Crops Research* 43: 77–86

Jamieson P D, Brooking I R, Semenov M A, Porter J R (1998a) Making sense of wheat development: a critique of methodology. *Field Crops Research* 55: 117–27

Jamieson P D, Semenov M A, Brooking I R, Francis G S (1998b) Sirius: a mechanistic model of wheat response to environmental variation. *European Journal of Agronomy* 8: 161–80

Jamieson P D, Porter J R, Goudriaan J, Ritchie J T, van Keulen H, Stoll W (1998c) A comparison of the models *AFRCWheat, CERES-Wheat, Sirius, SUCROS2* and *SWHEAT* with measurements from wheat grown under drought. *Field Crops Research* 55: 23–44

Jamieson P D, Stone P J, Semenov M A (2001) Towards modelling quality in wheat – from grain nitrogen concentration to protein composition. In Gooding M J, Barton S A, Smith G P (eds) *Wheat Quality. Aspects of Applied Biology* 64: 111–26

Jarvis A J, Mansfield T A, Davies W J (1999) Stomatal behaviour, photosynthesis and transpiration under rising CO_2. *Plant, Cell and Environment* 22: 639–48

Jefferies R A (1993) Responses of potato genotypes to drought. 1. Expansion of individual leaves and osmotic adjustment. *Annals of Applied Biology* 122: 93–104

Jensen C R, Mogensen V O, Mortensen G, Fieldsen J K, Milford G F J, Andersen M N, Thage J H (1996) Seed glucosinolate, oil and protein contents of field-grown rape (*Brassica napus* L.) affected by soil drying and evaporative demand. *Field Crops Research* 47: 93–105

Jeuffroy M-H, Devienne F (1995) A simulation model for assimilate partitioning between pods in pea (*Pisum sativum* L.) during the period of seed set; validation in field conditions. *Field Crops Research* 41: 79–89

Jeuffroy M-H, Ney B (1997) Crop physiology and productivity. *Field Crops Research* 53: 3–16

Jiang C-Z, Rodermel S R, Shibles R M (1993) Photosynthesis, rubisco activity and amount, and their regulation by transcription in senescing soybean leaves. *Plant Physiology* 101: 105–12

Johnson E C, Fischer K S, Edmeades G O, Palmer A F E (1986) Recurrent selection for reduced plant height in lowland tropical maize. *Crop Science* 26: 253–60

Johnson R, Frey N M, Moss D N (1974) Effect of water stress on photosynthesis and transpiration of flag leaves and spikes of barley and wheat. *Crop Science* 14: 728–31

Jones C A, Kiniry J R (1986) *CERES-Maize: A Simulation Model of Maize Growth and Development.* Texas A&M University Press, USA

Jones H G (1983) *Plants and Microclimate.* Cambridge University Press

Jones J L, Allen E J (1983) Effects of date of planting on plant emergence, leaf growth and yield in contrasting potato varieties. *Journal of Agricultural Science, Cambridge* 101: 81–95

Jones J W, Boote K J, Jagtap S S, Hoogenboom G, Wilkerson G G (1987) *SOYGRO V5.4, Soybean Crop Growth Model, User's Guide.* Florida Agricultural Experimental Station, University of Florida, USA

Jones J W, Boote K J, Jagtap S S, Mishoe J W (1991) Soybean development. In Ritchie J T, Hanks R J (eds) *Modeling Plant and Soil Systems.* American Agronomy Society, Agronomy Monograph 31, pp 71–90

Jones M B (1981) A comparison of sward development under cutting and continuous grazing management. In Wright C E (ed) *Plant Physiology and Herbage Production.* British Grassland Society Occasional Symposium 13: 63–7

Jones M B, Collett B, Brown S (1982) Sward growth under cutting and continuous stocking managements: sward canopy structure, tiller density and leaf turnover. *Grass and Forage Science* 37: 67–73

Jones R J, Schreiber B M N, Roessler J A (1996) Kernel sink capacity in maize: genotypic and maternal regulation. *Crop Science* **36**: 301–6

Jordan D B, Ogren W L (1984) The CO_2/O_2 specificity of ribulose 1,5 bisphosphate carboxylase/oxgenase. *Planta* **161**: 308–13

Justes E, Mary B, Meynard J M, Machet J M, Thelier-Huches L (1994) Determination of a critical nitrogen dilution curve for winter wheat crops. *Annals of Botany* **74**: 397–407

Kavakli I H, Slattery C J, Ito H, Okita T W (2000) The conversion of carbon and nitrogen into starch and storage proteins in developing storage organs. *Australian Journal of Plant Physiology* **27**: 561–70

Kemp D R (1980) The location and size of the extension zone of emerging wheat leaves. *The New Phytologist* **84**: 729–37

Kettlewell P S, Sothern R B, Koukkari W L (1999) UK wheat quality and economic value are dependent on the North Atlantic oscillation. *Journal of Cereal Science* **29**: 205–9

Khurana S C, McLaren J S (1982) The influence of leaf area, light interception and season on potato growth and yield. *Potato Research* **25**: 329–42

Killick R J, Simmonds N W (1974) Specific gravity of potato tubers as a character showing small genotype–environment interactions. *Heredity* **32**: 109–112

Kimball B A, Kobyashi K, Bindi M (2002) Responses of agricultural crops to free air CO_2 enrichment. *Advances in Agronomy* **77**: 293–368

Kiniry J R (1991) Maize phasic development. In Ritchie J T, Hanks R J (eds) *Modeling Plant and Soil Systems.* American Agronomy Society. Agronomy Monograph 31, pp 55–69

Kiniry J R, Otegui M E (2000) Processes affecting maize grain yield potential in temperate conditions. In Otegui M E, Slafer G A (eds) *Physiological Bases for Maize Improvement.* Food Products Press, pp 31–46

Kiniry J R, Ritchie J T (1985) Shade-sensitivity interval of kernel number of maize. *Agronomy Journal* **77**: 711–5

Kiniry J R, Ritchie J T, Musser R L (1983) Dynamic nature of the photoperiod response in maize. *Agronomy Journal* **75**: 700–3

Kiniry J R, Williams J R, Vanderlip R L, Atwood J D, Reicosky D C, Mulliken J, Cox W J, Mascagni H J, Hollinger S E, Wiebold W J (1997) Evaluation of two maize models for nine U.S. locations. *Agronomy Journal* **89**: 421–6

Kirby E J M (1967) The effect of plant density upon the growth and yield of barley. *Journal of Agricultural Science, Cambridge* **68**: 317–24

Kirby E J M, Appleyard M (1984) *Cereal Development Guide* 2nd edition. Arable Unit, National Agricultural Centre, Coventry

Kirby E J M, Appleyard M, Fellowes G (1982) Effect of sowing date on the temperature response of leaf emergence and leaf size in barley. *Plant, Cell and Environment* **5**: 477–84

Kirby E J M, Appleyard M, Fellowes G (1985) Leaf emergence and tillering in barley and wheat. *Agronomie* **5**: 193–200

Kirby E J M, Spink J H, Frost D L, Sylvester-Bradley R, Scott R K, Foulkes M J, Clare R W, Evans E J (1999) A study of wheat development in the field: analysis by phases. *European Journal of Agronomy* **11**: 63–82

Kirk W W, Marshall B (1992) The influence of temperature on leaf development and growth in potatoes in controlled environments. *Annals of Applied Biology* **120**: 511–25

Klepper B, Belford R K, Rickman R W (1984) Root and shoot development in winter wheat. *Agronomy Journal* **76**: 117–22

Knight J D, Livingston N J, van Kessel C (1994) Carbon isotope discrimination and water use efficiency of six crops grown under wet and dryland conditions. *Plant, Cell and Environment* 17: 173–9

Kokubun M, Shimada S, Takahashi M (2001) Flower abortion caused by preanthesis water deficit is not attributable to impairment of pollen in soybean. *Crop Science* 41: 1517–21

Komor E (2000) Source physiology and assimilate transport: the interaction of sucrose metabolism, starch storage and phloem export in source leaves and effects on sugar status in phloem. *Australian Journal of Plant Physiology* 27: 497–505

Krampitz M J, Klug K, Fock H P (1984) Rates of photosynthetic CO_2 uptake, photo-respiratory CO_2 evolution and dark respiration in water-stressed sunflower and bean leaves. *Photosynthetica* 18: 322–8

Kreig D R, Hutmacher R B (1986) Photosynthetic rate control in sorghum: stomatal and non-stomatal factors. *Crop Science* 26: 112–7

Krenzer E G, Nipp T L, McNew R W (1991) Winter wheat mainstem leaf appearance and tiller formation vs. moisture treatment. *Agronomy Journal* 83: 663–7

Kropff M J, van Laar H H (1993) *Modelling Crop–Weed Interactions*. CABI, UK

Lafitte H R, Edmeades G O (1994) Improvement for tolerance to low soil nitrogen in tropical maize. 2. Grain yield, biomass production, and N accumulation. *Field Crops Research* 39: 15–25

Lafitte H R, Edmeades G O, Taba S (1997) Adaptive strategies identified among tropical maize landraces for nitrogen-limited environments. *Field Crops Research* 49: 187–204

Lambers H (1997) Respiration and the alternative oxidase. In Foyer C H, Quick P (eds) *A Molecular Approach to Primary Metabolism in Plants*. Taylor and Francis, pp 295–309

Lambers H, Atkin O K, Scheurwater I (1996) Respiratory patterns in roots in relation to their functioning. In Waisel Y, Eshel A, Kafkaki U (eds) *Plant Roots: The Hidden Half*. Marcel Dekker, pp 323–62

Lambers H, Chapin F S, Pons T L (1998) *Plant Physiological Ecology*. Springer Verlag

Lambert R J, Johnson R R (1978) Leaf angle, tassel morphology, and the performance of maize hybrids. *Crop Science* 18: 499–502

Landes A, Porter J R (1989) Comparison of scales used for categorising the development of wheat, barley, rye and oats. *Annals of Applied Biology* 115: 343–60

Landsberg J J (1975) Temperature effects and plant response. *Progress in Biometeorology* C1: 86–107

Landsberg J J (1977) Some useful equations for biological studies. *Experimental Agriculture* 13: 273–86

Lantinga E A, Nasssiri M, Kropff M J (1999) Modelling and measuring vertical light absorption within grass–clover mixtures. *Agricultural and Forest Meteorology* 96: 71–83

Lauer J G, Coors J G, Flannery P J (2001) Forage yield and quality of corn cultivars developed in different eras. *Crop Science* 41: 1449–55

Law R D, Crafts-Brandner S J (1999) Inhibition and acclimation of photosynthesis to heat stress is closely correlated with activation of ribulose-1,5-bisphosphate carboxylase/oxygenase. *Plant Physiology* 120: 173–81

Lawless C, Semenov M A, Jamieson P D (2005) A wheat canopy model linking leaf area and phenology. *European Journal of Agronomy* 22: 19–32

Lawlor D W (1993) *Photosynthesis*. 2nd edition. Longman Scientific and Technical

Lawlor D W (2001) *Photosynthesis*. BIOS Scientific Publishers

Lemaire G, Gastal F (1997) N uptake and distribution in plant canopies. In Lemaire G (ed) *Diagnosis of the Nitrogen Status in Crops*. Springer Verlag, pp 3–43

Lemaire G, Salette J (1984) Relation entre dynamique de croissance et dynamique de prélèvement d'azote pour un peuplement de graminées fourragères. 2. Etude de la variabilité entre génotypes. *Agronomie* 4: 431–6

Lindquist J L (2001) Performance of INTERCOM for predicting corn-velvetleaf interference across north-central United States. *Weed Science* 49: 195–201

Löffler C M, Busch R H (1982) Selection for grain protein, grain yield, and nitrogen partitioning efficiency in hard red spring wheat. *Crop Science* 22: 591–5

Long S P (1991) Modification of the response of photosynthetic productivity to rising temperature by atmospheric CO_2 concentrations. Has its importance been underestimated? *Plant, Cell and Environment* 14: 729–39

Long S P (1994) Resource capture by single leaves. In Monteith J L, Scott R K, Unsworth M H (eds) *Resource Capture by Crops*. Nottingham University Press, pp 17–34

Long S P, Ainsworth E A, Rogers A, Ort D R (2004) Rising atmospheric carbon dioxide: plants FACE the future. *Annual Review of Plant Biology* 55: 591–628

Longnecker N, Robson A (1994) Leaf emergence of spring wheat receiving varying nitrogen supply at different stages of development. *Annals of Botany* 74: 1–7

Loreto F, Harley P C, Di Marco G, Sharkey T D (1992) Estimation of mesophyll conductance to CO_2 flux by three different methods. *Plant Physiology* 98: 1437–43

Loreto F, Di Marco G, Tricoli D, Sharkey T D (1994) Measurement of mesophyll conductance, photosynthetic electron transport and alternative electron sinks of field grown wheat leaves. *Photosynthesis Research* 41: 397–403

Lu C H, van Ittersum M K, Rabbinge R (2004) A scenario exploration of strategic land use options for the Loess Plateau in Northern China. *Agricultural Systems* 79: 145–70

Ludlow M M, Muchow R C (1990) A critical evaluation of traits for improving crop yields in water-limited environments. *Advances in Agronomy* 43: 107–153

Ludlow M M, Ng T T (1976) Effects of water deficit on carbon dioxide exchange and leaf elongation rate of *Panicum maximum* var *trichloglume*. *Australian Journal of Plant Physiology* 3: 401–13

Lueschen W E, Hicks D R (1977) Influence of plant population on field performance of three soybean cultivars. *Agronomy Journal* 69: 390–3

Lühs W W, Voss A, Seyis F, Friedt W (1999) Molecular genetics of erucic acid content in the genus *Brassica*. In Wratten N, Salisbury P A (eds) *New Horizons for an Old Crop. Proceedings of the 10th International Rapeseed Congress, Canberra, Australia.* http://www.regional.org.au/au/gcirc/4/442.htm

Lunn G D, Major B J, Kettlewell P S, Scott R K (2001) Mechanisms leading to excess *alpha*-amylase activity in wheat (*Triticum aestivum* L.) grain in the U.K. *Journal of Cereal Science* 33: 313–329

Lüscher A, Stäheli B, Braun R, Nösberger J (2001) Leaf area, competition with grass, and clover cultivar: key factors to successful overwintering and fast regrowth of white clover (*Trifolium repens* L.) in spring. *Annals of Botany* 88: 725–35

Lyndon R F (1994) Control of organogenesis at the shoot apex. *The New Phytologist* 128: 1–18

Maas E V, Grieve C M (1990) Spike and leaf development in salt-stressed wheat. *Crop Science* 30: 1309–13

MacKerron D K L, Jefferies R A (1988) The distribution of tuber sizes in droughted and irrigated crops of potatoes. 1. Observations on the effect of water stress on graded yields from different cultivars. *Potato Research* 31: 269–78

Maddonni G A, Chelle M, Drouet J-L, Andrieu B (2001) Light interception of contrasting azimuth canopies under square and rectangular plant spatial distributions: simulations and crop measurements. *Field Crops Research* 70: 1–13

Magliulo V, Bindi M, Rana G (2003) Water use of irrigated potato (*Solanum tuberosum* L.) grown under free air carbon dioxide enrichment in central Italy. *Agricultural Ecosystems & Environment* 97: 65–80

Mann M E, Bradley R S, Hughes M K (1999) Northern Hemisphere temperatures during the past millennium: inferences, uncertainties and limitations. *Geophysical Research Letters* 26: 759–62

Marshall B (1978) *Leaf and Ear Photosynthesis of Winter Wheat Crops*. PhD Thesis, University of Nottingham

Marshall B, Biscoe P V (1980) A model for C$_3$ leaves describing the dependence of net photosynthesis on irradiance. 1. Derivation. *Journal of Experimental Botany* 31: 29–39

Marshall B, Porter J R (1991) Concepts of nutritional and environmental interactions determining plant productivity. In Porter J R, Lawlor D W (eds) *Plant Growth: Interactions with Nutrition and Environment*. Cambridge University Press, pp 99–124

Martre P, Porter J R, Jamieson P D, Triboï E (2003) Modeling grain nitrogen accumulation and protein composition to understand the sink/source regulation of nitrogen remobilization for wheat. *Plant Physiology* 133: 1959–67

McCree K J (1970) An equation for the rate of respiration of white clover plants grown under controlled conditions. In Setlik I (ed) *Prediction and Measurement of Photosynthetic Productivity*. Pudoc, the Netherlands, pp 221–9

McCree K J (1974) Equations for the rate of dark respiration of white clover and grain sorghum, as functions of dry weight, photosynthetic rate and temperature. *Crop Science* 14: 509–14

McCree K J, Troughton J H (1966) Non-existence of an optimum leaf area index for the production rate of white clover grown under constant conditions. *Plant Physiology* 41: 559–66

McGarrahan J P, Dale R F (1984) A trend towards a longer grain-filling period for corn. A case study in Indiana. *Agronomy Journal* 76: 518–22

McMaster G S (1997) Phenology, development and growth of the wheat (*Triticum aestivum* L.) shoot apex: a review. *Advances in Agronomy* 59: 63–118

McMaster G S, Wilhelm W W (1995) Accuracy of equations predicting the phyllochron of wheat. *Crop Science* 35: 30–6

McMaster G S, Wilhelm W W, Palic D B, Porter J R, Jamieson P D (2003) Spring wheat leaf appearance and temperature: extending the paradigm? *Annals of Botany* 91: 697–705

Meier U (ed) (1997) *Growth Stages of Plants*. Blackwell Scientific Publications, Berlin

Milford G F J, Pocock T O, Riley J (1985a) An analysis of leaf growth in sugar beet. II. Leaf appearance in field crops. *Annals of Applied Biology* 106: 173–85

Milford G F J, Pocock T O, Riley J, Messem A B (1985b) An analysis of leaf growth in sugar beet. III. Leaf expansion in field crops. *Annals of Applied Biology* 106: 187–203

Millard P (1988) The accumulation and storage of nitrogen by herbaceous plants. *Plant, Cell and Environment* 11: 1–8

Millard P, MacKerron D K L (1986) The effects of nitrogen application on growth and nitrogen distribution within the potato canopy. *Annals of Applied Biology* 109: 427–37

Miralles D J, Slafer G A (1995) Individual grain weight responses to genetic reduction in culm length in wheat as affected by source-sink manipulations. *Field Crops Research* 43: 55–66

Miralles D J, Katz S D, Colloca A, Slafer G A (1998) Floret development in near isogenic wheat lines differing in plant height. *Field Crops Research* 59: 21–30

Monsi M, Saeki T (1953) Über der Lichtfaktor in den Pflanzengesellschaften und seine Bedeutung für die Stoffproduktion. *Japanese Journal of Botany* **14**: 22–52

Monteith J L (1965) Evaporation and environment. *Symposium of the Society of Experimental Biology* **19**: 205–34

Monteith J L (ed) (1976) *Vegetation and the Atmosphere,* Vol 2. Academic Press

Monteith J L (1977) Climate and the efficiency of crop production in Britain. *Philosophical Transactions of the Royal Society of London* **B281**: 277–94

Monteith J L (1978) Reassessment of maximum growth rates for C_3 and C_4 crops. *Experimental Agriculture* **14**: 1–5

Monteith J L (1981a) Does light limit crop production? In Johnson C B (ed) *Physiological Processes Limiting Plant Productivity*. Butterworths, pp 23–38

Monteith J L (1981b) Presidential address to the Royal Meteorological Society. *Quarterly Journal of the Royal Meteorological Society* **107**: 749–74

Monteith J L, Elston J (1983) Performance and productivity in the field. In Dale J E, Milthorpe F L (eds) *The Growth and Functioning of Leaves*. Cambridge University Press, pp 499–518

Monteith J L, Unsworth M H (1990) *Principles of Environmental Physics*, 2nd edition. Edward Arnold

Moorby J (1978) The physiology of growth and tuber yield. In Harris P M (ed) *The Potato Crop*. Chapman and Hall

Moorby J, Milthorpe F L (1975) Potato. In Evans L T (ed) *Crop Physiology*. Cambridge University Press, pp 225–57

Moot D J, Jamieson P D, Henderson A I, Ford M A, Porter J R (1996) Rate of change in harvest index during grain-filling of wheat. *Journal of Agricultural Science, Cambridge* **126**: 387–95

Morgan J M, Condon A G (1986) Water use, grain yield, and osmoregulation in wheat. *Australian Journal of Plant Physiology* **13**: 523–32

Morrison M J, McVetty P B E (1991) Leaf appearance rate of summer rape. *Canadian Journal of Plant Science* **71**: 405–12

Morrison M J, Voldeng H D, Cober E R (2000) Agronomic changes from 58 years of genetic improvement of short-season soybean cultivars in Canada. *Agronomy Journal* **92**: 780–4

Morton A G, Watson D J (1948) A physiological study of leaf growth. *Annals of Botany* **47**: 281–310

Mosaad M G, Ortiz-Ferrara G, Mahalakshmi V, Fischer R A (1995) Phyllochron response to vernalization and photoperiod in spring wheat. *Crop Science* **35**: 168–71

Moulia B, Loup C, Chartier M, Allirand J M, Edelin C (1999) Dynamics of architectural development of isolated plants of maize (*Zea mays* L.), in a non-limiting environment: the branching potential of modern maize. *Annals of Botany* **84**: 645–56

Muchow R C (1989) Effect of high temperature on grain growth in field grown maize. *Field Crops Research* **23**: 145–58

Muchow R C (1994) Effect of nitrogen on yield determination in irrigated maize in tropical and subtropical environments. *Field Crops Research* **38**: 1–13

Muchow R C, Carberry P S (1989) Environmental control of phenology and leaf growth in a tropically adapted maize. *Field Crops Research* **20**: 221–36

Muchow R C, Carberry P S (1990) Phenology and leaf area development in a tropical grain sorghum. *Field Crops Research* **23**: 221–37

Muchow R C, Sinclair T R (1991) Water deficit effects on maize yields modeled under current and 'greenhouse' climates. *Agronomy Journal* **83**: 1052–9

Muchow R C, Sinclair T R, Bennett J M (1990) Temperature and solar radiation effects on potential maize yield across locations. *Agronomy Journal* 82: 338–43

Mulholland B J, Craigon J, Black C R, Colls J J, Atherton J, Landon G (1998) Growth, light interception and yield responses of spring wheat (*Triticum aestivum* L.) grown under elevated CO_2 and O_3 in open-top chambers. *Global Change Biology* 4: 121–30

Muller B, Reymond M, Tardieu F (2001) The elongation rate at the base of a maize leaf shows an invariant pattern during both the steady-state elongation and the establishment of the elongation zone. *Journal of Experimental Botany* 52: 1259–68

Nanda R, Bhargava S C, Rawson H M (1995) Effect of sowing date on rates of leaf appearance, final leaf numbers and areas in *Brassica campestris, B. juncea, B. napus* and *B. carinata*. *Field Crops Research* 42: 125–34

Navratil R J, Burris J S (1980) Predictive equations for maize inbred emergence. *Crop Science* 20: 567–70

Nobel P S (1974) *Introduction to Biophysical Plant Physiology.* W H Freeman

O'Brien P J, Allen E J, Bean J N, Griffith R L, Jones S A, Jones J L (1983) Accumulated day degrees as a measure of physiological age and the relationships with growth and yield in early potatoes. *Journal of Agricultural Science, Cambridge* 101: 613–31

O'Leary M H (1988) Carbon isotopes in photosynthesis. *BioScience* 38: 328–36

Olesen J E, Bindi M (2002) Consequences of climate change for European agricultural productivity, land use and policy. *European Journal of Agronomy* 16: 239–62

Oparka K J, Davies H V (1985) Translocation of assimilates within and between potato stems. *Annals of Botany* 56: 45–54

Oparka K J, Viola R, Wright K M, Prior D A M (1992) Sugar transport and metabolism in the potato tuber. In Pollock C J, Farrar J F, Gordon A J (eds) *Carbon Partitioning within and between Organisms.* Bios Scientific Publications, pp 91–114

Ortiz-Monasterio J I, Sayre K D, Rajaram S, McMahon (1997) Genetic progress in wheat yield and nitrogen use efficiency under four nitrogen rates. *Crop Science* 37: 898–904

Otegui M E, Bonhomme R (1998) Grain yield components in maize. 1. Ear growth and kernel set. *Field Crops Research* 56: 247–56

Otegui M E, Andrade F H, Suero E E (1995) Growth, water use, and kernel abortion of maize subjected to drought at silking. *Field Crops Research* 40: 87–94

Parkhurst D (1994) Diffusion of CO_2 and other gases inside leaves. *New Phytologist* 126: 449–79

Parsons A J, Chapman D F (2000) The principles of pasture growth and utilization. In Hopkins A (ed) *Grass: Its Production and Utilization.* 3rd edition. Blackwell Science, pp 31–89

Parsons A J, Robson M J (1981a) Seasonal changes in the physiology of S24 perennial ryegrass (*Lolium perenne* L.). 2. Potential leaf and canopy photosynthesis during the transition from vegetative to reproductive growth. *Annals of Botany* 47: 249–58

Parsons A J, Robson M J (1981b) Seasonal changes in the physiology of S24 perennial ryegrass (*Lolium perenne* L.). 3. Partition of assimilates between root and shoot during the transition from vegetative to reproductive growth. *Annals of Botany* 48: 733–44

Parsons A J, Leafe E L, Collett B, Penning P D, Lewis J (1983) The physiology of grass production under grazing. 2. Photosynthesis, crop growth and animal intake of continuously-grazed swards. *Journal of Applied Ecology* 20: 127–39

Parsons A J, Johnson I R, Williams J H H (1988) Leaf age structure and canopy photosynthesis in rotationally and continuously grazed swards. *Grass and Forage Science* 43: 1–14

Patrick J W, Offler C E (1995) Post-seive element transport of sucrose in developing seeds. *Australian Journal of Plant Physiology* 22: 681–702

Patterson D T (1992) Temperature and canopy development of velvetleaf (*Abutilon theophrasti*) and soybean (*Glycine max*). *Weed Technology* 6: 68–76

Pearman I, Thomas S M, Thorne G N (1978) Effect of nitrogen fertiliser on growth and yield of semi-dwarf and tall varieties of winter wheat. *Journal of Agricultural Science, Cambridge* 91: 31–45

Pellny T K, Ghannoum O, Conroy J P, Schluepmann H, Smeekens S, Andralojc J, Krause K P, Goddijn O, Paul M J (2004) Genetic modification of photosynthesis with *E. coli* genes for trehalose synthesis. *Plant Biotechnology Journal* 2: 71–82

Penman H L (1948) Natural evaporation from open water, bare soil and grass. *Proceedings of the Royal Society of London* A193: 120–45

Penning de Vries F W T (1972) Respiration and growth. In Rees A R, Cockshull K E, Hand D W, Hurd R G (eds) *Crop Processes in Controlled Environments*. Academic Press, pp 327–47

Penning de Vries F W T (1974) Substrate utilisation and respiration in relation to growth and maintenance in higher plants. *Netherlands Journal of Agricultural Science* 22: 40–44

Penning de Vries F W T (1975) The cost of maintenance processes in plant cells. *Annals of Botany* 39: 77–92

Penning de Vries F W T, Brunsting A H M, van Laar H H (1974) Products, requirements and efficiency of biosynthesis: a quantitative approach. *Journal of Theoretical Biology* 45: 339–77

Penning de Vries F W T, van Laar H H, Chardon M C M (1983) Bioenergetics of growth of seeds, fruits and storage organs. In Smith W H, Banata S J (eds) *Potential Production of Field Crops under Different Environments*. International Rice Research Institute, the Philippines, pp 37–59

Perry M W, D'Antuono M F (1989) Yield improvement and associated characteristics of some Australian spring wheat cultivars introduced between 1860 and 1982. *Australian Journal of Agricultural Research* 40: 57–72

Pimental D, Pimental M (1979) *Food, Energy and Society*. Edward Arnold

Piper E L, Boote K J (1999) Temperature and cultivar effects on soybean seed oil and protein concentrations. *Journal of the American Oil Chemists' Society* 76: 1233–41

Plénet D, Etchebest S, Mollier A, Pellerin S (2000) Growth analysis of maize field crops under phosphorus deficiency. 1. Leaf growth. *Plant and Soil* 223: 117–30

Pollock C J, Cairns A J (1991) Fructan metabolism in grasses and cereals. *Annual Review of Plant Physiology and Plant Molecular Biology* 42: 77–101

Pollock C J, Jones T (1979) Seasonal patterns of fructan metabolism in forage grasses. *New Phytologist* 83: 9–15

Poorter H (1994) Construction costs and payback time of biomass: a whole plant perspective. In Roy J, Garnier E (eds) *A Whole Plant Perspective on Carbon–Nitrogen Interactions*. SPB Academic Publishing, the Netherlands, pp 111–127

Porter J R (1984) A model of canopy development in winter wheat. *Journal of Agricultural Science, Cambridge* 102: 383–92

Porter J R (1985a) Models and mechanisms in the growth and development of wheat. *Outlook on Agriculture* 14: 190–6

Porter J R (1985b) Approaches to modelling canopy development in wheat. In Day W, Atkin R T (eds) *Wheat Growth and Modelling*. Plenum Press, pp 69–81

Porter J R (1993) AFRCWHEAT2: A model of the growth and development of wheat incorporating responses to water and nitrogen. *European Journal of Agronomy* 2: 64–77

Porter J R, Gawith M (1999) Temperatures and the growth and development of wheat: a review. *European Journal of Agronomy* 10: 23–36

Porter J R, Leigh R A, Semenov M A, Miglietta F (1995) Modelling the effects of climatic change and genetic modification on nitrogen use by wheat. *European Journal of Agronomy* 4: 419–29

Prášil I T, Prášilová P, Pánková K (2004) Relationships among vernalization, shoot apex development and frost tolerance in wheat. *Annals of Botany* 94: 413–18

Presterl T, Seitz G, Landbeck M, Thiemt E M, Schmidt W, Geiger H H (2003) Improving nitrogen-use efficiency in European maize: estimation of quantitative genetic parameters. *Crop Science* 43: 1259–65

Priestley C H B, Taylor R J (1972) On the assessment of surface heat flux and evaporation using large-scale parameters. *Monthly Weather Review* 100: 81–92

Purcell L C (2003) Comparison of thermal units derived from daily and hourly temperatures. *Crop Science* 43: 1874–9

Rachidi F, Kirkham M B, Stone L R, Kanemasu E T (1993) Use of photosynthetically active radiation by sunflower and sorghum. *European Journal of Agronomy* 2: 131–9

Rahman S M, Kinoshita T, Anai T, Takagi Y (2001) Combining ability in loci for high oleic and low linolenic acids in soybean. *Crop Science* 41: 26–9

Rakocevic M, Sinoquet H, Christophe A, Varlet-Grancher C (2000) Assessing the geometric structure of a white clover (*Trifolium repens* L.) canopy using 3-D digitising. *Annals of Botany* 86: 519–26

Raschke K (1976) How stomata resolve the dimemma of opposing priorities. *Philosophical Transactions of the Royal Society of London* B273: 551–60

Ray J D, Sinclair T R (1997) Stomatal closure of maize hybrids in response to drying soil. *Crop Science* 37: 803–7

Reinhardt D, Pesce E-R, Steiger P, Mandel T, Baltensperger K, Bennett M, Traas J, Frimi J, Kuhlemeier B (2003) Regulation of phyllotaxis by polar auxin transport. *Nature* 426: 255–60

Rennie W J (2001) Seed potato production in Scotland. In Hay R K M (ed) *Scottish Agricultural Science Agency, Scientific Review 1997–2000*. ISBN 0 7559 0287 4. SASA, pp 42–4

Reynolds M P, van Ginkel M, Ribaut J M (2000) Avenues for genetic modification of radiation use efficiency in wheat. *Journal of Experimental Botany* 51: 459–73

Richards J H (1993) Physiology of plants recovering from defoliation. *Proceedings of the XVII International Grassland Congress* 85–94

Richards R A (1987) Physiology and the breeding of winter-grown cereals for dry areas. In Srivastava J P, Porceddu E, Acevedo E, Varma S (eds) *Drought Tolerance in Winter Cereals*. Wiley, pp 133–50

Richards R A (2000) Selectable traits to increase crop photosynthesis and yield of grain crops. *Journal of Experimental Botany* 51: 447–58

Richards R A, Rebetzke G J, Condon A G, van Herwaarden A F (2002) Breeding opportunities for increasing the efficiency of water use and crop yield in temperate cereals. *Crop Science* 42: 111–21

Ridge P E, Foale M A, Cox P G, Carberry P S (1996) Interpretation and value of soil nitrate nitrogen at depth. *Proceedings of the 8th Australian Agronomy Conference*. http://www.regional.org.au/au/asa/1996/contributed/478ridge.htm

Riggs T J, Hansen P R, Start N D, Miles D M, Morgan C L, Ford M A (1981) Comparison of spring barley varieties grown in England and Wales between 1880 and 1980. *Journal of Agricultural Science, Cambridge* 97: 599–610

Ritchie J T, Alagarswamy G (2003) Model concepts to express genetic differences in maize yield components. *Agronomy Journal* 95: 4–9

Ritchie J T, Otter S (1985) *Description and Performance of CERES-Wheat: A User-oriented Wheat Yield Model.* US Department of Agriculture, Agricultural Research Service

Robertson G P, Paul E A, Harwood R R (2000) Greenhouse gases in intensive agriculture: contributions of individual gases to the radiative forcing of the atmosphere. *Science* 289: 1992–5

Robertson M J, Brooking I R, Ritchie J T (1996) Temperature response of vernalization in wheat: modelling the effect on the final number of mainstem leaves. *Annals of Botany* 78: 371–81

Robinson D (1994) Resource capture by single roots. In Monteith J L, Scott R K, Unsworth M H (eds) *Resource Capture by Crops.* Nottingham University Press, pp 53–76

Robson M J (1981) Potential production – what is it, and can we increase it? In Wright C E (ed) *Plant Physiology and Herbage Production. Occasional Symposia of the British Grassland Society* 13: 5–18

Robson M J, Parsons A J (1978) Nitrogen deficiency in small closed canopies of S24 ryegrass. 1. Photosynthesis, respiration and dry matter production and partition. *Annals of Botany* 42: 1185–97

Rooney J M (1989) Biomass production and gas exchange of winter wheat as affected by flag leaf inoculations with *Septoria nodorum. New Phytologist* 112: 229–34

Rooney J M, Hoad G V (1989) Compensation in growth and photosynthesis of wheat (*Triticum aestivum* L.) following early inoculations with *Septoria nodorum* (Berk.) Berk. *New Phytologist* 113: 513–21

Rosegrant M W, Paisner M S, Meijer S, Witcover J (2001) *Global Food Projections to 2020: Emerging Trends and Alternative Futures.* International Food Policy Research Institute, Washington, USA

Rosenzweig C, Parry M L (1994) Potential impact of climate change on world food supply. *Nature* 337: 133–8

Rossing W A H, van Oijen M, van der Werf W, Bastiaans L, Rabbinge R (1992) Modelling the effects of foliar pests and pathogens on light interception, photosynthesis, growth rate and yield of field crops. In Ayres P G (ed) *Pests and Pathogens – Plant Responses to Foliar Attack.* Bios Scientific Publishers pp 161–80

Royal Society of London (2001) *The Role of Land Carbon Sinks in Mitigating Global Climate Change.* Royal Society of London

Russell G (1988) Physiological restraints on the economic viability of the evening primrose crop in Eastern Scotland. *Crop Research* 28: 25–33

Russell W A (1991) Genetic improvement of maize yields. *Advances in Agronomy* 46: 245–98

Sadras V O, Trápani N (1999) Leaf expansion and phenological development: key determinants of sunflower plasticity, growth and yield. In Smith D L, Hamel C (eds) *Crop Yield, Physiology and Processes.* Springer Verlag, pp 205–33

Saini H S, Aspinall D (1982) Abnormal sporogenesis in wheat (*Triticum aestivum* L.) induced by short periods of high temperature. *Annals of Botany* 49: 835–46

Sale P J M (1974) Productivity of vegetable crops in a region of high solar input. 3. Carbon balance of potato crops. *Australian Journal of Plant Physiology* 1: 283–96

Salih A A, Ali I A, Lux A, Luxova M, Cohen Y, Sugimoto Y, Inanaga S (1999) Rooting, water uptake and xylem structure adaptation to drought of two sorghum cultivars. *Crop Science* 39: 168–73

Scarisbrick D H, Ferguson A J (eds) (1995) *New Horizons for Oilseed Rape.* Wye College, London

Schippers P A (1968) The influence of rates of nitrogen and potassium application on the yield and specific gravity of four potato varieties. *European Potato Journal* 11: 23–33

Scholes J D, Lee P J, Horton P, Lewis D H (1994) Invertase: understanding changes in the photosynthetic and carbohydrate metabolism of barley leaves infected with powdery mildew. *New Phytologist* 126: 213–22

Schussler J R, Westgate M E (1991) Maize kernel set at low water potential. 2. Sensitivity to reduced assimilates at pollination. *Crop Science* 31: 1196–203

Scott R K, Wilcockson S J (1978) Aplication of physiological and agronomic principles to the development of the potato industry. In Harris P M (ed) *The Potato Crop*. Chapman and Hall, pp 678–704

Scott R K, Jaggard K W, Sylvester-Bradley R (1994) Resource capture by arable crops. In Monteith J L, Scott R K, Unsworth M H (eds) *Resource Capture by Crops*. Nottingham University Press, pp 279–302

Secor J, Shibles R, Stewart C R (1983) Metabolic changes in senescing soybean leaves of similar plant ontogeny. *Crop Science* 23: 106–10

Semenov M A, Porter J R (1995) Climatic variability and the modelling of crop yields. *Agricultural and Forest Meteorology* 73: 265–83

Shantz H L, Piemesal L N (1927) The water requirements at Akron, Colorado. *Journal of Agricultural Research* 34: 1093–190

Shaykewich C F (1995) An appraisal of cereal crop phenology modelling. *Canadian Journal of Plant Science* 75: 329–41

Shibles R M, Weber C R (1966) Interception of solar radiation and dry matter production by various soybean planting patterns. *Crop Science* 6: 55–9

Shibles R, Anderson I C, Gibson A H (1975) Soybean. In Evans L T (ed) *Crop Physiology*. Cambridge University Press, pp 151–89

Siddique K H M, Kirby E J M, Perry M W (1989) Ear:stem ratio in old and modern wheat varieties; relationships with improvement in number of grains per ear and yield. *Field Crops Research* 21: 59–78

Siddique K H M, Tennant D, Perry M W, Belford R K (1990) Water use and water use efficiency of old and modern wheat cultivars in a mediterranean-type environment. *Australian Journal of Agricultural Research* 41: 431–47

Simmonds N W (1995) The relation between yield and protein in cereal grain. *Journal of the Science of Food and Agriculture* 67: 309–15

Sinclair T R (1984) Cessation of leaf emergence in indeterminate soybeans. *Crop Science* 24: 483–6

Sinclair T R (1992) Mineral nutrition and plant growth response to climate change. *Journal of Experimental Botany* 43: 1141–6

Sinclair T R, Amir J (1992) A model to assess nitrogen limitations on the growth and yield of spring wheat. *Field Crops Research* 30: 63–78

Sinclair T R, Horie T (1989) Leaf nitrogen, photosynthesis and crop radiation use efficiency: a review. *Crop Science* 29: 90–8

Sinclair T R, Ludlow M M (1986) Influence of soil water supply on the plant water balance of four tropical grain legumes. *Australian Journal of Plant Physiology* 13: 329–41

Sinclair T R, Muchow R C (1995) Effect of nitrogen supply on maize yield. 1. Modeling physiological responses. *Agronomy Journal* 87: 632–41

Sinclair T R, Muchow R C (1999) Radiation use efficiency. *Advances in Agronomy* 65: 215–65

Sinclair T R, Muchow (2001) Systems analysis of plant traits to increase grain yield on limited water supplies. *Agronomy Journal* 93: 263–70

Sinclair T R, Shiraiwa T (1993) Soybean radiation-use efficiency as influenced by nonuniform specific leaf nitrogen distribution and diffuse radiation. *Crop Science* 33: 808–12

Sinclair T R, Purcell L C, Sneller C H (2004) Crop transformation and the challenge to increase yield potential. *Trends in Plant Science* 9: 70–5

Sinha N (1999) Leaf development in angiosperms. *Annual Review of Plant Physiology and Plant Molecular Biology* 50: 419–46

Sinha S K, Bhargave S C, Goel A (1982) Energy as the basis of harvest index. *Journal of Agricultural Science, Cambridge* 99: 237–8

Sivakumar M V K, Shaw R H (1979) Attenuation of radiation in moisture-stressed and unstressed soybeans. *Iowa State Journal of Research* 53: 251–7

Skinner R H, Nelson C J (1994) Epidermal cell division and the coordination of leaf and tiller development. *Annals of Botany* 74: 9–15

Slafer G A (2003) Genetic basis of yield as viewed from a crop physiologist's perspective. *Annals of Applied Biology* 142: 117–28

Slafer G A, Rawson H M (1997) Phyllochron in wheat as affected by photoperiod under two temperature regimes. *Australian Journal of Plant Physiology* 24: 151–8

Slafer G A, Andrade F H, Satorre E H (1990) Genetic-improvement effects on pre-anthesis physiological attributes related to wheat grain-yield. *Field Crops Research* 23: 255–63

Slatyer R O (1967) *Plant–Water Relationships*. Academic Press, London

Smil V (2003) *Energy at the Crossroads*. MIT Press, Massachusetts, USA

Soussana J F, Besle J M, Chabaux I, Loiseau P (1997) Long-term effects of CO_2 enrichment and temperature on forage quality in a temperate grassland. In Buchanan-Smith J G, Bailey L D, McCaughey P (eds) *Proceedings of the XVIII Grassland Congress*. International Grassland Society, Saskatoon, Canada, pp 23–4

Spedding C R W (1971) *Grassland Ecology*. Clarendon Press, Oxford

Spitters C J T (1986) Separating the diffuse and direct components of global radiation and its implications for modelling canopy photosynthesis. 2. Calculation of canopy photosynthesis. *Agricultural and Forest Meteorology* 38: 231–42

Spitters C J T, Toussaint H A J M, Goudriaan J (1986) Separating the diffuse and direct components of global radiation and its implications for modelling canopy photosynthesis. 1. Components of incoming radiation. *Agricultural and Forest Meteorology* 38: 217–29

Spollen W G, Sharp R E, Saab I N, Wu Y (1993) Regulation of cell expansion in roots and shoots at low water potentials. In Smith J A C, Griffiths H (eds) *Water Deficits*. BIOS Scientific Publishers, pp 37–52

Stern K R, Jansky S, Bidlack J E (2003) *Introductory Plant Biology*. International Edition. McGraw Hill Higher Education

Stern W R, Donald C M (1962) Light relationships in grass–clover swards. *Australian Journal of Agricultural Research* 13: 599–614

Stern W R, Kirby E J M (1979) Primordium initiation at the shoot apex of four contrasting varieties of spring wheat in response to sowing date. *Journal of Agricultural Science, Cambridge* 93: 203–15

Stone L R, Goodrum D E, Schlegel A J, Jaafar M N, Khan A H (2002) Water depletion depth of grain sorghum and sunflower in the Central High Plains. *Agronomy Journal* 94: 936–43

Storey R M J, Davies H V (1992) Tuber quality. In Harris P M (ed) *The Potato Crop*. 2nd edition. Chapman and Hall, pp 507–69

Stoy V (1980) Grain filling and the properties of the sink. In *Physiological Aspects of Crop Productivity*. International Potash Institute, Basel, Switzerland

Streit L G, Fehr W R, Welke G A, Hammond E G, Cianzio S R (2001) Family and line selection for reduced palmitate, saturates, and linolenate of soybean. *Crop Science* **41**: 63–7

Struik P C, van Heusden E, Burger-Meijer K (1988) Effects of short periods of long days on the development, yield and size distribution of potato tubers. *Netherlands Journal of Agricultural Science* **36**: 11–22

Subedi K D, Gregory P J, Summerfield R J, Gooding M J (1998) Cold temperatures and boron deficiency caused grain set failure in spring wheat (*Triticum aestivum* L.). *Field Crops Research* **57**: 277–88

Subedi K D, Gooding M J, Gregory P J (2001) Cultivar variation in boron accumulation and grain set in wheat under the influence of cold temperature. *Annals of Applied Biology* **138**: 97–101

Summerfield R J, Asumadu H, Ellis R H, Qi A (1998) Characterization of the photoperiodic response of post-flowering development in maturity isolines of soyabean [*Glycine max* (L.) Merrill] 'Clark'. *Annals of Botany* **82**: 765–71

Summerfield R J, Roberts E H, Ellis R H, Lawn R J (1991) Towards the reliable prediction of time to flowering in six annual crops. 1. The development of simple models for fluctuating field environments. *Experimental Agriculture* **27**: 11–31

Sunderland N (1960) Cell division and expansion in the growth of the leaf. *Journal of Experimental Botany* **11**: 68–80

Sung S, Amasino R M (2004) Vernalisation in *Arabidopsis thaliana* is mediated by the PHD finger protein V1N3. *Nature* **427**: 159–64

Svenning M M, Røsnes K, Junttila O (1997) Frost tolerance and biochemical changes during hardening and dehardening in contrasting white clover populations. *Physiologia Plantarum* **101**: 31–7

Swaminathan M S (1998) Crop production and sustainable food security. *Proceedings of the Second International Crop Science Congress, Delhi, 1996*, pp 3–17

Szabo L J, Bushnell W R (2001) Hidden robbers: the role of fungal haustoria in parasitism of plants. *Proceedings of the National Academy of Sciences* **98**: 7654–5

Takami S, Turner N C, Rawson H M (1981) Leaf expansion of four sunflower (*Helianthus annuus* L.) cultivars in relation to water deficits. 1. Patterns during plant development. *Plant, Cell and Environment* **4**: 399–407

Takami S, Rawson H M, Turner N C (1982) Leaf expansion in four sunflower (*Helianthus annuus* L.) cultivars in relation to water deficits. 2. Diurnal patterns during stress and recovery. *Plant, Cell and Environment* **5**: 279–86

Tanner C B, Jury W A (1976) Estimating evaporation and transpiration from a row crop during incomplete cover. *Agronomy Journal* **68**: 239–43

Tanner C B, Sinclair T R (1983) Efficient water use in crop production: research or re-search? In Taylor H M, Jordan W R, Sinclair T R (eds) *Limitations to Efficient Water Use in Crop Production*. American Society of Agronomy, pp 1–27

Tasma I M, Shoemaker R C (2003) Mapping flowering time gene homologs in soybean and their association with maturity (*E*) loci. *Crop Science* **43**: 319–28

Taylor H M, Mason W K, Bennie A T P, Rowse H R (1982) Responses of soybeans to two row spacings and two soil water levels. 1. An analysis of biomass accumulation, canopy development, solar radiation interception and components of seed yield. *Field Crops Research* **5**: 1–14

Thomas H, Howarth C J (2000) Five ways to stay green *Journal of Experimental Botany* **51**: 329–37

Thompson L (1993) The influence of the radiation environment around the node on morphogenesis and growth of white clover (*Trifolium repens*). *Grass and Forage Science* **48**: 271–8

Thorne G N and Taylor P J (1980) Factors limiting yields of winter wheat. *Rothamsted Experimental Station, Report for 1979*, Part 1, 18–19

Thorne G N, Wood D W (1987) Effects of radiation and temperature on tiller survival, grain number and grain yield in winter wheat. *Annals of Botany* 59: 413–26

Thorne G N, Pearman I, Day W, Todd A D (1988) Estimation of radiation interception by winter wheat from measurements of leaf area. *Journal of Agricultural Science, Cambridge* 110: 101–8

Thornley J H M, Cannell M G R (2000) Modelling the components of plant respiration: representation and realism. *Annals of Botany* 85: 55–67

Thornley J H M, Johnson I R (1990) *Plant and Crop Modelling. A Mathematical Approach to Plant and Crop Physiology.* Oxford University Press

Timsina J, Connor D J (2001) Productivity and management of rice – wheat cropping systems: issues and challenges. *Field Crops Research* 69: 93–132

Tollenaar M (1991) Physiological basis of genetic improvement of maize hybrids in Ontario from 1959 to 1988. *Crop Science* 31: 119–24

Tollenaar M, Aguilera A (1992) Radiation use efficiency of an old and a new maize hybrid. *Agronomy Journal* 84: 536–41

Tollenaar M, Hunter R B (1983) A photoperiod and temperature sensitive period for leaf number of maize. *Crop Science* 23: 457–60

Tollenaar M, Dwyer L M, Stewart D W (1992) Ear and kernel formation in maize hybrids representing three decades of grain yield improvement in Ontario. *Crop Science* 32: 432–8

Tottman D R, Broad H (1987) The decimal code for the growth stages of cereals, with illustrations. *Annals of Applied Biology* 110: 441–54

Trenbath B R, Angus J F (1975) Leaf inclination and crop production. *Field Crops Abstracts* 28: 231–44

Triboï E, Triboï-Blondel A M (2002) Productivity and grain or seed composition: a new approach to an old problem. *European Journal of Agronomy* 16: 163–86

Troughton J H (1969) Plant water stress and carbon dioxide exchange of cotton leaves. *Australian Journal of Biological Sciences* 22: 289–302

Truong H-T, Duthion C (1993) Time of flowering of pea (*Pisum sativum* L.) as a function of leaf appearance rate and node of first flower. *Annals of Botany* 72: 133–42

Turgeon R (1989) The sink-source transition in leaves. *Annual Review of Plant Physiology and Plant Molecular Biology* 40: 119–38

Turgeon R, Webb J A (1973) Leaf development and phloem transport in *Cucurbita pepo*: transition from import to export. *Planta* 113: 179–91

Turgeon R, Webb J A (1975) Leaf development and phloem transport in *Cucurbita pepo*: carbon economy. *Planta* 123: 53–62

Turner L B, Pollock C J (1998) Changes in stolon carbohydrates during the winter in four varieties of white clover (*Trifolium repens* L.) with contrasting hardiness. *Annals of Botany* 81: 97–107

Uchijima Z (1970) Carbon dioxide environment and flux within a corn crop canopy. In Setlik I (ed) *Prediction and Measurement of Photosynthetic Productivity.* Pudoc, the Netherlands

Unkovich M J, Pate J S (2000) An appraisal of recent field measurements of symbiotic N_2 fixation by annual legumes. *Field Crops Research* 65: 211–28

Unsworth M H, Colls J J, Sanders G E (1994) Air pollutants as constraints for crop yields. In Boote K J, Bennett J M, Sinclair T R, Paulsen G M (eds) *Physiology and Determination of Crop Yield.* Crop Science Society of America, pp 467–86

Upadhyay A P, Ellis R H, Summerfield R J, Roberts E H, Qi A (1994a) Characterization of photothermal flowering responses in maturity isolines of soyabean [*Glycine max* (L.) Merrill] cv. Clark. *Annals of Botany* 74: 87–96

Upadhyay A P, Summerfield R J, Ellis R H, Roberts E H, Qi A (1994b) Variation in the durations of the photoperiod-sensitive and photoperiod-insensitive phases of development to flowering among eight maturity isolines of soyabean [*Glycine max* (L.) Merrill]. *Annals of Botany* 74: 97–101

Uribelarrea M, Cárcova J, Otegui M E, Westgate M E (2002). Pollen production, pollination dynamics, and kernel set in maize. *Crop Science* 42: 1910–8

Vaidyanathan L V (1984) Soil and fertiliser nitrogen use by winter wheat. In *The Nitrogen Requirements of Cereals. MAFF Reference Book* 385. HMSO, London, pp 107–18

Valentinuz O R, Tollenaar M (2004) Vertical profile of leaf senescence during the grain-filling period in older and newer maize hybrids. *Crop Science* 44: 827–34

van Beem J, Smith M E (1997) Variation in nitrogen use efficiency and root system size in temperate maize genotypes. In Edmeades G O, Bänziger M, Mickelson H R, Peña-Valdivia C B (eds) *Developing Drought- and Low N-Tolerant Maize*. CIMMYT, Mexico D.F., pp 241–4

van der Werf A, Poorter H, Lambers H (1994) Respiration as dependent on a species' inherent growth rate and on the nitrogen supply to the plant. In Roy J, Garnier E (eds) *A Whole Plant Perspective of Carbon–Nitrogen Interactions*. SPB Academic Publishing, the Netherlands, pp 61–77

van Heemst H D J (1986) The distribution of dry matter during growth of a potato crop. *Potato Research* 29: 55–66

van Ittersum M K, Rabbinge R (1997) Concepts in production ecology for analysis and quantification of agricultural input–output combinations. *Field Crops Research* 52: 197–208

Varlet-Grancher C, Gosse G, Chartier M, Sinoquet H, Bonhomme R, Allirand J M (1989) Mise au point: rayonnement solaire absorbé ou intercepté par un couvert végétal. *Agronomie* 9: 419–39

Vega C R C, Sadras V O, Andrade F H, Uhart S A (2000) Reproductive allometry in soybean, maize and sunflower. *Annals of Botany* 85: 461–8

Veit B, Briggs S P, Schmidt R J, Yanofsky M F, Hake S (1998) Regulation of leaf initiation by the *terminal ear 1* gene of maize. *Nature* 393: 166–8

Villalobos F J, Ritchie J T (1992) The effect of temperature on leaf emergence rates of sunflower genotypes. *Field Crops Research* 29: 37–46

Vinocur M G, Ritchie J T (2001) Maize leaf development biases caused by air-apex temperature differences. *Agronomy Journal* 93: 767–72

Viola R, Taylor M A, Oparka K J (2001a) Meristem activation in potato: impact on tuber formation, development and dormancy. *Annual Report of the Scottish Crop Research Institute for 2000/01*, pp 99–102

Viola R, Roberts A G, Haupt S, Gazzani S, Hancock D, Marmiroli N, Machray G C, Oparka K J (2001b) Tuberization in potato involves a switch from apoplastic to symplastic phloem unloading. *The Plant Cell* 13: 385–98

Vos J (1981) *Effects of Temperature and Nitrogen Supply on Post-floral Growth of Wheat: Measurements and Simulations*. Pudoc, Wageningen, the Netherlands

Vos J, Biemond H (1992) Effects of nitrogen on the development and growth of the potato plant. 1. Leaf appearance, expansion growth, life spans of leaves and stem branching. *Annals of Botany* 70: 27–35

Vos J, Groenwold J (1989) Genetic differences in water-use efficiency, stomatal conductance and carbon isotope fractionation in potato. *Potato Research* 32: 113–21

Vos J, van der Putten P E L (1998) Effect of nitrogen supply on leaf growth, leaf nitrogen economy and photosynthetic capacity in potato. *Field Crops Research* 59: 63–72

Vouillot M O, Devienne-Barret F (1999) Accumulation and remobilization of nitrogen in a vegetative winter wheat crop during or following nitrogen deficiency. *Annals of Botany* 83: 569–75

Wachendorf M *et al.* (2001a) Overwintering of *Trifolium repens* L. and succeeding growth: results from a common protocol carried out at twelve European sites. *Annals of Botany* 88: 669–82

Wachendorf M *et al.* (2001b) Overwintering and growing season dynamics of *Trifolium repens* L. in mixture with *Lolium perenne* L.: a model approach to plant–environment interactions. *Annals of Botany* 88: 683–702

Walker K C, Booth E J (2003) Sulphur nutrition and oilseed quality. In Abrol Y P, Ahmad A (eds) *Sulphur in Plants*. Kluwer Academic Publishers, pp 323–39

Walters D R, Ayres P G (1983) Changes in nitrogen utilization and enzyme activities associated with CO_2 exchanges in healthy leaves of powdery mildew infected barley. *Physiological Plant Pathology* 23: 447–59

Wardlaw I F (1990) The control of carbon partitioning in plants. *New Phytologist* 116: 341–81

Wardlaw I F, Willenbrink J (2000) Mobilization of fructan reserves and changes in enzyme activities in wheat stems correlate with water stress during kernel filling. *New Phytologist* 148: 413–22

Warren Wilson J (1959) Analysis of the distribution of foliage area in grassland. In Ivins J D (ed) *The Measurement of Grassland Productivity*. Butterworths, pp 51–61

Warrington I J, Kanemasu E T (1983) Corn growth response to temperature and photoperiod. II. Leaf-initiation and leaf appearance rates. *Agronomy Journal* 75: 755–61

Watson D J (1947) Comparative physiological studies on the growth of field crops. 1. Variation in net assimilation rate and leaf area between species and varieties and within and between years. *Annals of Botany* 11: 41–76

Watson D J (1952) The physiological basis of variation in yield. *Advances in Agronomy* 4: 101–45

Weir A H, Barraclough P B (1986) The effect of drought on the root growth of winter wheat and its uptake from a deep loam. *Soil Use and Management* 4: 33–40

Weir A H, Bragg P L, Porter J R, Rayner J H (1984) A winter wheat crop simulation model without water or nutrient limitation. *Journal of Agricultural Science, Cambridge* 102: 371–82

Weir A H, Day W, Sastry T G (1985) Using a whole crop model. In Day W, Atkin R T (eds) *Wheat Growth and Modelling*. Plenum Press, pp 339–55

Wells R (1991) Soybean growth responses to plant density: relationships among canopy photosynthesis, leaf area, and light interception. *Crop Science* 31: 755–61

Westgate M E (2000) Strategies to maintain ovary and kernel growth during drought. In Otegui M E, Slafer G A (eds) *Physiological Bases for Maize Improvement*. Food Products Press, pp 113–37

Westgate M E, Lizaso J, Batchelor W (2003) Quantitative relationships between pollen shed density and grain yield in maize. *Crop Science* 43: 934–42

Whaley J M, Sparkes D L, Foulkes M J, Spink J H, Semere T, Scott R K (2000) The physiological response of winter wheat to reductions in plant density. *Annals of Applied Biology* 137: 165–77

Whigham D K (1983) Soybean. In *Potential Productivity of Field Crops under Different Environments*. International Rice Research Institute, the Phillippines, pp 205–25

Whitfield D M, Smith C J, Gyles O A, Wright G C (1989) Effects of irrigation, nitrogen and gypsum on yield, nitrogen accumulation and water use by wheat. *Field Crops Research* 20: 261–77

Wiebold W J, Shibles R, Green D E (1981) Selection for apparent photosynthesis and related leaf traits in early generations of soybeans. *Crop Science* 21: 969–73

Wilcox J R, Cavins J F (1995) Backcrossing high seed protein to a soybean cultivar. *Crop Science* 35: 1036–41

Wilkerson G G, Jones J W, Boote K J, Ingram K T, Mishoe J W (1983) Modeling soybean growth for crop management. *Transactions of the American Society of Agricultural Engineers* 26: 63–73

Willcott J, Herbert S J, Zhi–Yi L (1984) Leaf area display and light interception in short-season soybeans. *Field Crops Research* 9: 173–82

Williams J (1954) *The Compleat Stratergist*. McGraw Hill

Williams R F (1975) *The Shoot Apex and Leaf Growth*. Cambridge University Press

Williams R F, Rijven A H G C (1965) The physiology of growth in the wheat plant. 2. The dynamics of leaf growth. *Australian Journal of Biological Sciences* 18: 721–43

Williams W A, Loomis R S, Lepley C R (1965) Vegetative growth of corn as affected by population density. 1. Productivity in relation to interception of solar radiation. *Crop Science* 5: 211–3

Williams W A, Loomis R S, Duncan W G, Dovrat A, Ninez A (1968) Canopy architecture at various population densities and the growth and grain yield of corn. *Crop Science* 8: 303–8

Wittenbach V A, Ackerson R C, Giaquinta R T, Hebert R R (1980) Changes in photosynthesis, ribulose bisphosphate carboxylase, proteolytic activity, and ultrastructure of soybean leaves during senescence. *Crop Science* 20: 225–31

Woledge J (1977) The effects of shading and cutting treatments on the photosynthetic rate of ryegrass leaves. *Annals of Botany* 41: 1279–86

Wolfe D W, Henderson D W, Hsiao T C, Alvino A (1988a, b) Interactive water and nitrogen effects on senescence of maize. 1. Leaf area duration, nitrogen distribution, and yield. 2. Photosynthetic decline and longevity of individual leaves. *Agronomy Journal* 80: 859–64, 865–70

Woodward F I, Sheehy J E (1983) *Principles and Measurements in Environmental Biology*. Butterworths

Wright D P, Baldwin B C, Shephard M C, Scholes J D (1995a, b) Source–sink relationships in wheat leaves infected with powdery mildew. 1. Alterations in carbohydrate metabolism. 2. Changes in the regulation of the Calvin cycle. *Physiological and Molecular Plant Pathology* 47: 237–53, 255–67

Wright G C, Hubick K T, Farquhar G D, Rao R C N (1993) Genetic and environmental variation in transpiration efficiency and its correlation with carbon isotope discrimination and specific leaf area in peanut. In Ehleringer J R, Hall A E, Farquhar G D (eds) *Stable Isotopes and Plant–Water Relations*. Academic Press, pp 247–67

Wuest S B, Cassman K G (1992) Fertilizer-nitrogen use efficiency of irrigated wheat. 2. Partitioning efficiency of preplant versus late-season application. *Agronomy Journal* 84: 689–94

Wurr D C E (1974) Some effects of seed size and spacing on the yield and grading of two maincrop potato varieties. *Journal of Agricultural Science, Cambridge* 82: 37–45, 47–52

Wurr D C E, Allen E J (1974) Some effects of planting density and variety on the relationship between tuber size and tuber dry matter percentage in potato. *Journal of Agricultural Science, Cambridge* 82: 277–82

Yaklich R W, Vinyard B, Camp M, Douglass S (2002) Analysis of seed protein and oil from soybean Northern and Southern Region Uniform Tests. *Crop Science* **42**: 1504–15

Youssefian S, Kirby E J M, Gale M D (1992) Pleiotropic effects of the GA-insensitive *Rht* dwarfing genes in wheat. 2. Effects on leaf, stem, ear and floret growth. *Field Crops Research* **28**: 191–210

Yunusa I A M, Siddique K H M, Belford R K, Karimi M M (1993) Effect of canopy structure on efficiency of radiation interception and use in spring wheat cultivars during the pre-anthesis period in a mediterranean-type environment. *Field Crops Research* **35**: 113–22

Zerihun A, McKenzie B A, Morton J D (1998) Photosynthetic costs associated with the utilisation of different nitrogen-forms: influence on the carbon balance of plants and shoot-root biomass partitioning. *New Phytologist* **138**: 1–11

Zhang H, Charles T C, Driscoll B T, Prithiviraj B, Smith D L (2002) Low temperature-tolerant *Bradyrhizobium japonicum* strains allowing improved soybean yield in short-season areas. *Agronomy Journal* **94**: 870–5

Zoebel D (1996) Controversies around resource use efficiency in agriculture: shadow or substance? Theories of C T de Wit (1924–1993). *Agricultural Systems* **50**: 415–24

Zrenner R, Salanoubat M, Willmitzer L, Sonnewald U (1995) Evidence of the crucial role of sucrose synthase for sink strength using transgenic potato plants (*Solanum tuberosum* L.). *The Plant Journal* **7**: 97–107

Index

starch, 146
stem population density, 170
transpiration efficiency, 187–8, 190
tuber initiation and filling, 31–2, 151, 167–71, 216
water stress, 45, 169
precision agriculture, 203
protein turnover, 137–9, 142

quality of crop products, 205–21, 269–70
α-amylase, 206
bread wheat, 205–209, 270
erucic acid, 214–15
forage digestibility, 176, 218–21, 270
glucosinolates, 212–15
glutenin, 206
inverse relationship with yield, 206–208, 212, 218–20
oilseed rape, oil and protein, 212
potato tubers, 215–17
soybean, oil and protein, 209–211

reflection of solar radiation, 66, 70, 78, 111
resource,
capture, 202–204
use, 266
respiration (crop), 62–4, 117–44
alternative pathways, 118, 119, 137
growth and maintenance respiration (McCree model), 118, 120, 121, 123–6, 130–34, 136–40, 142, 236, 246, 250
mitochondrial (dark), 74, 77–8, 102, 119, 120, 126
pasture, 172–3
respiration of substrates, 126–30
respiratory costs of biosynthesis, 118, 126, 129, 135–8, 236, 245, 250
respiratory demands in winter, 176
responses to temperature, 120, 124, 141
root, 130–31
water potential, 141
rice (*Oryza sativa*),
C$_3$ physiology, 104
domestication and range, 7
modelling, 226
quality, 205
radiation use efficiency, 78, 96–7
respiration, 120–21
Rubisco (ribulose bisphosphate carboxylase-oxygenase), 73–116, 197, 199, 246, 273–4

senescence, 8–12, 40, 50–52, 54, 70, 105, 108, 111, 114, 126, 152, 171–3, 178–9, 192, 201, 207, 223, 240, 243, 253, 262
shading and shade tolerance, 33–4, 46
Sorghum spp.,
C$_4$ physiology, 99, 104, 108, 109, 188
photosynthesis, 110
protein turnover, 139
radiation use efficiency, 78
transpiration efficiency, 188, 190
water relations, 182–3, 185, 187, 189, 190–91
sources and sinks, 8, 25, 60, 61, 105, 147–8, 148–51, 191, 208, 226, 261–2
limitation of yield by sink, 157–62
limitation of yield by source, 153–7, 161
ontogeny, 148–51, 171
soybean (*Glycine max*),
apical development, 36
assimilate partitioning, 151–2, 155
branching, 50
C$_3$ physiology, 141
CROPGRO soybean model, 243–53
determinate and indeterminate types, 20
development, 10, 11, 18–21, 26, 28–9, 41, 42, 244–5, 243–4
domestication and range, 7
drought tolerance, 192
extinction coefficient, 69
growth (development) stages, 19
growth and maintenance respiration, 125
harvest index, 153
interception of solar radiation, 61, 72
juvenility (pre-inductive phase), 20, 26–9, 42, 244, 261
leaf angle, 66
leaf area index, 53–4, 59
maturity group, 28, 29, 210
nitrogen fixation, 194
ozone, 113–14
phasic development, 21, 26, 28
photoperiod, 20, 26, 28–9, 261
photosynthesis, 63–4, 246–8, 273
protein turnover, 130
quality of products, 209–211
radiation use efficiency, 78–9, 96–7
respiration, 126–30
sucrose, 145
transpiration efficiency, 188
squash (*Curcubita pepo*), 148–9
'stay green' character, 51–2, 199